T0325330

Respiratory Physiology
of Newborn Mammals

Respiratory Physiology of Newborn Mammals

A Comparative Perspective

Jacopo P. Mortola, M.D.
Professor, Department of Physiology
McGill University
Montreal, Quebec, Canada

The Johns Hopkins University Press
Baltimore and London

The Johns Hopkins University Press
2715 North Charles Street
Baltimore, Maryland 21218-4363
www.press.jhu.edu

Library of Congress Cataloging-in-Publication Data

Mortola, Jacopo P.
  Respiratory physiology of newborn mammals : a comparative
perspective / Jacopo P. Mortola.
    p.   cm.
  Includes bibliographical references (p.   ) and index.
    ISBN 0-8018-6497-6 (acid-free paper)
    1. Respiratory organs—Growth.   2. Infants
    Newborn)—Physiology.   3. Physiology, Comparative.
    4. Developmental biology.   I. Title.
    QP121.M65 2001
    573.2'19—dc21
                                            00-009834

A catalog record for this book is available from the British Library.

To Cosimo, Ginni, and Giudi

Contents

Preface

This book began as a few notes and schemas that I had prepared for some lectures on neonatal respiration. At the time, my objective was to make extensive use of the comparative approach in discussing the main aspects of the respiratory physiology of the developing mammal. In due course, that objective took on added flesh, and expanding and transforming the original notes into a book required that I include a substantial amount of detailed information and source citation. Still, I would like to think that these additions do not seriously detract from the strength of the original approach, in which comparative analysis across species, including the human infant, represented the starting point for further discussion.

Practicality and a measure of conformity with traditional approaches to the study of respiratory physiology have suggested the separation of the manuscript into five main chapters: Gestation and Birth, Metabolic and Ventilatory Requirements, Mechanical Behavior of the Respiratory Pump, Reflex Control of the Breathing Pattern, and Changes in Temperature and Respiratory Gases. Technical matters are usually not discussed, the chief exception being certain aspects of respiratory mechanics; the methodological details offered in Appendix A should facilitate an understanding of this topic. A brief discussion of the problems related to comparisons and normalization procedures is offered in Appendix B. Those readers less familiar with the jargon of respiratory physiology may find it useful to consult the Glossary or the Abbreviations list from time to time.

Each chapter concludes with two very brief sections, Interspecies Comparisons and Clinical Implications. These sections, which emphasize some of the comparative and pathological aspects mentioned in the chapters, are offered mainly for the use of readers with interests in comparative zoology or medicine. Indeed, it is my hope that this book, though intended primarily for students in developmental physiology, or comparative biology or zoology, will also be found useful by neonatologists and pediatric pulmonologists keen on refreshing their appreciation of the basic physiological concepts of their clinical practice, by pursuing an approach quite different from what is adopted in medical classes.

Many societies, organizations, and publishers have kindly granted me permission to reproduce materials from their publications. These include Academic Press, the American Pediatric Society, the American Physiological Society, the American Thoracic Society, the British Thoracic Society, Cambridge University Press, the Canadian National Research Council, CSIRO Australia, Elsevier, Growth Publishing Company, Marcel Dekker, Mosby, the National Geographic Society, *Nature*, Oxford University Press, Portland Press, the Royal Society of Medicine, Saunders, *Scientific American*, and John Wiley & Sons.

Erin Seifert and Teresa Trippenbach of McGill University, Montreal, and Peter Frappell of La Trobe University, Melbourne, have critically read and given constructive suggestions about various sections of the manuscript. Jay P. Farber of the University of Oklahoma, Oklahoma City, contributed the photo of the newborn opossum, and Pat Woolley of La Trobe University, the photo of the pouch-young dunnart. Domnica Marghescu has helped with the preparation of the figures. Wendy Harris of the Johns Hopkins University Press and Bill Carver, the editor commissioned by the Press, have provided immensely useful suggestions on the layout of the book and the editing of the text. I owe special thanks to all the students, visitors, and colleagues with whom I have been fortunate to collaborate over the past twenty-five years. To all, I am very grateful.

Abbreviations

See also the Glossary following the text.

a	arterial
A	alveolar
ab	abdomen
ao	airway opening
ATP	adenosine triphosphate
BAT	brown adipose tissue
BPD	bronchopulmonary dysplasia
BTPS	body temperature, pressure and saturation
C	compliance ($= 1$/elastance)
CAT	catalase
CH	carbohydrate
CI	confidence interval
Cl^-	chlorine ion
C_L	lung compliance
$C_L$dyn	dynamic lung compliance
CO_2	carbon dioxide
Crs	respiratory system compliance
CSF	cerebrospinal fluid
Cw	chest wall compliance
di	diaphragm
DNA	deoxyribonucleic acid
DPG	diphosphoglycerate
E	expiratory
E	elastance ($= 1$/compliance)

el	elastic
EMG	electromyogram, electromyographic
EMGdi	diaphragm electromyogram
es	esophageal
f	breathing frequency
F	fractional concentration
FBM	fetal breathing movements
F_{EO_2}	fractional expired concentration of oxygen
F_{ICO_2}	fractional inspired concentration of carbon dioxide
F_{IO_2}	fractional inspired concentration of oxygen
FRC	functional residual capacity (or end-expiratory level)
	surface tension
GP	glutathione peroxidase
H^+	hydrogen ion
Hb	hemoglobin
HCl	hydrochloric acid
Hct	hematocrit
H_2O	water molecule
I	inspiratory
K^+	potassium ion
L	lung
L0	optimal muscle fiber length
LW	lung weight
mus	muscles
Mb	myoglobin
Na^+	sodium ion
Nb	newborn
NEEP	negative end-expiratory pressure
NST	nonshivering thermogenesis
O_2	oxygen
P	pressure
Pab	abdominal pressure
$PaCO_2$	arterial partial pressure of carbon dioxide
P_{ACO_2}	alveolar partial pressure of carbon dioxide
Pao	pressure at the airway opening
PaO_2	arterial partial pressure of oxygen
Pb	barometric pressure

PCA	posterior cricoarytenoid (muscle)
PCO_2	partial pressure of carbon dioxide
Pdi	transdiaphragmatic pressure
PEEP	positive end-expiratory pressure
Pel(rs)	elastic pressure of the respiratory system
Pes	esophageal pressure
PEG_2	prostaglandin E_2
pH	acidity. *See also* Glossary
PIO_2	partial pressure of the inspired oxygen
PL	transpulmonary pressure
PL_{FRC}	transpulmonary pressure at FRC
pl	pleural
Pmus(I)	muscle activity during inspiration
PO_2	partial pressure of oxygen
Ppl	pleural pressure (or pleural *surface* pressure)
Ppl(liq)	pleural liquid pressure
Prc	pressure across t he rib cage
Prs	transrespiratory system pressure
PTE	recoil pressure at any time during expiration
Pw	pressure across the chest wall
P_{50}	partial pressure of O_2 at 50% hemoglobin saturation
Q_{10}	an index of temperature sensitivity. *See also* Glossary
R	resistance
RAR	rapidly adapting receptor
rc	rib cage
RDS	respiratory distress syndrome
REM	rapid-eye-movement (sleep)
RER	respiratory exchange ratio
RL	total pulmonary resistance
RQ	respiratory quotient ($\dot{V}CO_2/\dot{V}O_2$)
Rrs	respiratory-system resistance
rs	respiratory system
SAR	slow-adapting receptor
SD	standard deviation
SDH	succinic dehydrogenase
SEM	standard error of the mean
SIDS	sudden infant death syndrome

SLN	superior laryngeal nerve
SO_2	sulphur dioxide
SOD	superoxide dismutase
STPD	standard temperature, pressure, and dry conditions
T	time; or temperature
Ta	ambient temperature
TA	thyroarytenoid (muscle)
Tb	body temperature
T_E	expiratory time
T_I	inspiratory time
TLC	total lung capacity
T_{TOT}	total time of the cycle
UCP	uncoupling protein
V	volume
$\dot{V}$	flow
$\dot{V}_A$	alveolar ventilation
$\dot{V}_{CO_2}$	carbon dioxide production
V_D	dead space
A	alveolar ventilation
$\dot{V}_E$	pulmonary ventilation
$\dot{V}_{O_2}$	oxygen consumption
$\dot{V}_{O_2}max$	maximal oxygen consumption
Vr	resting volume
V_T	tidal volume
V_{TE}	volume at any time during expiration
V_T/T_I	mean inspiratory flow
w	chest wall
W	body weight
γ	surface force (*also* surface tension)
τ	time constant
τ_L	time constant of the lungs
τ_{rs}	time constant of the respiratory system
$\tau_{rs}(E)$	expiratory time constant

Respiratory Physiology
of Newborn Mammals

Introduction

With very few exceptions—those few to be found among the smallest newborn marsupials—all mammals exchange oxygen and carbon dioxide through the lungs. Hence, in newborns, as in adults, changes in metabolic rate are likely to be accompanied by changes in pulmonary air convection. Indeed, the blood gases are an excellent indicator of the accuracy by which pulmonary ventilation ($\dot{V}E$) tracks metabolic needs; and during growth, even during the early postnatal hours, the blood gases remain within a narrow range of concentration and exhibit quite similar values among animals of very different dimensions, life modes, and age.

Despite the fact that age and body size have profound effects on the functional demands of the organism and the structural characteristics of the respiratory apparatus, the ventilatory function remains consistently adequate during growth—powerful validation of the evolutionary success of the mechanism by which air convection through the lungs ensures stable gas exchange through a lifetime.

The interaction between body size and age is an issue that will emerge frequently in the discussions of neonatal respiratory physiology in the chapters that follow, and comparative analysis among species is a convenient tool in addressing it. By comparing a given structural or functional parameter among newborns of different species one can gain some appreciation of the role of geometrical and physical factors in determining such variables. Similarly, from comparisons of newborns with adults of different species but of the same body mass, the role of growth and development, independent of body size, can be considered. The choice of comparing animals grouped by *species* rather than as individuals or by other taxonomic or evolutionary criteria such as their membership in various orders or families is a convenient and time-honored practice. Nevertheless, it is an arbitrary decision, one that should not be allowed to distract us from the view of the animal kingdom as an uninterrupted continuum or steadily elaborated shrub.

As an example, consider body temperature, which in adult mammals is maintained around values that are approximately constant and typically higher than ambient temperature. The smaller body mass of the neonate, coupled with its consequent higher surface-to-volume ratio, reduces its heat-sustaining capacity, thus tending to reduce body temperature. Hence, simply because of geometrical considerations, one would expect that a newborn mammal would have greater difficulties in maintaining a constant body temperature (homeothermy) than would an adult of the same species, and that the effect would be more pronounced in newborns of small species than in those of large species. Interspecies comparisons confirm this prediction, but they also indicate that a newborn is often more prone to hypothermia even than an adult of a different species but with a body size similar to its own, because in the neonatal period the mechanisms for the control of heat loss are still very limited. Hence, the propensity of the neonate to a low body temperature is the combined result of body size and age.

Interspecies analysis will also indicate that most newborns, in attempting to achieve homeothermy, consume relatively more oxygen than do the adults, therefore generating higher levels of $\dot{V}E$. The question arises: What are the mechanisms permitting these solutions? But again, even before examining the data one can put forward some hypotheses. A large pulmonary surface area and therefore high levels of gas diffusion could be achieved by designing big lungs, relative to body mass, containing large numbers of alveoli of small size. Comparative analysis will reveal that, generally, lung mass in newborns is in fact large, although the alveoli are usually not small, presumably because smaller dimensions would imply a greater tendency for alveolar collapse.

With respect to the mechanisms permitting high levels of $\dot{V}E$, one may hypothesize that if the newborn is to produce greater tidal volumes it may need bigger or stronger respiratory muscles per unit of body mass, compared to those of the adult, or larger and more compliant lungs, or perhaps longer inspiratory times. One could also hypothesize that the greater $\dot{V}E$ results from a higher breathing rate (which would require a more rapid central generation of neural output), a low mechanical resistance, and a short time constant. A lower resistance could be obtained by building a lung with disproportionately large airways, although this would mean an unusually large dead space, one that compromises the gas-exchange efficiency of $\dot{V}E$. Again, comparative analysis will help to sort out these possibilities, pointing toward the most commonly adopted solutions. It will then be our task to speculate on why evolution has favored some of these options but not others, and to understand the mechanisms that permit them.

In essence, therefore, the chief goal of this book is to identify and analyze the solutions and mechanisms that have evolved to guarantee adequate pulmonary ventilation in the neonatal mammal. Our attention will be focused principally on the mechanical, metabolic, and neural regulatory aspects of pulmonary ventilation, which are addressed in the central chapters. But because the phenomena occurring during gestation and the events surrounding birth are important determinants of neonatal respiration and survival, they will be reviewed first.

The neonate mammal's responses to sudden or sustained variations in temperature and/or respiratory gases will be discussed last. These responses offer an important opportunity to study the animal's strategies for survival, and to make use of the information on the metabolic, mechanical, and neural mechanisms controlling breathing discussed in the preceding chapters. Changes in oxygenation and temperature have undoubtedly been among the main environmental stimuli in the evolution of the ventilatory control mechanisms. Hence, the responses to temperature and oxygenation in the neonatal period seen today are likely to present some similarities to the less differentiated responses that occurred along the way during evolution, and their progressive development in the life of the individual is probably a good example of ontogeny recapitulating phylogeny.

Chapter 1

Gestation and Birth

The development of the mammalian embryo inside the mother probably represents an adaptive strategy, driven by evolution, of defense against predators. With the sole exception of the egg-laying monotremes[1] (Fig. 1.1), live bearing is common to all extant mammalian species (see Appendix C). The moment of birth can arrive as soon as 2 weeks after conception, as in the shrews and some marsupials (Fig. 1.2), or up to 2 years later, as in the blue whale and the African elephant. As a first approximation, the duration of gestation, defined as the length of time from fertilization until the delivery of the fetus, is proportional to the body mass of the adult, but the variability is great (Fig. 1.3). In marsupials, gestation can be very short. In fact, in most marsupials the intrauterine nourishment of the embryo is provided only by a yolk-sac structure (the choriovitelline placenta), which can sustain the embryo's needs for a limited period of time. The marsupial offspring is therefore born at a very immature, or altricial,[2] stage and continues its development confined to the maternal pouch for considerable additional time, as a *pouch young.* Some true-placental mammals (i.e., mammals with chorioallantoic placenta, or *eutherians*) regularly give birth to altricial newborns, but these are never,

[1] This order comprises only three species: the duck-billed platypus (*Ornithorhynchus anatinus*) and the short-beaked echidna (*Tachyglossus aculeatus*), both of Australia and the long-beaked echidna (*Zaglossus bruijnii*) of Papua and New Guinea. (See Appendix C.)

[2] The term *altricial* implies birth at an immature stage of development, as regards general appearance and response to the environment. Its opposite is *precocial.* Most insectivores, rodents, and carnivores are altricial, whereas artiodactyls, perissodactyls, and cetaceans generally have precocial offspring.

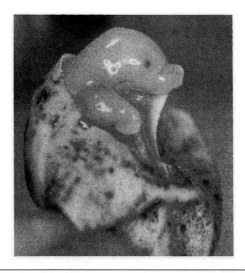

Figure 1.1. The emergence of the echidna from its leathery, oval egg. The tough-skinned egg is slit open by an enamel-covered egg tooth, as in some reptiles and birds. At hatching, this monotreme has a jellylike appearance. The spines characteristic of the adult will begin to appear a few weeks later. From Rismiller and Seymour, 1991.

at birth, as immature as marsupial newborns. Relative to their size, cetaceans (whales, dolphins, and allies) tend to have short gestations. Primates, especially the anthropomorphic primates (apes), and rodents of several families, have longer gestations than would be indicated by their size alone. Among common laboratory species, a remarkable deviation is offered by the guinea pig (*Cavia porcellus*), which has a gestation lasting more than 2 months (i.e., more than twice the average for a rodent of its size) and bears its pups at an unusually precocial stage.

In some species, such as rats and rabbits, mating stimulates ovulation, and the presence of sperm in the vagina is an almost certain sign of fertilization. In several others, notably hibernating bats, delayed fertilization is common; in these species, after mating, which occurs in late summer, spermatozoa are retained on one side of the uterus until the following spring, when fertilization of the ovum occurs. In many others, implantation of the embryo to the uterine wall occurs within days of fertilization, but there are some in which implantation is delayed, perhaps for months,

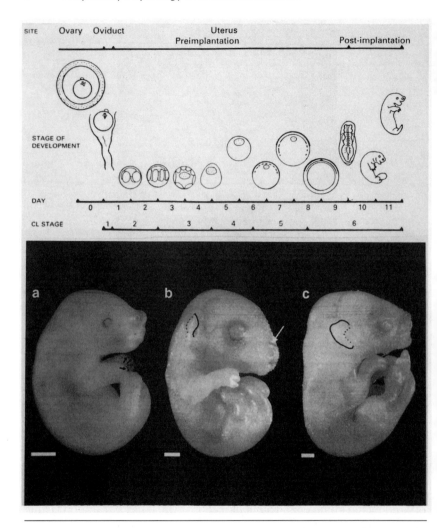

Figure 1.2. The gestational development of the stripe-faced dunnart, a marsupial born after a gestation of only 11 days. Note the very brief postimplantation phase. At birth, the skeleton is entirely cartilaginous. Development continues in the pouch, three phases of which are shown in the photos at the bottom. Photos *a, b,* and *c* are of pouch young at postnatal days 5, 20, and 25, respectively. Scale bars, 1 mm. *Above:* from Selwood and Woolley, 1991; *below:* from Frigo and Woolley, 1997.

yielding unusually long gestations. Delayed implantations are known to occur in many Mustelidae (pine marten, sea otter), seals and sea lions, bears, and armadillos. Lactation of a young from a previous litter can also cause delayed implantation, as for example in the embryonic diapause of marsupials. And even after implantation has been effected, there are opportunities to prolong gestation by slowing down the development of the embryo, as occurs in some fruit bats, which aim to perfectly coincide the time of birth with the abundance of fruit.

Across mammalian species, the weight of the pup at birth is approximately proportional to the adult's body mass, although, as is true for the duration of gestation, the variability is great, even when monotremes and marsupials are considered separately from eutherian (placental) mammals (Fig. 1.4). The very immature pups of marsupials usually have well-developed forelimbs suited for grasping and crawling toward the marsupium, an anatomical structure (permanent or transitory) that opens ventrally, transversely, or even posteriorly, as in the koala, the wombat, and other burrowers. Once a nipple in the marsupium has been located, the pup draws it into its mouth. The nipple will then swell, securing the pup to the mother so firmly that, if forcibly pulled out, the pup will no longer be able to reconnect. In some marsupials the milk delivered to the pouch young differs in content from that produced for an older, independent joey; thus the female can simultaneously produce two types of milk, and as the pouch young grows, it will remain constantly attached to the nipple for weeks or months. The echidna, a monotreme, has a transitory pouch where the pup feeds and grows for a couple of months after hatching, until its developing spines force the mother to eject it and place it in a nest.

Because medium-size and large mammals usually give birth to a single offspring, and large litters are more common in the smaller species, the relationship between the mass of the whole litter and the adult's body weight has an exponent slightly lower than that of the newborn's body weight (0.75 instead of 0.86). In any case, exceptions to these generalities are numerous, and variability even among species of the same body size is great. Bears, for example, have great body mass, yet most often give birth not to one but to two or three pups of remarkably small size. The armadillo consistently gives birth to a quadruplet of identical offsprings (Fig. 1.5). And in the smaller mammals, the size of the litter can

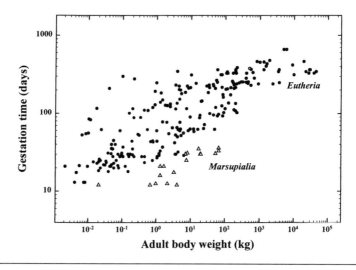

Figure 1.3. Duration of gestation as a function of adult weight. Each point represents a different species, and the species represented are from virtually all mammalian orders. Marsupials (△) are born after a very short gestation. Among eutherians (●), only some insectivores have a gestation similarly short. The egg-laying monotremes are not included; in the platypus, for example, the incubation of the egg lasts only 10–12 days. Data compiled mostly from Stonehouse, 1985; Forsyth, 1985; Tyndale-Biscoe and Renfree, 1987; Haltenorth and Diller, 1988; Eisenberg, 1989; Redford and Eisenberg, 1992; publications of Torstar Books and the Audubon Society; and the *Complete Book of Australian Mammals,* published by the Australian Museum, Sydney.

vary enormously around the mean value. Within a species, litters are usually larger in bigger females, or in circumstances of abundant food. In species producing multiple litters per breeding season, the size of the second and following litters of the year progressively decreases. The issues of the reproductive values of large versus small litters and the role of evolution and environmental factors in these matters are still lively and unresolved topics in mammalian ecology.

The Respiratory System before Birth

In the seventeenth century, Harvey asked how the fetus could survive in utero, considering that, when it is prematurely born, breathing is imme-

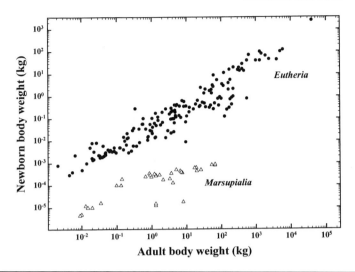

Figure 1.4. Body weight of the newborn as a function of adult weight. Each point (●) represents a different species, and the species represented are from all mammalian orders. Marsupials, which are born at an early stage of development, have far lower body weight at birth than same-size eutherian mammals. The exponent of the allometric equation for eutherians (164 species) was 0.86; that for marsupials (36 species) was 0.50. Data compiled mostly from Stonehouse, 1985; Forsyth, 1985; Tyndale-Biscoe and Renfree, 1987; Haltenorth and Diller, 1988; Eisenberg, 1989; Redford and Eisenberg, 1992; publications of Torstar Books and the Audubon Society; and the *Complete Book of Australian Mammals,* published by the Australian Museum, Sydney.

diately essential to its survival, and indeed begins immediately. The answer came more than two centuries later, having had to wait for the discovery of oxygen and the understanding of gas diffusion (Barron, 1976). Before birth, in mammals, gas exchange is provided by the placenta (or, in marsupials, the choriovitelline membrane), and the lungs do not contribute to the transfer of O_2 and CO_2. We also know now that, because of the very low blood perfusion in fetal life, the metabolic functions of the lungs during that time are probably of negligible importance.

Hence, before birth, the lung is a nonessential organ, and its prenatal growth is designed for its postnatal tasks. Two major phenomena contribute to fetal lung growth, the production of pulmonary fluid and the onset of incipient and intermittent breathing movements. Alterations in

Figure 1.5. The genetically identical four offspring of the nine-banded armadillo. Storrs and Lavies, 1982.

either of these processes do not interfere with the survival of the fetus in
utero, but do compromise the prenatal development of the lung, with potentially catastrophic consequences for the viability of the newborn.

Fetal Pulmonary Fluid

It was established many years ago that during prenatal development the
lungs contribute substantially to the production of the amniotic fluid, owing to their having a large surface area perfused with a hydraulic pressure
greater than the protein osmotic pressure (Setnikar et al., 1959). It is now
clear that the processes of fluid secretion and its postnatal reabsorption
are finely dependent on a system of active ion transport across the pulmonary epithelium, namely, of chlorine and sodium ions (Strang, 1991).

Fluid Production. The rate of production of the pulmonary fluid has
been measured in a few species; in the lamb it amounts to $\sim$2 ml $\cdot$ kg^{-1} $\cdot$
h^{-1} in the second third of gestation, increasing to almost twice as much
in the final third (Normand et al., 1971; Olver et al., 1981; Harding and
Hooper, 1996). Experimental closure of the trachea or congenital abnormalities with bronchial atresia during fetal life cause accumulation of
fluid in the airways and lung overexpansion (Potter and Bohlender, 1941;

Jost and Policard, 1948; Adams et al., 1963). Conversely, reduction in fluid leads to pulmonary hypoplasia, which in the fetal lamb can be corrected by tracheal obstruction, with good recovery of the postnatal respiratory function (Davey et al., 1999). In part, the pulmonary fluid is swallowed; in part it flows into the amniotic liquid, the latter explaining why pulmonary surfactant (a lipoprotein complex that functions to lower surface tension) can be detected in samples of the amniotic liquid (Gluck et al., 1971). At the end of pregnancy the total pulmonary fluid in the lung of the fetal lamb was estimated to be ~30 ml/kg (Normand et al., 1971; Olver et al., 1981), which would be similar to the quantity of air present a few hours after birth. A more recent series of studies, however, by Harding, Hooper, and co-workers (Harding and Hooper, 1996) on unanesthetized ovine fetuses, has indicated that the total pulmonary fluid continues to increase until the end of gestation, reaching values considerably above the end-expiratory volume of the postnatal air-filled lungs (Fig. 1.6). Because upper-airway resistance is low during periods of fetal breathing movements (see below), and high in their absence, lung fluid volume oscillates around the mean value inversely with the breathing

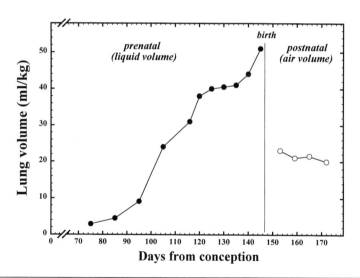

Figure 1.6. Lung volume of ovine fetus (●) and newborn (○). At term, the volume of the fetus's liquid-filled lung is greater than the functional residual capacity of the newborn's air-filled lung. From data compiled by Harding and Hooper, 1996.

movements (Dickson et al., 1987; Harding and Hooper, 1996). The absence of air, and therefore of surface tension phenomena, reduces lung recoil, and this, coupled with the high chest wall compliance, explains the very low, almost nil, value of end-expiratory transpulmonary pressure in the fetus (Agostoni, 1959; Avery and Cook, 1961; Vilos and Liggins, 1982).

Fluid Removal. During labor, pulmonary filtration rapidly ceases and absorption begins, both mechanisms probably related to the level of circulating epinephrine and its action on epithelial ion transport (Walters and Olver, 1978; Strang, 1991). Indeed, there is an enormous surge of catecholamines during parturition, a surge possibly responsible for the animal's alertness and arousal state immediately after birth (Lagercrantz, 1996). Other factors, including production of the hormone vasopressin (Perks and Cassin, 1989; Wallace et al., 1990), are likely to contribute to the perinatal switch from fluid secretion to fluid reabsorption, and to the long-term maintenance of the balance in favor of reabsorption. The greater part of the fluid remaining in the lung after delivery, in the mature lamb, is eventually absorbed by pulmonary circulation, and the remainder (less than one-third) by the pulmonary lymphatics or by way of the pleural space (Humphreys et al., 1967; Gonzalez-Crussi and Boston, 1972; Bland et al., 1982; Cummings et al., 1993). In the lamb, measurements of lymph flow from the lung (Humphreys et al., 1967), as well as radiological observations (Fletcher et al., 1970), suggest that most of the fluid removal occurs in a couple of hours. And in other species, when changes in lung weight or in lung wet-to-dry ratio are examined, or changes in extravascular lung water are specifically considered, fluid reabsorption seems to be complete in a few hours (Fauré-Fremiet and Dragoiu, 1923; Aherne and Dawkins, 1964; Adams et al., 1971; Bland et al., 1980). Because different and often indirect methodologies are used to obtain data of this sort, it is difficult to compare rates of fluid clearance among different species. However, since the Starling equilibrium[3] and vascular transmural pressure are likely to be species-size-invariant, the reabsorption time should be proportional to the ratio of capillary surface

[3] The Starling equilibrium for capillary exchange is the balance of forces involved in fluid filtering outward from the capillary and in fluid absorption into the capillary. These forces include the capillary and interstitial fluid pressures and the osmotic pressures of plasma and interstitial proteins.

area to pulmonary fluid volume, which is greater in small than in large mammals. If these are indeed the main factors involved, a shorter clearance time may be expected in smaller species.

Effects of the Mode of Delivery. During delivery, some of the fetal lung fluid is eliminated through the upper airways during the squeezing of the thorax in the pelvic canal (Borell and Fernstrom, 1962; Karlberg et al., 1962a; Saunders and Milner, 1978), although the removal realized in this phase is probably not an essential requirement for adequate onset of breathing and lung expansion, nor is it necessary for the cardiovascular adaptation at birth (Agata et al., 1995). In fact, in many small mammalian species, deliveries with breech presentation are common, and they are probably the rule in dolphins and other marine mammals (Scheffer, 1976) (Fig. 1.7). Further, the onset of breathing in babies delivered by cesarean section does not seem to differ appreciably from that of vaginally delivered babies (Vyas et al., 1981; Mortola et al., 1982a). One can argue that the absence of squeezing through the pelvic canal in artificially delivered newborns leaves a somewhat larger amount of fluid in their airways, as suggested by measurements of extravascular lung water in rabbit pups (Bland et al., 1980), and is therefore the cause of the delay in their establishment of functional residual capacity (FRC) (Klaus et al., 1962a; Milner et al., 1978a; Boon et al., 1979; Vyas et al., 1981).

As mentioned above, however, epinephrine concentration is a key causative event in the inhibition of pulmonary fluid production and the beginning of its absorption (Walters and Olver, 1978). It is therefore the *occurrence* of labor, rather than the mode of delivery, that chiefly determines the amount of fluid retained at birth (Bland et al., 1980). This probably explains the finding that respiratory problems are more likely to occur in human infants delivered by elective cesarean section than in those delivered by emergency C-section after the onset of labor (Faxelius et al., 1982; Hales et al., 1993; Morrison et al., 1995). Births by C-section are often elective deliveries opted for before labor begins, hence before the level of endogenous epinephrine rises. In lambs, artificial reduction of the pulmonary fluid before elective C-section delivery has a major favorable effect on postnatal gas exchange (Berger et al., 1996). Rats born before term by C-section did not adapt to the first day of extrauterine life as well as rats that, in addition to C-section, experienced anoxia, presumably because in the latter group the concentration of plasma cate-

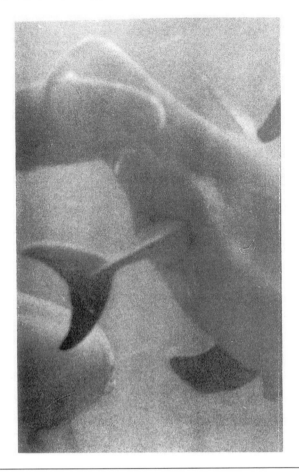

Figure 1.7. The birth of a bottlenose dolphin, tail first. Scheffer, 1976.

cholamines was higher (El-Khodor and Boksa, 1997). Rabbits delivered operatively after labor have no more water in their lungs than rabbits born vaginally (Bland et al., 1979). Differences in the establishment of FRC between infants born by natural delivery and those born by C-section delivery, therefore, should not necessarily be attributed to differences in the mechanics of delivery. The surge of fetal plasma catecholamines during labor and delivery is likely to be important not only in the respiratory adaptation at birth (surfactant production and fluid absorption) but also in several metabolic functions of the newborn, possibly including thermogenesis (Christensson et al., 1993), and in the stress response to hypoxia and asphyxia (Lagercrantz and Slotkin, 1986).

During delivery, irrespective of its modalities, a portion of the pulmonary fluid fills the newborn's upper airway. Animals are known to lick their offspring extensively at birth, but whether or not this includes cleaning of the upper airway, as is claimed to be the case for rats (Faridy, 1983), is difficult to establish. Oronasopharyngeal suction is a common routine in the modern human neonatal setting, but whether this practice is really essential for the establishment of extrauterine life, at least in healthy infants, has been questioned (Estol et al., 1992).

Fetal Breathing Movements

Periods of rhythmic contractions of the respiratory muscles and decreased pleural pressure have been recognized in fetuses of several species (Jansen and Chernick, 1983), although by far the most abundant information comes from observations of human and sheep fetuses (Jansen and Chernick, 1991). These fetal breathing movements (FBM), possibly already present by the end of the first third of gestation, occur almost exclusively during a sleep phase (low voltage, high frequency), and are behaviorally reminiscent of the adult's rapid-eye-movement (REM) sleep, whereas they are absent during high-voltage, slow-wave sleep (Dawes et al., 1972). FBM are characterized by contraction of the diaphragm (Clewlow et al., 1983), sometimes accompanied by activation of the intercostals, as well as activity of the laryngeal abductors (Harding et al., 1980) (Fig. 1.8). The adductor thyroarytenoid muscle, so active postnatally, is rarely active prior to birth, but it is recruited intermittently in the periods between bursts of FBM (Harding, 1980; Harding et al., 1980).

Deep, abrupt inspiratory efforts have also been described in fetuses of humans and several other nonruminant species; they have been compared to the postnatal hiccup, although the latter is characterized by a diaphragmatic contraction against upper-airway closure, rather than opening. Their incidence decreases with gestation, and eventually, toward term, is about 2–4% of the total period of FBM (Pillai and James, 1990), a value similar to the occurrence of postnatal hiccups in the premature infant (Brouillette et al., 1980).

The incidence of FBM increases with gestation, and toward the last

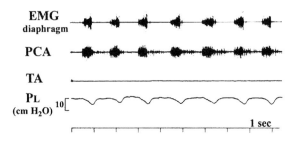

Figure 1.8. Recordings of fetal breathing of sheep fetus during low-voltage electrocortical state. From *top to bottom:* electromyographic activity of the diaphragm, posterior cricoarytenoid (PCA, laryngeal abductor), thyroarytenoid muscle (TA, laryngeal adductor), and changes in transpulmonary pressure (PL). During fetal breathing movements, TA is usually silent. Modified from Harding, 1980.

few weeks the movements occur about 30% of the time, at a rate not too different from that of postnatal breathing. Only a few days before term the FBM decrease, and they cease totally during labor (Richardson et al., 1979; Berger et al., 1986), although some sporadic diaphragmatic bursts have been recorded even during labor (Berger et al., 1990). In occurrence, frequency, and amplitude, variation in FBM is great. Just why that should be so is not clear, nor is the functional basis for this fetal activity yet understood, although the possibility of a control by prostaglandins, namely PGE_2, has much experimental support (Thorburn, 1992). Numerous factors affect FBM, including time of day, levels of corticoids and glucose in maternal plasma (Patrick et al., 1978), and the mother's use of drugs, alcohol, or tobacco, but the mechanisms of their influence on FBM are not known.

Maternal, and therefore fetal, hypercapnia and hypocapnia respectively increase and decrease FBM in both humans and other animals (Van Weering et al., 1979; Ritchie and Lakhani, 1980; Jansen et al., 1982; Ioffe et al., 1987; Connors et al., 1988). In the lamb fetus, combined hypercapnia and central hypothermia, both applied directly to the fetus, can stimulate FBM even during high-voltage, slow-wave sleep, thus overriding the inhibition of FBM characteristic of this sleep phase (Kuipers et al., 1997a). Hypercapnia alone does not stimulate FBM during high-voltage sleep, unless the lateral pons is lesioned (Johnston and Gluckman, 1989). Conversely, hypocapnia reduces FBM (Kuipers et al., 1994). All these observations, as well as additional considerations, have led to the hypothesis that the level of fetal CO_2 represents an important background "drive" for the occurrence of FBM, against which are inhibitory inputs related to sleep state (Blanco, 1994). Acute hypoxia or asphyxia, in the sheep fetus, reduces both pulmonary liquid production and FBM (Bocking et al., 1988; Hooper et al., 1988; Koos and Sameshima, 1988; Koos et al., 1988; Hooper and Harding, 1990), but eventually the fetus appears to adapt, resuming its normal rate of FBM. Very severe hypoxia, by contrast, triggers powerful gasps and upper-airway dilation; and bilateral lesions of the upper pons in lamb fetuses reverse the hypoxic depression into stimulation, suggesting that during hypoxia, as for hypercapnia, this brain region inhibits FBM (Gluckman and Johnston, 1987).

Prenatal lung growth is known to be highly influenced by the degree of fetal lung expansion (Alcorn et al., 1977; Moessinger et al., 1990; Hooper et al., 1993; Harding and Hooper, 1996), which depends on the balance between pulmonary fluid secretion and upper-airway resistance. The volume distention of the lung, rather than the intra-airway pressure, is the important parameter determining tissue hyperplasia (Nardo et al., 1998). The exact role of FBM in this interplay of factors determining lung growth is not completely clear, but it is well documented that the abolition of FBM, whether by phrenicotomy or upper cervical-cord transection, results in a hypoplastic lung. In infants with severe lesions of the central nervous system and an absence of FBM, or in cases of congenital absence of the phrenic nerves, the lungs are hypoplastic (Dornan et al., 1984; Goldstein and Reid, 1980). Harding (1994), in summarizing the results of several studies pertinent to this topic, concluded that FBM counteract the lung recoil, which, albeit very small, favors the elimination of lung fluid; hence, their absence would lead to lung hypoplasia because of the reduced lung expansion. The potentially contributing role of the glottis and of the muscles controlling upper-airway resistance, in cases of pulmonary hypoplasia, has not yet been thoroughly evaluated. In addition to their role in normal lung development, FBM are likely to influence the development of the inspiratory muscles and their neural pathways, as well as, perhaps, cardiac output and blood-flow distribution.

Birth

At birth, the newborn's respiratory apparatus must be mature enough to take over the gas-exchange function previously provided by the placenta. This role requires periodic back-and-forth convection of air. The beginning of an adequate external ventilation, in place of the irregular fetal breathing movements (FBM), and an appropriately large pulmonary perfusion, are therefore obvious urgencies. But they are not the only ones. For gas exchange to occur properly, it is equally important that the pulmonary fluid be rapidly absorbed and replaced by air. Hence, one may reasonably expect that the major priorities of the respiratory adaptation at birth include the following: onset of continuous external ventilation, pulmonary-fluid elimination and reabsorption with establishment of a functional residual capacity, and adequate perfusion. All these requirements have been extensively studied, although in very few species, namely, infants and lambs. These studies have yielded a large body of information on the adjustments of the pulmonary circulation, the regulation of the pulmonary fluid, and the mechanical aspects of the first breath. But the answer to the key question—what are the factors prompting the onset of breathing and the maintenance of continued ventilation?—remains elusive.

What Controls the Onset of Continuous Breathing at Birth?

What triggers breathing at birth is a fascinating question, one that has intrigued researchers for a long time. Its answer remains vague, although more recent knowledge of FBM has somewhat focused the way the old question was phrased. In fact, rather than questioning what initiates regular breathing at birth, researchers now, more specifically, ask what are the factors that transform the irregular and periodic FBM into regular postnatal breathing, and conversely, what are the mechanisms that periodically inhibit fetal breathing. Experimental answers have been sought either by reproducing the postnatal conditions with the fetus still in utero, or by mimicking the prenatal conditions on the newly born after birth.

In addition to lung expansion and oxygenation and the number of inputs associated with the cardiovascular adaptation (Serwer, 1992) and changes in the endocrine systems, a number of stimuli—tactile, pressure, thermal, and, in some species, visual and acoustic—also accompany birth. It is conceivable that all these variables participate in the establishment of continuous breathing, any one of them assuming a dominant role in a given situation (Jansen and Chernick, 1991). In particular, the cold stimulus has often been cited as an important candidate for triggering continuous respiration at birth (Harned et al., 1970); indeed, in the lamb fetus, stimulation of cutaneous thermoreceptors induces FBM and shivering (Gluckman et al., 1983). This response to inputs from the thermoreceptors could mediate the changes in the state of arousal, in oxygenation, and in metabolic rate (specifically CO_2 production) that accompany the cold-induced thermogenesis. Oxygenation by itself would increase FBM, possibly because of arousal (Baier et al., 1990; Hasan and Rigaux, 1991, 1992a). It is of interest that the increase in PaO_2 (arterial partial pressure of oxygen) in the hyperoxic fetus is accompanied by an increase in $PaCO_2$ (corresponding pressure of carbon dioxide), a phenomenon that could indicate a hyperoxic-induced increase in metabolic rate, as is known to occur postnatally in newborns of numerous species (Dawes and Mott, 1959; Mortola and Gautier, 1995; see Chapter 5, under "Hyperoxia"). Hence, the possibility exists that $PaCO_2$ may have been the respiratory stimulant during the hyperoxic increase in FBM (Blanco et al., 1987a, 1987b; Kuipers et al., 1997b).

The time-honored notion that occlusion of the umbilical cord initiates breathing, and the more recent documentation that reestablishment of the placental circulation abruptly *stops* breathing, even under controlled fetal lung ventilation and oxygenation, are both suggestive that fetal inhibition of breathing is of placental origin (Adamson et al., 1987). In those experiments, however, $PaCO_2$ was not controlled, and in fact rose substantially when the chord was clamped. In a later study, the fetal breathing response to hypercapnia was found to increase when the cord was clamped; Adamson et al. (1991) concluded by reaffirming the importance of the placental inhibition of breathing, possibly induced by a prostaglandin (Kitterman, 1987; Thorburn, 1992), but also postulated that its role is played in conjunction with other factors, perhaps $PaCO_2$ itself. In fetal rats exteriorized just before term, umbilical-cord occlusion, whether in warm or cold environment, was not sufficient to successfully establish a sustained breathing pattern, whereas it was when combined with rhythmic compressions of the chest at pressures comparable to those produced by uterine contractions (Ronca and Alberts, 1995).

Kuipers et al. (1992, 1994, 1997b) addressed the potentially crucial role of CO_2 in driving FBM by studying fetal lambs with extracorporeal membrane oxygenation, in which arterial $PaCO_2$ could be manipulated at constant PaO_2 and temperature. Their finding that breathing did not originate in lambs within 5 minutes of umbilical-chord occlusion, so long as CO_2 was not allowed to rise, contradicts the idea that the initiation of breathing at birth is linked to the abrupt elimination of an inhibitory factor of placental origin; at the same time, it emphasizes the possible role of endogenous CO_2, and therefore metabolic rate, in maintaining rhythmic breathing at birth. Similarly, goat fetuses given extracorporeal circulation and thus disconnected from the original placenta did not shift their intermittent breathing into a continuous pattern (Kozuma et al., 1999). The possibility that an increase in metabolic rate is the underlying drive to a sustained breathing pattern at birth, one in which neuroendocrine factors may act as modulators, seems an appealing one. It could explain the facilitative effect of seemingly unrelated variables, such as cold (which could also lower the threshold to $PaCO_2$, [Moss, 1983]), arousal by tactile or pressure stimuli, and hyperoxia, all of which are known to increase oxygen consumption in the newborn, and the inhibition by hypoxia, which in newborns decreases metabolic rate. Unfortunately, measurements of oxygen consumption have rarely been performed in studies concerned with the mechanisms of the initiation of breathing.

As is the case for the onset of breathing at birth in mammals, the mechanisms of the initiation of breathing at hatching time in birds remain mysterious. Avian embryos seem to make respiratory movements before the breaking of the air cell membrane ("internal pipping") and the onset of lung ventilation (Vince and Tolhurst, 1975). It is known that the gradual increases in hypoxia and hypercapnia toward the end of incubation, as the result of the mismatch between the embryo's metabolic rate and the gas conductance of the eggshell and membranes, are key parameters in the timing of the hatching processes (Visschedijk, 1968a, 1968b, 1968c; Pettit and Whittow, 1982). The reversal of hypoxemia as soon as lung ventilation begins, and probably the hypoxic release of catecholamines (Wittmann and Prechtl, 1991), rapidly raise metabolic rate, which is considered an essential step for the successful completion of hatching.

Some of the hypotheses about the mechanisms of the onset of regular breathing could be more strongly supported or refuted if we had more details of the events surrounding birth. For example, the deep breaths at birth have been interpreted as gasping activity accompanying the transient hypoxia and hypercapnia, in support of the idea that birth-related asphyxia is an important event in the onset of breathing (Bryan et al., 1986). But recent recordings of electromyographic (EMG) activity of the lamb's diaphragm during birth have indicated that the early breathing acts do not have the EMG characteristics of gasps, suggesting to the authors that asphyxia is not a primary mechanism in the initiation of breathing at birth (Berger et al., 1990). Further, a mismatch between the establishment of diaphragmatic activity and the changes in arterial pH would not support acidemia as a key variable in the process. Present information is essentially confined to observations on humans and lambs. A broader view to other species could be illuminating. For example, information on the characteristics of the onset of breathing in monotremes or in marsupials, which, unlike eutherian mammals, have yolk-sac or primitive chorioallantoic placentas, could shed some light on the putative role of placental factors on the initiation of continuous breathing at birth.

Birth, a Hyperoxic Event

The toxic effects of oxygen have been suspected almost since its discovery. More recent, however, is the notion that free radicals of O_2[4] are produced as part of the normal aerobic metabolism, and are increased in conditions of hyperoxia. The lungs, an internal organ in contact with the O_2-rich external environment, are highly susceptible to the cytotoxic effects of O_2 free radicals. In fact, O_2-toxicity is thought to be involved in the pathophysiology of several pulmonary diseases, including neonatal bronchopulmonary dysplasia and the adult respiratory distress syndrome. Normally, a battery of antioxidant defense systems controls cellular O_2 radicals, and among the better known of these are the various species of superoxide dismutase (SOD), catalase (CAT), and glutathione peroxidase (GP) (Fisher, 1988; Sosenko and Frank, 1991).

---

[4] *Free radical* is any species of element capable of independent existence with at least one unpaired electron. In the case of O_2, the most abundant free radicals are superoxide $O_2^{\cdot-}$, hydroxyl radical $HO^{\cdot}$, and hydrogen peroxide (H_2O_2). It is estimated that the mitochondrial production of O_2 free radicals is about 1–2% of oxygen consumption in normoxia, but it can increase severalfold in hyperoxia (Turrens et al., 1982).

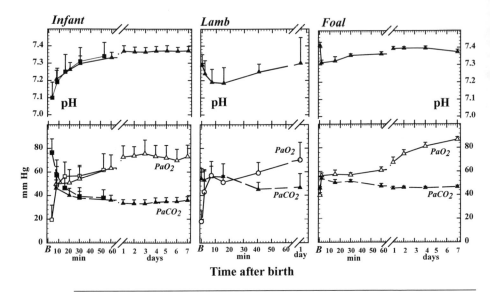

Figure 1.9. Arterial values of the partial pressures of oxygen (PaO$_2$) and carbon dioxide (PaCO$_2$) (*bottom panels, open and filled symbols,* respectively), and pH (*top panels*) in normal infants, lambs, and foals during the first hours and days after birth (*B*). Symbols are mean values; bars are 1 SD. Data for infants from Oliver et al., 1961, and Polgar and Weng, 1979; for lambs, from Berger et al., 1990; for foals, from Stewart et al., 1984.

Birth in mammals, like hatching in birds, is an example of physiological hyperoxia. In fact, within hours, arterial PaO$_2$ rises three to four times the fetal values of ~20–25 mm Hg (Fig. 1.9). It is therefore perhaps not too surprising that in preparation for birth the levels of the antioxidant enzymes increase (Fig. 1.10), as a result of gene upregulation (Clerch and Massaro, 1992). The oxygen consumption of the lung tissue itself does not appreciably increase during the last phase of gestation (Frank and Groseclose, 1984). The rise in the pulmonary antioxidant enzymes, which begins in the last 10–15% of gestation, almost coincides temporally with the development of the surfactant system, and it is possible that controlling mechanisms not now known are common to both systems (Sosenko and Frank, 1991). But the gender difference in the development of surfactants, which appear earlier in females, does not apply to the antioxidant enzymes (Sosenko et al., 1989).

The earlier observations on fetal rabbits and rats (Frank and Grose-

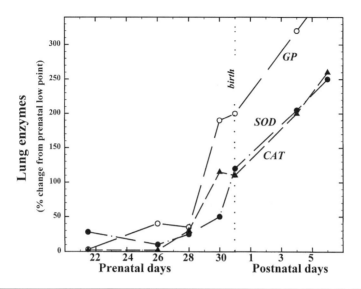

Figure 1.10. Developmental changes in the activity of the primary antioxidant enzymes from the lungs of fetal and newborn rabbits. The enzymes are CAT, catalase; SOD, superoxide dismutase; and GP, glutathione peroxidase. Values are expressed as a percentage of the lowest measured fetal value. Slightly modified from Frank and Groseclose, 1984.

close, 1984; Tanswell and Freeman, 1984; Gerdin et al., 1985) have since been extended to hamsters, guinea pigs, and lambs (Frank and Sosenko, 1987; Walther et al., 1989). In the hamster the prenatal rise of SOD, CAT, and GP is less than that in other newborns, and in the fetal guinea pig it peaks at an unusually early time (Fig. 1.11). To what extent these differences have some relation to the natural history of these species, the hamster being semifossorial and the guinea pig notoriously precocial, is only a matter of speculation. It is nevertheless a curious coincidence that the newborns of these two species, contrary to the newborns of many others, are as susceptible to hyperoxia as adult mammals are, being unable to induce antioxidants (Frank et al., 1978) (Fig. 1.12). Before speculating on mechanisms, it would be informative to extend further these comparative observations to newborns of other species that develop in hypoxic or otherwise unusual environments, such as burrowers, high-altitude mammals, deep-diving mammals, and various marsupials.

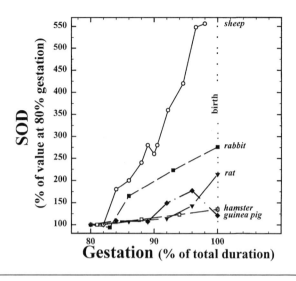

Figure 1.11. Changes in the activity of superoxide dismutase (SOD) in some species during the last portion of gestation, expressed as a percentage of the value at 80% gestation. From data of Frank and Sosenko, 1987, and Walther et al., 1991.

Whether the low antioxidant activity of the premature makes it more susceptible to O_2 toxicity is not necessarily borne out from the above data. In fact, it is known that oxygen tolerance is closely related to the ability to induce antioxidant activity, rather than to absolute normoxic level. Currently, the only data on O_2 tolerance in prematures are from guinea pigs and rabbits, and the results are discordant (Sosenko and Frank, 1987, 1990). In the guinea pig, unlike the rabbit, premature pups are more O_2-tolerant and more capable of elevating antioxidant enzymes than full-term newborns are. Incidentally, this emphasizes once more the peculiarity of those precocial pups, which at birth have the appearance of miniaturized adults.

Mechanics of the First Breath

In the early 1960s, studies at the Pediatric Clinic of the University of Stockholm provided very valuable information regarding the mechanics of the first breath. Fawcitt, Lind, and Wegelius (1960) took serial roentgenograms of full-term infants. They reported that both diaphragm

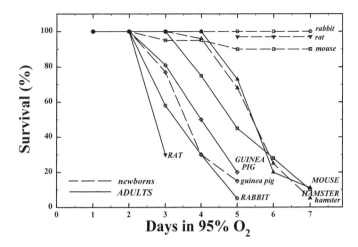

Figure 1.12. Survival rate in hyperoxia in newborns and adults of five species. Newborn rats, mice, and rabbits have an extraordinary resistance to oxygen toxicity, whereas newborn guinea pigs and hamsters are as sensitive as adults. Slightly modified from Frank et al., 1978.

and upper-airway muscles seemed to be involved in the initial air infla-tion of the lung. Before lung inflation, the rib cage is sucked inward, par-ticularly in its upper portion. During a successful inspiration the in-trathoracic trachea dilates, and air then fills first the posterior portions of the lung bases. One or more lobes of the lung may fail to aerate with the first breath. During the first expiration a high degree of de-aeration oc-curs, although some air remains in the lung and variable closure of the pharynx-larynx may be seen.

At about the same time, Karlberg and associates (Karlberg et al., 1962b) attempted to record the changes in lung volume (with a reverse plethysmograph) and in pleural pressure (with a water-filled catheter placed in the esophagus) during the onset of breathing of full-term in-fants born vaginally. They succeeded in 11 cases out of 79 attempts. From these and subsequent measurements (Milner and Saunders, 1977; Saun-ders and Milner, 1978; Vyas et al., 1981; Mortola et al., 1982a) emerged the following picture of the first P_L-V loop (transpulmonary pressure and volume, shown schematically in Fig. 1.13): (1) changes in P_L and V usu-ally begin at the same time; (2) peak P_L is elevated, between 30 and up

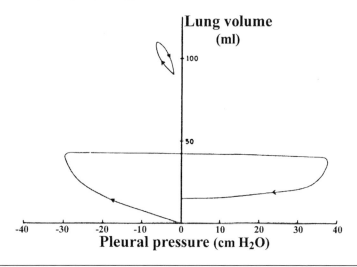

Figure 1.13. Schematic representation of pleural pressure-volume loop during first postnatal breath. Note the large negative and positive pleural pressure swings and the establishment of functional residual capacity at zero pressure. For purposes of comparison, a breathing cycle at one day after birth is also represented (*upper loop*). From Mortola, 1987.

to 100 cm H_2O, which is much higher than the 5–7 cm H_2O P_L swing of the few-days-old infant (Cook et al., 1955; Cook et al., 1957; Dinwiddie and Russel, 1972; Asher et al., 1982); (3) V_T (tidal volume) averages ~35–45 ml (i.e., less than the infant's later vital capacity during crying but almost twice its V_T during resting breathing); (4) in the first phase of expiration, a large positive airway pressure is generated; and (5) a substantial amount of air, on average 15 ml, or ~40% of the inhaled air, is retained in the lungs at the end of the first expiration.

Inspiration. In human infants, the amount of air inhaled with the first inspiratory effort can vary considerably; on average, it is slightly above 40 ml (Karlberg, 1960; Karlberg et al., 1962b; Milner and Saunders, 1977; Saunders and Milner, 1978; Mortola et al., 1982a), that is, greater than the resting V_T seen after a few days, and probably one of the deepest breaths of the whole neonatal period (Fig. 1.14). Why this is so, despite the obvious mechanical constraints, is unclear. One may speculate, with Cross (1961), that it represents a manifestation of Head's paradoxi-

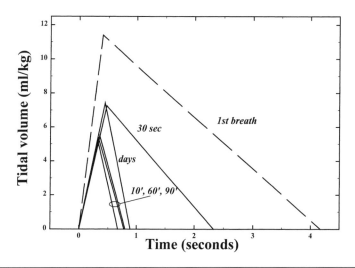

Figure 1.14. Schematic representation of the average spirogram of the first breath (*broken line*), and of those at 30 seconds, at 10, 60, and 90 min, and at a few days after birth. The first breath is commonly one of the deepest and slowest of the whole neonatal period, whereas at 10–90 min the breathing rate tends to be high.

cal reflex,[5] whereby a sudden inflation of the lungs promotes further inspiratory activity. It could also be hypothesized that during the first lung inflation the respiratory controller does not yet fully recognize the inspiratory-inhibitory inputs from the airway's slowly adapting receptors (see Chapter 4, under "Airway Receptors"), since it had never previously been exposed to such a stimulus. This may seem to contradict with the recent notion that the fetal lung volume exceeds the postnatal lung volume (see Fig. 1.6, above), but airway stretch receptors are activated by transpulmonary pressure, not lung volume, and the former is undoubtedly much greater postnatally than it is prenatally.

Muscle pressure during the first inspiration is directed chiefly against overcoming the elastance of the lung tissue, the surface forces, and the frictional resistance to the movement of air and fluid in the tracheobronchial tree. Tissue resistance is probably of relatively minor impor-

---

[5] A vagal reflex of pulmonary origin first described by Head (1889) (see Chapter 4, under "Vagal Reflexes").

tance, as shown by the fact that the pressure required to expand the lungs of human fetuses (Gruenwald, 1947) or animal fetuses (Agostoni et al., 1958) with liquid is only a few centimeters of H_2O. The surface tension (γ) of plasma is about 55 dyne/cm, and the maximal γ during the expansion of fluids washed out from the peripheral always approaches this value (Clements, 1957). Hence, during inspiration, the role of the pulmonary surfactant[6] in reducing γ is likely to be small, although by no means negligible. For example, full-term rabbits (having passed 29–30 days of gestation) can inhale, during the first inspiratory effort, about 0.6 ml of air by generating 35 cm H_2O of negative esophageal pressure, whereas slightly premature rabbits (28 days) can inhale only one-tenth of that volume for similar PL (Lachmann et al., 1979). Similarly, passive inflation of paralyzed rabbit fetuses at term requires only a fraction of the pressure needed to aerate the lungs of premature pups (Humphreys and Strang, 1967; Nilsson, 1979).

The flow resistance of the first inspiration must in any case be much higher than that faced during resting air breathing in later days, because of the high viscosity of the column of pulmonary fluid filling the airways. Agostoni and co-workers (1958) calculated that in the guinea pig fetus at flow rates of 0.7 ml/s (which are lower but within the same order of magnitude, on a per kg basis, of those recorded in infants) the flow-resistive pressure is ~60 cm H_2O. These authors pointed out that, although the pressure losses due to γ and those due to flow resistance are additive, their maximal values do not occur in the same regions of the tracheobronchial tree. During the progression of the air-liquid column toward the lung periphery, the pressure drop across the meniscus increases progressively, since the airway radius decreases,[7] whereas flow resistance decreases, since the total sectional area of the airways increases and the flow velocity of the pulmonary fluid drops (Fig. 1.15). Hence, the total pressure needed during inspiration as the air-liquid meniscus moves toward the lung periphery may remain within narrow limits. But this conclusion

[6] An informative and entertaining historical review of the discovery of the role of surfactants in pulmonary physiology was written by J. H. Comroe (1977a, 1977b, 1977c).

[7] According to the Young-Laplace relationship, P is proportional to γ/r, r being the radius of curvature of the air-liquid interface and P the pressure generated by the surface tension γ.

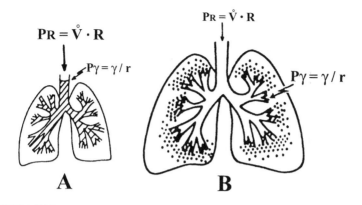

Figure 1.15. Before the newborn's first inspiration (*A*), the air-liquid interface is in the large airways. Hence, the flow-resistive pressure (P$_R$ = airflow $\dot{V}$ · resistance R) is high, whereas the pressure Pγ determined by the surface tension γ is low because the radius of curvature at the interface (r) is large. As the fluid moves toward the smallest airways (*B*) and eventually into the interstitium (*dotted area*), Pγ progressively increases and P$_R$ decreases.

does not account for additional sources of energy dissipation, such as those caused by the shape distortion of the rib cage, the viscous resistance of the chest, the decreased lung compliance due to lung distortion, and the inertia of the pulmonary fluid.

Simple physical considerations can explain how the small inspiratory muscle mass of the newborn can generate the large pressures required for the first inspiration (see Chapter 3, under "Mechanical Properties"). As in infants, also in lambs large negative pleural pressures have been measured during the onset of breathing at birth (Berger et al., 1990). It would be of interest to extend these observations to very small altricial species, such as the marsupials Virginia opossum or Julia Creek dunnart (see Figs. 2.1 and 3.14), which are fully capable of air breathing after only 12–13 days of gestation; such measurements, however, are at present confronted by insurmountable technical challenges. Interspecies comparisons would probably permit some generalizations about the key features of the establishment of breathing at birth, information that should also help us to understand of the mechanisms responsible for the establishment of a steady ventilation.

The uneven distribution of air during the first lung expansion has been noted in isolated lungs from human infants or animal fetuses (Agostoni et al., 1958; Gruenwald, 1947, 1963; Scarpelli et al., 1981) and in infants during radiographic observations (Fawcitt et al., 1960). Expansion of the rabbit fetal lungs with positive airway pressure has revealed that some areas pop open as others remain unexpanded (Scarpelli et al., 1981), a pattern similar to that observed during air expansion of gas-free adult lungs and not seen during liquid filling (Mead, 1961). A sudden expansion is the expected behavior once the radius of the air-liquid interface of the region (e.g., an alveolar duct) exceeds that of the airway leading to it (e.g., a terminal bronchiole). In this transition the air space passes from a situation of geometric stability (in which the radius decreases with the increase in lung volume) to an, initially, unstable situation (Mead, 1961) (Fig. 1.16).

Some calculations based on the PL-V loop would indicate that the work involved in the first expansion of the lungs is not extraordinarily high, corresponding to approximately the work of a cry at a few days of life (Karlberg et al., 1962b). This information, however, is probably misleading to the extent that it represents only the external ventilatory work; the internal work done in distorting the rib cage, not apparent on the PL-V diagram, is probably a large fraction of the total work performed by the inspiratory muscles.

What other factors, besides the contraction of the inspiratory muscles, could help the first inspiration? Karlberg and co-workers (Karlberg, 1960; Karlberg et al., 1962a) thought that the outward recoil of the chest immediately after its squeezing through the pelvic canal may favor the passive entrance of some air into the lungs even before the first actual breath. This possibility seemed also supported by high-speed radiographic studies during the establishment of ventilation (Bosma and Lind, 1960) and by the observation, in two infants, that the amount of air exhaled after the first breath exceeded the volume inhaled (Karlberg et al., 1962b). Although in the large majority of infants studied by Karlberg and co-workers (Karlberg et al., 1962b), and as also seen in subsequent measurements (Milner and Saunders, 1977; Saunders and Milner, 1978; Vyas et al., 1981; Mortola et al., 1982a), the amount of air exhaled never exceeded that previously inhaled, it is not inconceivable that the chest wall outward recoil may contribute to some passive entrance of air into the upper airways. For this component to become a relevant aspect of the

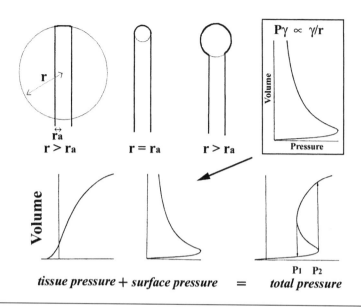

$$P\gamma \propto \gamma/r$$

tissue pressure + surface pressure   =   total pressure

Figure 1.16. *Top:* Pressure-volume relationship of an idealized alveolus, in which surface tension is the determinant of the component of recoil pressure due to surface forces (Pγ). At first, with the increase in air volume, the radius of curvature of the alveolus r decreases, which implies an increase in Pγ according to the Young-Laplace relationship. Once r reaches its minimal value (i.e., r equals the radius of the airways ra), further increases in volume are accompanied by an increase in r and a decrease in Pγ. *Bottom:* The surface phenomena described in the top panels (*panel at bottom center*) and the elasticity of the tissue (*panel at bottom left*) combine in determining the pressure-volume curve of the alveolus (*panel at bottom right*). A unit of this type would be unstable, since during inflation (at P2) it can suddenly "pop open," and during deflation (at P1) it can suddenly collapse. Surfactants prevent this instability by varying γ according to r.

first lung expansion, however, the chest must be squeezed to very low volumes, since at birth it is a highly compliant structure with an undefined resting position, whereas the pressure required to inhale, as discussed previously (see Fig. 1.13), is great.

Additional mechanisms proposed as possible contributors to the entrance of air into the lungs include an active inflation through contraction of the pharyngeal muscles, resembling the swallowing-like respiration of the frog (Bosma et al., 1959), and the erection of the pulmonary capillaries as lung volume expands and pulmonary vascular resistance falls (Jaykka, 1958). Glossopharyngeal respiration has been observed in patients with respiratory poliomyelitis (Dail et al., 1955; Burke, 1957), but it seems to be a rather slow and inefficient means of achieving sizable effects. The inspiratory pressure generated by the erection of the pulmonary capillaries with the increase in blood flow is less than 5 cm H_2O (Avery et al., 1959), which is too little to be of any practical significance.

In fact, modifications in the cardiovascular system have little bearing on the establishment of ventilation at birth. Rather, the contrary is true, that is, the onset of ventilation per se, even without the associated increase in oxygen level, has remarkable effects on the adjustments of the pulmonary circulation (see "Implications for Pulmonary Circulation," below). If a sizable amount of air was present in the lung before the first inspiration, the air-liquid interface would be placed in small airways and would thus generate some recoil pressure, in accordance with the Young-Laplace relationship (see preceding footnote). Hence, one would expect some pressure change before any change in volume, that is, an opening pressure, but during the first inspiratory effort in most of the cases the onset of volume change did not lag behind that of PL (Karlberg et al., 1962b; Milner and Saunders, 1977; Saunders and Milner, 1978; Vyas et al., 1981). Neglecting flow resistance, one would expect peak PL to be achieved only when the air-liquid interface is approaching the first gas-exchange units, hence when a volume approximately equivalent to the dead space volume has been inhaled. Since the flow-resistive pressure is considerable and progressively decreases as the column of fluid moves toward the lung periphery (see Fig. 1.15, above), the peak pressure may be expected to occur after a change in lung volume somewhat smaller than the dead space volume. This may explain why the inflation of isolated lungs, in which the dead space is much reduced, may yield higher values of peak-inflation pressures (Agostoni et al., 1958). These pressure changes, therefore, rather than true opening pressures (Milner and Vyas, 1982; Faridy, 1987) probably represent the sum of elastic and resistive pressures, the former including the pressure due to surface-tension phenomena.

Expiration. The expiration following the first inspiration is often slow and protracted (Karlberg et al., 1962b; Milner and Saunders, 1977; Mortola et al., 1982a) (see Fig. 1.14, above). It may be accompanied by a narrowing, or occasionally a total closure, of the upper airway passage, a pattern clearly documented for the first hours of extrauterine life in several newborn species (see Chapter 2, under "The First Hours after Birth"). Consistent with these reports are radiological observations (Fawcitt et al., 1960), and high positive values of esophageal pressure have been recorded (see Fig. 1.13, above), values much higher than expected from the passive recoil of the chest wall against the increased upper airway resistance. Similarly, high positive values of pleural pressure were recorded in lambs (Berger et al., 1990). This finding indicates active recruitment of the expiratory muscles in generating a positive airway pressure that can promote clearing of the fluid from the lung and perhaps a more even lung expansion. In tracheostomized rabbits, during the first expiratory phase no positive esophageal pressures were recorded (Lachmann et al., 1979), presumably because an increase in upper airway resistance is required for the generation of these pressures.

Establishment of the Functional Residual Capacity

The amount of air retained in the infant lung after the first expiration (Fig. 1.17) averages, in different studies, ~11–19 ml (Karlberg et al., 1962b; Milner and Saunders, 1977; Saunders and Milner, 1978; Vyas et al., 1981; Mortola et al., 1982a). This is a considerable amount, since it represents 10–20% of the functional residual capacity (or end-expiratory level, FRC) of the infant a few days old. Although the variability among the infants studied is rather great, there does not seem to be a systematic difference between infants delivered vaginally and those by cesarean section; if present, the difference is small (Vyas et al., 1981; Mortola et al., 1982a).

The fact that some air remains in the lung after the first breath is an expression of alveolar stability mostly attributable to the presence of surfactants. As lung deflation progresses and the tissue recoil diminishes, the mechanical behavior of the peripheral lung units (or of those spaces where surface tension phenomena occur) becomes more and more dependent on the properties of the air-liquid surface. Since at intermediate lung volumes the component of lung elastance due to surface tension γ has a negative sign (i.e., an increase in volume, hence in surface radius, is accompanied by a decrease in the surface tension component of the

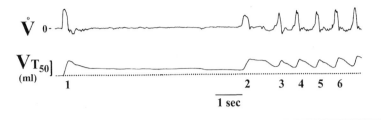

Figure 1.17. Recordings of the change in lung volume (VT, *bottom*) and corresponding airflow (V̊ *top*) of the first six breaths of a human infant after birth. Note the difference between the end-expiratory level and the starting lung volume (*dotted line*), which represents the early phases in the process of establishing a functional residual capacity.

transpulmonary pressure), in this region of the lung's P-V curve the air spaces are in an unstable state, tending either to collapse or to expand toward stable conditions (see Fig. 1.16, above). Clements and associates (Clements, 1957; Clements et al., 1961) demonstrated that the surface-active properties of lung surfactant depend not only on surface area but also on the direction of its changes, γ being markedly reduced during surface area reduction. The decreased surface tension during lung deflation not only decreases the lung recoil pressure at any given lung volume but also reduces the nonstable region of the P-V curve. Both mechanisms reduce air space stresses and favor retention of air during expiration. Hence, the absence of surfactant not only decreases lung compliance and alveolar stability, but also compromises air-retention in deflation (Fig. 1.18), and therefore the establishment of FRC.

It was later emphasized (Scarpelli, 1978) that during the first breath the classical air-bulk liquid model is a too-simplistic view of the air-pulmonary fluid interaction in the fetus. In fact, the dispersion of air in liquid leads to the formation of foam, a process similar to that previously observed during ventilation of excised lungs of adult dogs and demonstrated to be dependent on the presence of surfactant (Faridy and Permutt, 1971). Stereomicroscopy of rabbit or lamb fetal lung during passive inflation from the natural liquid-filled state indicated that foam is already formed in the peripheral airways with the first inflation. Foam was also found shortly after birth in the airways of spontaneously breath-

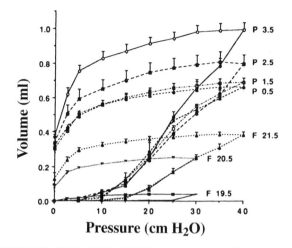

Figure 1.18. Static pressure-volume relationship of the respiratory system in fetal (F) and postnatal (P) rats of different ages. Numbers refer to the day of fetal (term = 22 days) or postnatal age. Error bars are 1 SEM. Note that on deflation, more air is retained in the lungs of the postnatal than in the fetal rats. Slightly modified from Suen et al., 1994.

ing pups (Scarpelli, 1978) (Fig. 1.19). The presence of air bubbles in the lung periphery explains why a rather large amount of air can be trapped in the lung at end-expiration. This, in turn, contributes to the ability of the newborn's lung to act as an effective gas exchanger following the first breaths (Oliver et al., 1961), despite the extended time required for the absorption of the pulmonary fluid (see "Fetal Pulmonary Fluid," above). In any case, immature lungs retain less air and produce less foam than do mature lungs (Humphreys and Strang, 1967; Kotas and Avery, 1971; Lachmann et al., 1979; Scarpelli et al., 1979, 1981).

The relationship between lack of surfactant and the lung's inability to retain air at end-expiration, as an important aspect of the pathophysiology of the infant's respiratory distress syndrome (RDS, formerly called hyaline membrane disease), is well established (Avery and Mead, 1959; Brumley et al., 1967; Gribetz et al., 1959; Gruenwald, 1963; Scarpelli et al., 1971). Conversely, intratracheal instillation of surfactant material to premature lambs restores their lungs' P-V curve, improving in particular the amount of air retained at deflation (Ikegami et al., 1980). The lungs

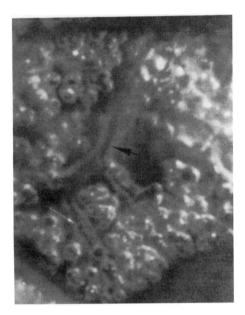 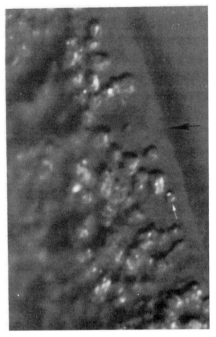

Figure 1.19. Stereomicroscopy of the lungs during inflation. *Left:* inflation of the lungs of a mature rabbit fetus: foam and air bubbles are mixed with some air columns (*indicated by arrows*). *Right:* lung surface at the resting volume of a newborn rabbit that had breathed spontaneously for 45 sec. The flat, homogeneous surface (*black arrow*) indicates areas still filled with liquid; the raised or distended surface is of areas that contain foam (*white arrow*) and air. From Scarpelli, 1978.

of rabbit pups that breathed spontaneously for a few minutes after birth had a smaller ratio of lung weight to body weight and a more pronounced alveolar stability than those of pups of the same gestational age examined before the onset of spontaneous breathing (Kotas et al., 1971; Taeusch et al., 1974). These results emphasize the fact that adequate foam formation requires not only lung distention, which is known to promote fluid clearance and the release of surfactant (Egan et al., 1975; Faridy, 1976; Lawson et al., 1979; Hildebran et al., 1981; Massaro and Massaro, 1983), but also the occurrence of labor itself, with its associated hormonal effects and perhaps other, yet undefined factors related to the onset of spontaneous respiration.

Since the newborn's lung tissue has pronounced viscous-elastic properties (see Chapter 3, under "Mechanical Properties of the Lungs"), one would expect the tissue hysteresis to favor a larger V, at any given PL, during deflation than during inflation, therefore contributing to the retention of air during expiration. The importance of this plastic behavior of the newborn lung in promoting the establishment of FRC at birth is controversial (Agostoni et al., 1958; Agostoni, 1959; Taeusch et al., 1974; Scarpelli et al., 1981) and rather difficult to assess in quantitative terms. Additional extrapulmonary factors that could contribute to the establishment of FRC may include an augmented outward recoil of the chest, possibly promoted by geometric changes attending lung expansion, and active control mechanisms aiming to keep mean lung volume elevated, such as the expiratory increase in upper airway resistance (see Chapter 2, under "The First Hours after Birth").

Implications for Pulmonary Circulation

Probably all vascular beds undergo a major change in the transition from fetal to neonatal life, and in the lungs these changes are among the most dramatic. Having received only a small blood flow before birth (less than 10% of cardiac output is needed to provide for the metabolic needs of the growing organ), after birth the lungs will be functionally in series with the systemic circulation, receiving almost the whole cardiac output. This shift is achieved through a major drop in vascular resistance accompanied by a parallel rapid decrease in pulmonary arterial pressure, which in infants is close to the adult value within 3 days after birth (Emmanouilides et al., 1964). In calves, pulmonary arterial pressure drops from 80 to 30–40 mm Hg within 9–10 hours after birth (Reeves and Leathers, 1964). Because the flow through the *ductus arteriosus* and, with it, the right-to-left shunt both decrease, blood flow toward the abdomen and lower extremities drops in favor of the upper body and brain (Fig. 1.20). Hence the changes in pulmonary blood flow at birth have implications for the redistribution of blood flow throughout the body.

The newborn lamb, goat, and pig have been, almost without exception, the preferred animal models for research studies in perinatal pulmonary circulation. The primary object of research has been the identi-

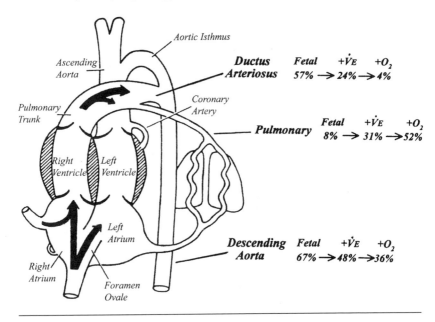

Figure 1.20. Diagrammatic representation of the fetal circulation, with direction of flow indicated by arrows. Numbers refer to the percentages of the combined ventricular output (left + right) in a sheep fetus before and during mechanical lung expansion, without changes in blood gases or combined with increased oxygenation. From data of Teitel et al., 1987.

fication of the factors involved in the vascular changes seen at birth, and their mechanisms of action. Some, like the lung expansion and onset of ventilation, the rise in oxygenation, the bypassing of the placental circulation and interruption of umbilical blood flow, have been known for a long time. Other factors, like the metabolic factors of pulmonary origin (e.g., vasoactive prostaglandins and endothelium-derived nitric oxide), have been appreciated and studied only recently, and their role is still under scrutiny (Tod and Cassin, 1997).

The onset of ventilation per se, even setting aside the associated increase in oxygen level, has remarkable effects on the adjustments of the pulmonary circulation, with consequences for the whole cardiovascular system (Adrian et al., 1952; Assali et al., 1963; Rudolph, 1979; Teitel et al., 1987, 1990; Serwer, 1992). Because the total volume of the liquid-

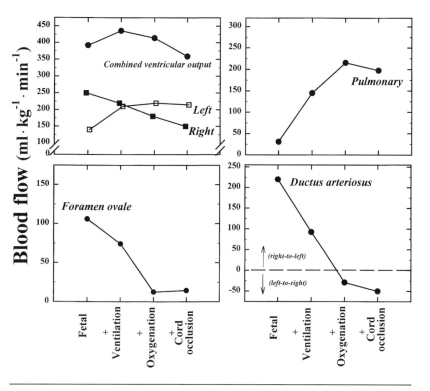

Figure 1.21. Blood flow to various segments of the vascular system of the sheep fetus as the result of ventilation only, ventilation with 100% O_2 (oxygenation), and oxygenation plus clamping of the umbilical cord. The onset of air ventilation is by itself a key factor in increasing pulmonary circulation, irrespective of the increase in oxygenation. Computations based on data of Teitel et al., 1987.

filled fetal lungs at term exceeds the lung volume of the newborn (see Fig. 1.6, above), it is conceivable that at birth the drop in pulmonary vascular resistance is due not to lung expansion per se, but rather to the increase in lung recoil pressure. With the air expansion of the lungs, the formation of surface tension and the increase in alveolar recoil reduce the interstitial pressure on the pulmonary capillaries; the effect, at least in the experimental lamb, is an astonishing ~fivefold increase in pulmonary blood flow (Fig. 1.21). In addition to the mechanical effect on the pulmonary vasculature, lung ventilation promotes the synthesis and release of prostacyclin and possibly other prostaglandins, with powerful vasodilating effects on the pulmonary vascular bed (Leffler et al., 1978, 1984a, 1984b; Velvis et al., 1991). The important role of oxygenation in decreasing pulmonary artery pressure and vascular resistance has long been appreciated (Cassin et al., 1964; Campbell et al., 1967; Clarke et al., 1990) (see Figs. 1.20 and 1.21). Hence, lung ventilation and oxygenation are key factors in the drop of the pulmonary vasculature's resistance, although the former would affect it chiefly by increasing blood flow, and the latter by decreasing pulmonary arterial pressure (Teitel et al., 1990). The opposite is equally true; that is, hypoxemia increases vascular resistance in both the fetus and the newborn.

The quantitative aspects of the pulmonary vascular response to hypoxia, which is qualitatively comparable to that in the adult, and its developmental changes have been addressed by several studies, with conflicting conclusions (Tod and Cassin, 1997). In calves, hypoxia could induce an important pulmonary hypertension during the first hours after birth, and a lesser effect at older postnatal ages (Reeves and Leathers, 1964). Acute acidosis and alkalosis provide, respectively, a synergistic and a blunting effect on hypoxic vasoconstriction (Lyrene et al., 1985; Schreiber et al., 1986), indicating that the acute alkalosis experienced at birth (see Fig. 1.9, above) is an additional factor in the pulmonary vasculature adjustments required at birth. The occlusion of the cord, however, adds little to the pulmonary vascular changes determined by ventilation and oxygenation (Teitel et al., 1987, 1990) (see Fig. 1.21). Finally, the role of airways and alveolar pressure is intriguing. On the one hand, an increased alveolar pressure, as during mechanical ventilation, can, by squeezing the interstitium and the capillaries, increase vascular resistance and arterial pressure. On the other hand, some positive airway pressure on the alveolar side could represent an additional back force against the pulmonary fluid in the direction of absorption. Experimental evidence suggests that indeed the latter occurs, as long as the lungs are in "functional zone 3" (i.e., when alveolar pressure is small and not higher than left atrial pressure) (Fike and Lai-Fook, 1990).

The grunting-like pattern in the first hours after birth (see "Mechanics of the First Breath," above, and Chapter 2, under "Mechanical Events in The First Hours after Birth") could favor liquid absorption by generating small positive airway pressures. Equally, the end-expiratory airway pressure could remain slightly elevated whenever breathing rate is high (see Chapter 3, under "Expiration"); in this respect, the rapid breathing in infants who experience a delay in lung fluid clearance ("transient tachypnea syndrome"), could favor liquid absorption. A grunting-like pattern has also been reported in foals with neonatal respiratory distress (Mahaffey and Rossdale, 1959; Rossdale et al., 1967). It is of interest that in reptiles, in which filtration of fluid into the air spaces is a constant menace, owing to their high pulmonary arterial pressure and low protein osmotic pressure, the breathing pattern is characterized by end-inspiratory glottis occlusions with low positive airway pressure (Bartlett et al., 1986); during this expiratory phase, the fluid that may have filtrated during inspiration is reabsorbed (Burggren, 1982).

In marsupials, despite their short gestation, light and electron microscopy observations have indicated that cardiovascular changes in the perinatal period are, in essence, similar to those of eutherian mammals (Runciman et al., 1995). Right-to-left shunts include the *ductus arteriosus*, despite assertions to the contrary (Bernard et al., 1996), and intraatrial communications. But in eutherian mammals the *foramen ovale* is closed very rapidly by the flap valve of the primary and secondary septum, whereas in marsupials the closure of the intra-atrial shunt requires tissue proliferation over a period of several days.

Interspecies Comparisons

The larger a species, the longer will be its gestation period and the bigger the pups at birth. These generalizations, however, have numerous exceptions, both among and within mammalian orders; furthermore, the definition of the gestation period is complicated, in some groups, by phenomena like the delay in egg fertilization or implantation. The mode of delivery (headfirst or tailfirst) is not important for the onset of breathing, and it is quite variable among species. But at the same time, labor and its associated hormonal changes are important for the establishment of adequate ventilation and normalization of blood gases. Before birth, in preparation for the acute hyperoxia that accompanies the onset of air breathing, most, but not all, species increase the antioxidant enzymes. Among the exceptions are the hamster and the guinea pig, which as newborns have low antioxidant levels and are very sensitive to oxygen toxicity. After birth, pulmonary arterial pressure in mammals decreases probably more slowly in the smaller species, whereas the absorbance of the fetal pulmonary fluid is likely to be faster in the smaller species. In marsupials, despite their dramatically altricial characteristics, the various processes surrounding the onset of breathing at birth are very similar to those of eutherian mammals; only the smallest newborn marsupials have an important level of gas exchange through the skin.

Clinical Implications

In the fetus, the production of pulmonary fluid and the rise of intermittent breathing movements are crucial for the normal prenatal growth of the lungs; abnormalities in the lungs, although inconsequential prenatally, can lead to lung hypoplasia, with catastrophic consequences for the viability of the newborn. The stress of labor and delivery, with the associated surge in catecholamines, is important for the adequate adaptation of the pulmonary and cardiovascular systems at birth, including the processes of fluid absorption and surfactant production, whereas the *mode* of delivery (head or breech presentation, vaginal or C-section) is of minor importance to its survival. In prematures, the immaturity of the surfactant system is responsible for low lung compliance, alveolar instability, and poor establishment of a functional residual capacity; these abnormalities eventually lead to hyaline membrane disease (or respiratory distress syndrome). In these cases, the mechanical trauma of prolonged artificial ventilation and the administration of oxygen can contribute to the development of a chronic lung injury, known as bronchopulmonary dysplasia (BPD). Oxygen toxicity is in fact a serious problem of oxygen therapy, probably even more in prematures because of the meagre antioxidant defense mechanisms of these newborns. Because lung expansion and oxygenation contribute to the postnatal fall in pulmonary vascular resistance, BPD is commonly associated with pulmonary hypertension and right-ventricular hypertrophy.

Summary

1. Both the duration of gestation and the body mass of the newborn at birth, in first approximation, are proportional to the size of the adult, although many exceptions are known. Marsupials differ from all other or-

ders of live-bearing mammals in their shorter gestations and the much smaller size of their offsprings at birth.

2. Before birth, the lung is a nonvital organ. Two fundamental prenatal mechanisms, however, fetal breathing movement and pulmonary fluid secretion, are essential for the normal development of the lung.

3. Fetal breathing movements occur only intermittently, depending on sleep state, and are sensitive to chemical inputs and other stimuli. The mechanisms responsible for their change in frequency during gestation, their reduction at term, and their complete cessation during labor are not yet understood. Experimental evidence suggests that CO_2 may represent an important tonic "drive" for fetal breathing, a drive modulated by factors related to sleep state. Prostaglandins of placental origin are thought to be involved in various aspects of cardiorespiratory control before birth.

4. Fetal pulmonary fluid production and reabsorption are finely controlled processes, the action of epinephrine on epithelial ion transport being the primary factor. The onset of labor, more than the route of delivery, is important in stimulating epinephrine release and in reversing the process of pulmonary fluid secretion in the fetus into reabsorption in the newborn.

5. What triggers the onset of regular respiration at birth remains one of the unsolved mysteries of the regulation of breathing. Some evidence favors the possibility of a metabolic role, possibly via changes in CO_2, but numerous other mechanisms, including that of a respiratory inhibition by a placental factor, have been proposed.

6. The first inspiration of the human infant requires high transpulmonary pressures to overcome mainly surface and viscous forces, and it is very deep, despite associated mechanical constraints.

7. The squeezing of the chest wall during delivery and the following outward recoil have minimal importance in the processes of lung aeration at birth, whereas labor itself is important because of the associated norepinephrine release. In many species, tail first is a common mode of delivery.

8. At birth, the expansion of the lungs and their ventilation by air play the main role in lowering pulmonary vascular resistance, resulting in a major increase in pulmonary blood flow. They are also a fundamental factor in the redistribution of systemic blood flow from that of the fetus to that of the neonatal condition.

9. The expiratory phase of the first breath is usually long, with positive airway pressure, probably because of the activation of the expiratory muscles against an increased upper airway resistance. The amount of air inhaled exceeds that exhaled, the difference representing the first important contribution to the establishment of the functional residual capacity. This process is crucially dependent on the function of surfactants. The dispersion of air into the surfactant-pulmonary liquid solution is the cause of intra-airway formation of air bubbles and foam, which are probably important for gas exchange during the transition from the liquid-filled to the air-filled state of the lungs.

10. Birth is the only hyperoxic exposure physiologically occurring during life. Numerous mechanisms are available following the last phase of gestation to counteract the potential risk of oxygen- toxicity associated with birth. The rapid induction of enzymes scavenging O_2 free radicals is one of the most important mechanisms in establishing the resistance of newborns to oxygen-toxicity, a resistance much greater than that of adults.

Chapter 2

Metabolic and Ventilatory Requirements

The design of the mammalian respiratory system, with the lungs representing the sole possible avenue for gas exchange, implies the existence of a close link between aerobic metabolism and pulmonary air convection. Indeed, in adults, changes in pulmonary ventilation follow closely the changes in metabolic rate determined, for example, by muscle exercise, cold exposure, aging, or pharmacological interventions. And at least during resting conditions the differences in metabolic requirements among adults of different mammalian species find a close correspondence in the interspecies differences in pulmonary ventilation.

In this chapter we will consider the metabolic needs of newborn mammals of various species, and to what extent these requirements are met by corresponding changes in pulmonary and alveolar ventilation. The effects wrought on these variables by changes in ambient temperature and oxygenation level will be addressed in Chapter 5.

Oxygen Consumption and Carbon Dioxide Production

The energy required by the newborn mammal, as for all animals, is derived chiefly from the aerobic transformation of food into catabolic products. For carbohydrates (CH), for example, the equation of full oxidation is

$$CH + O_2 = H_2O + CO_2 + Energy,$$

(Equation 2.1)

the stoichiometry of which depends on the type of carbohydrate substrate in question. This equation gives a general overview of the processes of energy production, but it can be misleading if taken in a strict biochemical or physiological sense. In fact, O_2 is not a true reagent interacting with CH; rather, it is the final acceptor of electrons in the process of oxidative phosphorylation. In addition, by presenting O_2 as a substrate, the equation may suggest that the rate of energy production depends on O_2 concentration, which, physiologically, would be true only in the total absence of regulatory mechanisms.

In conditions of steady state, the rate of change of any of the parameters of Equation 2.1, whether products or substrates, provides equivalent information about the rates of the metabolic processes. The exchange of oxygen and carbon dioxide between the organism and the environment, respectively oxygen uptake ($\dot{V}O_2$) and carbon dioxide production ($\dot{V}CO_2$), are easy to measure and therefore represent convenient and very common parameters for the assessment of the metabolic status of the whole organism.[1] During transient or nonsteady-state conditions, however, pulmonary gas exchange may differ substantially from the total gas exchange occurring at the tissue level, because of the O_2 and CO_2 body stores; since the former are much smaller than the latter, $\dot{V}O_2$ measured at the airway opening in nonsteady-state conditions can represent the rate of gas exchange at the tissue level more accurately than does $\dot{V}CO_2$.

As are those of the adult, the newborn's metabolic processes are mostly aerobic. Transfer of oxygen and carbon dioxide occurs almost exclusively through the lungs, although in very small marsupials the skin can provide an important route for gas exchange (Figs. 2.1 and 2.2). An equilibrium exists between the rate of the metabolic processes or energy production, on the one hand, and the whole-body gaseous metabolism, on the other hand, and these two terms, *metabolic rate* and *oxygen uptake* or *consumption* ($\dot{V}O_2$), are often used interchangeably, an approximation that is valid in conditions of steady state.

The lines in Figure 2.3 represent the relationships between $\dot{V}O_2$ and body mass in adult eutherian mammals. Since the time of the first measurements it has been apparent that mammals of various species, even after taking into account their differences in body mass, do not consume the same quantities of oxygen. In fact, $\dot{V}O_2$/kg is less in larger species than

---

[1] The energy equivalent of 1 liter (STPD) of oxygen (or ~45 mMol) is ~4.8 Cal.

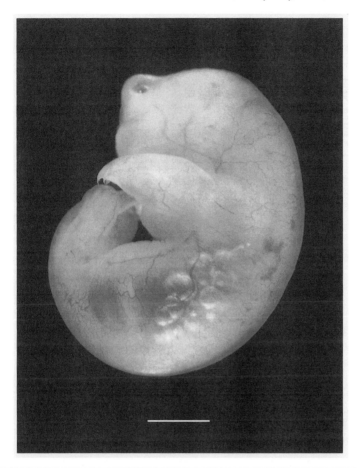

Figure 2.1. One-day old Julia Creek dunnart (*Sminthopsis douglasi*), a dasyurid marsupial of northeast Australia, born after about 13 days of gestation. The lungs are visible through the air sacs on each side of the heart. Despite the presence of cartilaginous ribs, no respiratory movements can be detected. Bar is 1 mm. Courtesy of P. A. Woolley, La Trobe University, Melbourne.

in smaller species (Fig. 2.3, *bottom panel*). In other words, if we could sample a 1-gram *average* flesh sample from a 50-gram mouse, a 500-gram rat, a 10-kg dog, and an 8-ton elephant, we would find that $\dot{V}o_2$ decreases progressively from the mouse sample to the rat, the dog, and the elephant. Hence, the exponent of the $\dot{V}o_2$ allometric curve is less than unity

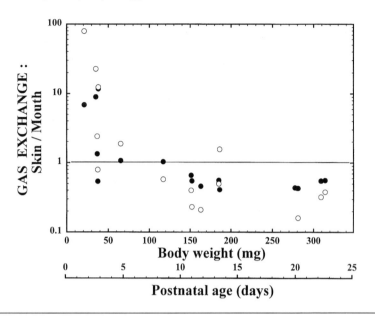

Figure 2.2. Relative importance of skin and pulmonary gas exchange in the newborn Julia Creek dunnart (*Sminthopsis douglasi*). A skin-to-mouth ratio of 1 (*horizontal line*) indicates an equal contribution between skin and lungs. ●, oxygen consumption; ○, carbon dioxide production. All data taken under ambient temperature of 36°C, which is approximately pouch temperature. At birth, gas exchange through the skin can be 10 to 80 times higher than through the lungs. By 20 days of age, skin gas exchange is approximately half that contributed by the lungs. Slightly modified from Mortola et al., 1999a.

(Appendix B provides further explanations of the allometric analysis). A value close to 0.75 (i.e., $\dot{V}_{O_2} \propto W^{0.75}$, where W is body weight) was originally proposed by Kleiber (1932) and Brody and Procter (1932), and has been essentially confirmed ever since (Fig. 2.3, *top panel*).

Figure 2.3. (*opposite*) The "mouse-to-elephant" oxygen consumption ($\dot{V}_{O_2}$) curve for adult mammalian species, over the range in body size between a few grams (approximately the size of a shrew) to about 900 kg. *Top:* Double-logarithmic plot with $\dot{V}_{O_2}$ expressed in ml/min. The *thin unbroken lines* indicate the expected relationship for direct proportionality ($\propto W^1$) and for a 2/3 proportionality ($\propto W^{0.67}$); the *dashed line* is the experimental curve ($\propto W^{0.75}$). *Bottom:* Representation of the "mouse-to-elephant" curve with $\dot{V}_{O_2}$ normalized by body weight (ml · kg^{-1} · min^{-1}).

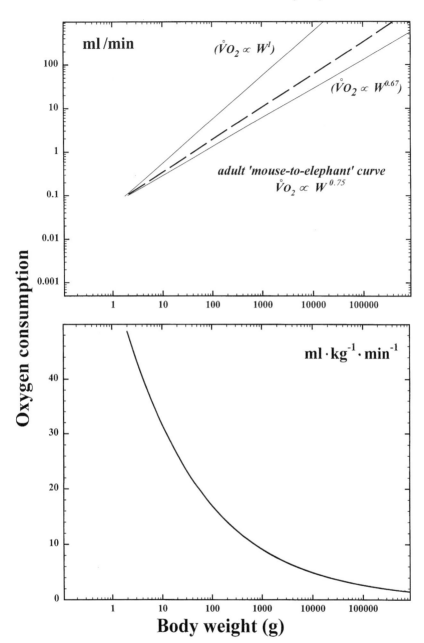

Oxygen consumption

ml /min

$(\overset{\circ}{V}O_2 \propto W^1)$

$(\overset{\circ}{V}O_2 \propto W^{0.67})$

adult 'mouse-to-elephant' curve
$\overset{\circ}{V}O_2 \propto W^{0.75}$

ml · kg^{-1} · min^{-1}

Body weight (g)

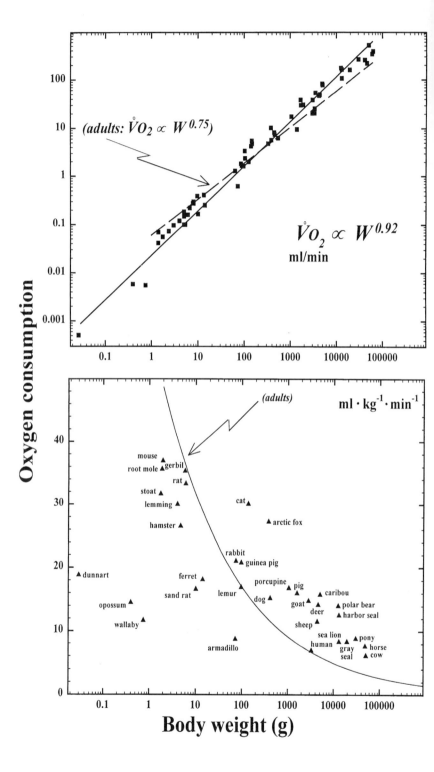

$(adults: \overset{\circ}{V}O_2 \propto W^{0.75})$

$\overset{\circ}{V}O_2 \propto W^{0.92}$
ml/min

Oxygen consumption

$(adults)$

$ml \cdot kg^{-1} \cdot min^{-1}$

mouse
root mole ▲ gerbil
rat ▲
stoat ▲
lemming ▲
hamster ▲
cat ▲
▲ arctic fox
rabbit
▲▲ guinea pig
▲ dunnart
ferret ▲
sand rat ▲
lemur
porcupine ▲ pig
▲ caribou
opossum ▲
dog ▲
goat ▲ ▲ ▲ polar bear
deer ▲ harbor seal
wallaby ▲
sheep ▲
sea lion ▲ pony
human ▲ ▲ horse
gray ▲ cow
armadillo ▲
seal

Body weight (g)

The Metabolic Curve in Newborns

One would predict that newborns, because of their smaller size and their greater propensity for heat loss, should have higher weight-specific resting metabolic rates than adults. Furthermore, newborns face the additional energetic cost of growth and development. In the early postnatal phases, the surge of catecholamines associated with the stress of parturition, and the sudden exposure to hyperoxia, are additional factors increasing neonatal metabolism. Within a species, the expectation of higher $\dot{V}O_2$/kg in newborns than in adults is essentially confirmed by numerous measurements. In fact, in many newborns $\dot{V}O_2$ is also larger than in the same-size adults of other species.

The data points of several neonatal species are plotted in Fig. 2.4. As is the case for the adults, among newborns mass-specific $\dot{V}O_2$ ($\dot{V}O_2$/kg) is not a constant; rather, it decreases progressively in the larger species. But

---

Figure 2.4. (*opposite*) Oxygen consumption in newborn mammals, plotted as a function of the newborn body mass. *Top:* Allometric relationship. The *dashed line* represents the adult curve (as in the top panel of Fig. 2.3). Although all data points are represented, the regression line was calculated after averaging the values pertinent to any particular species. *Bottom:* Average values of $\dot{V}O_2$/kg of newborns of several species. *Continuous line* is the adult line, as in the bottom panel of Fig. 2.3. The three species at lower left are all marsupials. Data for armadillo are from[1]; rat[2-7]; sand rat[3]; rabbit[8,9]; mouse[3,9,10,39]; hamster[3,4,9,11]; opossum[12]; cat[13-17]; guinea pig[3,17]; gerbil[3,9]; dog[1,3,7,18,19]; pig[3,9,19]; deer[3,9]; ferret[3]; lemur[3]; sheep[20,40,41]; porcupine[9]; wallaby[21]; horse[22,38]; pony[22]; cow[23-25,36]; human[26-28,42]; sea lion[29]; gray seal[31]; harbor seal[30]; caribou[32]; lemming[33]; polar bear[34]; stoat[35]; goat[3,9,22]; arctic fox[9]; dunnart[37]; root vole.[43] [1]Mortola, unpublished; [2]Taylor, 1960b; [3]Mortola et al., 1989; [4]Mortola, 1991; [5]Mortola and Dotta, 1992; [6]Matsuoka and Mortola, 1995; [7]Saiki and Mortola, 1996; [8]Hull, 1965; [9]Mortola and Lanthier, 1996; [10]Fitzgerald, 1953; [11]Bartlett and Areson, 1977; [12]Farber et al., 1972; [13]Hill, 1959; [14]Mortola and Rezzonico, 1988; [15]Frappell et al., 1991; [16]Mortola and Matsuoka, 1993; [17]Adamsons et al., 1969; [18]Crighton and Pownall, 1974; [19]Mount and Rowell, 1960; [20]Blackmore, 1969; [21]Baudinette et al., 1988; [22]Brody, 1945; [23]Stowe and Good, 1961; [24]Kiorpes et al., 1978; [25]Bisgard et al., 1974; [26]Talbot, 1938; [27]Mortola et al., 1992a; [28]Mortola et al., 1992b; [29]Iversen and Krog, 1973; [30]Miller et al., 1976; [31]Mortola and Lanthier, 1989; [32]Hart et al., 1961; [33]Hissa, 1968; [34]Blix and Lentfer, 1979; [35]Segal, 1975; [36]Reeves and Leathers, 1964; [37]Mortola et al., 1999a; [38]Stewart et al., 1984; [39]Mortola and Tenney, 1986; [40]Fahey and Lister, 1989; [41]Moss et al., 1987; [42]Polgar and Weng, 1979; [43]Ru-Yung and Jinxiang, 1987.

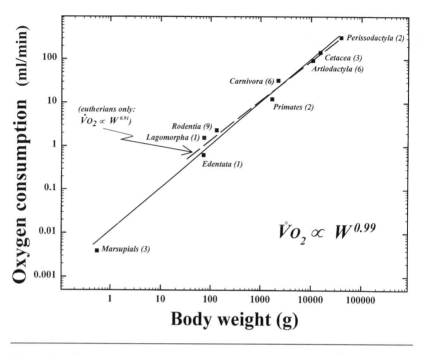

Figure 2.5. Allometric curve of neonatal oxygen consumption ($\dot{V}o_2$). Values of species of the same mammalian order have been averaged. The *dashed line* shows the relationship without marsupials. The numbers in parentheses are numbers of species. Data from sources indicated in legend of Fig. 2.4.

the interspecies differences are less pronounced than those among adults. In fact, the allometric exponent of the curve in newborns is ~0.92, hence much higher than that of the adults, and close to unity. The difference between the allometric relationship of adults and that of newborns can also be appreciated when, instead of individual species, analysis is performed on the average data of mammalian orders (Fig. 2.5). In general, in newborns of species whose newborns have body weight more than ~100 grams the mass-specific $\dot{V}o_2$ falls above the adult's curve (Fig. 2.4), indicating that the newborn's $\dot{V}o_2$/kg is not only higher than that of the adult of its own species but also higher than that of adults of different species with similar body weight. For example, the $\dot{V}o_2$/kg of a 500-gram puppy is not only higher than that of an adult dog, but also higher than that of a 500-gram adult rat. These results and the differences between the adult and newborn functions are open to various interpretations, some of which we turn to now.

The "Mouse-to-Elephant" Metabolic Curve: What Does It Really Mean?

The Adult Curve. The metabolic "mouse-to elephant" curve is often the subject of debate, with respect to both its analytical aspects (Heusner, 1982; Feldman and McMahon, 1983; Smith, 1984) and its physiological, ecological, and evolutionary meaning (Schmidt-Nielsen, 1984). Although values are collected under similar conditions for all species, usually a resting state, one cannot totally eliminate the possibility of a systematic bias. For example, if the metabolic cost against gravity were an important component of resting metabolism (Economos, 1979), aquatic mammals, many of which are large (Cetacea, Sirenia), could bias the curve. Further, measurements obtained at the same time of day may be confounded by the different circadian oscillations of diurnal and nocturnal animals, since diurnal species tend to be relatively large (Prothero, 1984). But notwithstanding the possibility of unaccounted sources of systematic errors, an exponent of, or very close to, 0.75 is the consistent finding, even when analyses have been limited to behaviorally or phylogenetically similar mammals (Peters, 1983). Furthermore, metabolic measurements of homologous tissues from different species and morphological analysis of their mitochondria have indicated a cellular basis for the size-dependency of endothermic metabolism (Krebs, 1950; Smith, 1956; Hulbert and Else, 1990).

The most commonly given interpretation of the adult mouse-to-elephant curve is the "surface law," proposed more than a century ago by Max Rubner (1883) in discussing the metabolic rates of dogs of different sizes.[2] In fact, the idea that in homeotherms heat production should be proportional to their surface area rather than their body mass dates back

[2] When animals of different size but the *same* species are compared, the allometric exponent is close to 0.67; it thus differs somewhat from the 0.75 exponent of the interspecies comparison. The difference (0.75 versus 0.67), albeit small, could have physiological significance (Heusner, 1982). An important exception could be that of pouch-young animals; in the marsupial wallaby, between birth and the time of first exit from the pouch (i.e., between 1 and 250 g), $\dot{V}_{O_2}$ was found to be directly proportional to body mass (Baudinette et al., 1988).

to Sarrus and Rameaux (1838). Because mammals are homeotherms (neglecting some special conditions like hibernation, estivation, and torpor), their body temperature thus maintained within a very narrow range (37–39°C), the huge differences in size across species imply that either thermodispersion, or thermoproduction, or both, must vary among species. From a geometrical perspective, thermodispersion is mostly determined by body surface, whereas heat production is determined by the metabolic activities of the cells, hence is proportional to body mass. If, for simplicity, we think of an animal as a sphere, then because a sphere's surface area is proportional to the square of its diameter and its volume is proportional to the third power of its diameter, it follows that the larger the animal the smaller its surface-to-volume ratio. Hence, if heat production were directly proportional to body size, then large mammals, which have relatively smaller surfaces for heat loss, should have higher values of body temperature. That this is not the case is because the overall metabolic activity of large animals is relatively less pronounced than that in the smaller species, $\dot{V}o_2/kg$ progressively decreasing with the increase in animal weight.

This interpretation of the 0.75 exponent of the adult metabolic curve as a function of the surface law, as appealing and accepted as it may be, still presents some difficulties. In a strict geometrical application of the surface law the exponent should be 0.67, not the consistently found 0.75. This apparent discrepancy could be due to factors that influence metabolic rate systematically, in a size-dependent fashion. As mentioned previously, the gravitational field could be one such factor, because of its effect on body structure and shape, and on the size-dependent gravitational component of metabolism (Economos, 1979). If so, the geometric interspecies similarity may not be the correct premise, and should be replaced by criteria of elastic similarity (McMahon, 1973); these criteria would indeed predict an exponent of 0.75, not 0.67. Systematic differences in the mechanisms of thermodispersion could also contribute to the difference between the experimentally found exponent (0.75) and that expected strictly on the basis of geometric similarity (0.67).

One compelling argument against the interpretation of the 0.75 exponent as an effect of surface law is that this exponent not only occurs among mammals, but also is the consistent result of interspecies comparisons of metabolic rate among lower vertebrates and invertebrates

(Schmidt-Nielsen, 1984) (i.e., of non-homeothermic animals). In fact, an exponent equal, or very close, to 0.75 has been found even in comparing metabolic rates of bacteria and other unicellular organisms (Hemmingsen, 1960; Schmidt-Nielsen, 1970). Indeed, this generality in the scaling processes of biological phenomena continues to stimulate the search for unifying principles and common mechanisms (Lindstedt and Calder, 1981; West et al., 1997).

The Newborn Curve. Given that the factors responsible for the size-dependency of metabolic rate among adult mammals are not completely clear, the reasons for the differences among neonates of various species also remain speculative. The newborn curve unequivocally differs from that of adults, its allometric exponent being closer to unity than to the classic 0.75. One factor underlying the high allometric exponent could be the major difference in development, and therefore metabolic status, among various species at birth. In fact (see Chapter 1, opening paragraphs), at one end of the spectrum are *precocial* species, the offspring born already standing, with eyes open, and active a few minutes after birth. At the opposite end are the *altricial* species, blind and almost helpless at birth. Although at birth metabolic rate is likely to increase in all species, the changes during the following hours and days can be quite varied. Many small altricial species have a low, mass-specific metabolic rate in the first postnatal days. Of these, some (e.g., marsupials) continue to have a low metabolic rate; others, like the mouse and many other rodents (Waldschmidt and Müller, 1988), substantially increase $\dot{V}O_2/kg$ during the first 1–2 weeks, in some cases even above the values of the adult allometric curve (Fig. 2.6). In the human infant, $\dot{V}O_2$ follows a similar pattern (Brück, 1998). Some of these changes seem to be related to changes in weight-specific liver metabolism, specifically in the energy required by the sodium pump (Else, 1991). But other species (e.g., the horse) drop $\dot{V}O_2$ during the first days (Fig. 2.6, bottom right). The fact that many altricial and precocial species belong to, respectively, the small and large species, tends to decrease the lower end and increase the upper end of the metabolic curve, therefore increasing its allometric exponent. In other words, the high allometric exponent of the neonatal metabolic curve could derive from the fact that differences in the developmental stage among species tend to have some degree of size-dependency.

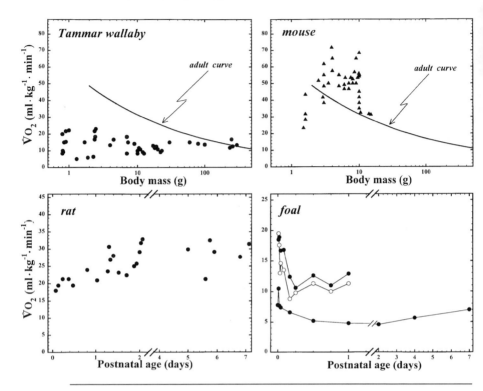

Figure 2.6. Oxygen consumption, normalized by the weight of the animal ($\dot{V}O_2$/kg) during the postnatal period in the Tammar wallaby, mouse, rat, and foal. In the *top panels,* data are presented as a function of body mass, in logarithmic scale; the continuous function represents the allometric relationship of adult species. In the *bottom panels,* data are presented as a function of postnatal age. For foals, *open symbols* are ponies; *filled symbols,* thoroughbred. Data for wallaby and mouse from Baudinette et al., 1988; for rat from Taylor, 1960b; for foal from Ousey et al., 1991, and Stewart et al., 1984.

An important fraction of the total metabolic energy is devoted to maintaining transmembrane ionic gradients; the greater tendency of mammalian cells to passively leak sodium and potassium ions, compared to that of reptilian cells, has been suggested as a factor in the differences in metabolic level between endotherms and ectotherms (Else and Hulbert, 1987; Else, 1991). Differences in the structural and functional characteristics of the mitochondria, including their enzymatic activity, could also be a contributing factor (Else and Hulbert, 1981, 1985a). Among adult mammals, the larger density of mitochondria and the greater mitochondrial leak of hydrogen ions across the inner membrane (Porter and Brand, 1993) probably account for the higher mass-specific $\dot{V}o_2$ of the smaller species.

Could, then, any of these factors also help to explain the differences in postnatal changes in $\dot{V}o_2$ among species? Data are very few, but some are consistent with the possibility that some of the variables considered responsible for the phylogenetic differences in $\dot{V}o_2$ may also explain the $\dot{V}o_2$ changes during ontogenesis. In studies of the Tammar wallaby, *Macropus eugenii*, the mitochondrial membrane surface area of various tissues was measured by electron microscopy in animals at different stages of postnatal development, from the 10-day-old pouch young to the 6-kg adult (Hulbert et al., 1991). Normalized by tissue weight, this area increased in heart, kidney, and brain. The total surface area gradually increased from the low "reptilian" value of the very young to the higher "mammalian" value of the adult (Fig. 2.7). In fact, in first approximation, the gradual development in the total "summed" mitochondrial surface

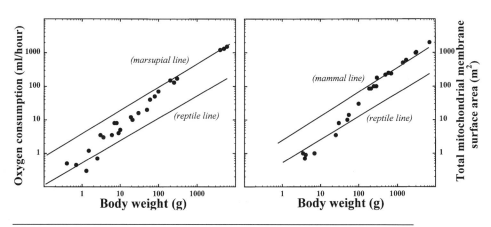

Figure 2.7. Total mitochondrial membrane area (*right panel*), calculated from liver, kidney, heart, and brain samples, and oxygen consumption (*left panel*) in the Tammar wallaby, *Macropus eugenii*, as a function of body mass. The *continuous lines* represent the corresponding allometric curves of reptiles, marsupials, and mammals, as indicated. (Modified from Hulbert et al., 1991, with data from Baudinette et al., 1988, and Else and Hulbert, 1985b.

area from the various organs appeared to have a developmental pattern similar to that of metabolic rate. In the rat brain, as well, the cellular density of mitochondria increases during postnatal development, in conjunction with an increase in the efficiency of oxidative phosphorylation, and in the concentration and activity of enzymes (Nioka and Chance, 1991). To what extent the differences in mitochondrial characteristics could account for the differences in metabolic rate between precocial and altricial species is difficult to estimate. Comparative data on postnatal mitochondrial development in both altricial and precocial species would be very helpful, but no information is currently available. Equally, a comparative analysis of the metabolic needs of fetuses close to term would be of interest, and would also shed some light on the interpretation of the neonatal allometric curve. In fact, in the fetus, thermal needs are not part of the metabolic requirements. Moreover, the weightless intrauterine environment eliminates the gravitational factor that among adult species, as mentioned earlier, is thought to be a potential contributor to the interspecies differences in body-weight-specific metabolism.

Pulmonary Ventilation

Pulmonary ventilation is the sole form of air convection, and therefore of oxygen convection, in mammals, whether newborns or adults, excepting only some extraordinarily small and altricial newborn marsupials in which the skin provides an important route for gas exchange. The mechanical changes accompanying the aeration of the lungs at birth, discussed in Chapter 1, and the necessity of clearing the pulmonary fluid from the peripheral airways, are important determinants of the very early breathing pattern. Later, pulmonary ventilation is expected to be influenced less by mechanical factors and more by the metabolic needs of the newborn.

Mechanical Events in the First Hours after Birth

Major Initial Changes. With the first breath, the three major changes that characterize the respiratory adaptation of mammals at birth (onset of external ventilation, clearing of the fetal pulmonary fluid, and establishment of a functional residual capacity) have begun. In infants, respiratory-system compliance is low and resistance is high in the first minutes after birth; in a few days the former increases by 80% and the latter decreases by 20% (Mortola et al., 1982b). Both changes (i.e., the increase in lung compliance and the drop in total pulmonary resistance) reflect mainly the changes in the mechanical properties of the lung (Geubelle et al., 1959; Karlberg et al., 1962b) that accompany the progressive clearing of the pulmonary fluid and the expansion of the lung. Most of these changes occur within a few hours, during which time functional residual capacity progresses, and the improvement in the matching between lung ventilation and blood perfusion leads to a normalization of the blood gases (see Fig. 1.9). The time required for clearance of the pulmonary fluid is probably of the same order of magnitude. The pressure needed to expand the lungs decreases progressively with time after birth in lambs (Dawes et al., 1953) and in infants (Hull, 1969), and increases in lung compliance similar to those observed in infants have been observed in rats, rabbits, and lambs. In rabbits, dynamic lung compliance increased by 80% within the first hour after birth (Lachmann et al., 1979). Mechanical ventilation of newborn rabbits increases both lung compliance (CL) and the compliance of the whole respiratory system (Crs) (Nilsson, 1979). The increase in Crs, however, was substantially less than that of CL, indicating a gradual stiffening of the chest wall. Serial measurements in rats have also indicated a drop in chest wall compliance (Cw) during the first hours.

Although a gradual decrease in Cw during growth is to be expected, as a result of the progressive stiffening of the cartilaginous structures (see Fig. 3.11), such rapid changes during the first hours are difficult to explain. They may reflect variations in chest wall geometry that accompany lung expansion and the entrance of air into the digestive tract. In lambs and rabbits, but not in human infants, the lung's airflow-resistive characteristics did not seem to change during the first hours (Shaffer et al., 1976; Lachmann et al., 1979). It should be noted, however, that experiments on animals have required both artificial delivery (to assure accurate timing of the first breath) and tracheostomy (for measurements of changes in lung volume by pneumotachography); both procedures make any comparison to the human data uncertain.

The Early Breathing Pattern. In the human infant, after a slow start the breathing rate increases (see Fig. 1.14); in the first 90 min it averages between 70 and 90 breaths/min, although the pattern is very irregular and short periods of apnea may alternate with bursts of extremely high breathing frequencies (Fisher et al., 1982). This pattern does not seem to be affected by the mode of delivery or type of maternal anesthesia employed (Fisher et al., 1983a). Rapid breathing, which usually subsides within one day, may last for several days in some infants; in these cases, presumably, low lung compliance derives from a delay in lung fluid clearance (transient tachypnea syndrome). This breathing pattern evidently minimizes the increase in respiratory work caused by the low compliance (Avery et al., 1966). Radiologic observations in lambs have indicated that the grade of lung aeration is inversely related to the wet-to-dry lung weight ratio and to the animal's resting breathing rate (Fletcher et al., 1970). The tachypnea of the first hours may be mediated by the activation of pulmonary vagal fibers, pulmonary irritant receptors, or nonmyelinated fibers ("J-receptors"), a pattern similar to that seen during lung congestion in the adult (Sellick and Widdicombe, 1969; Paintal, 1973) (see Chapter 4, under "Airway Receptors").

A characteristic feature of the early breathing pattern is the frequent interruption of the expiratory flow (Fig. 2.8), with lung volume consequently maintained above the passive resting volume (Milner et al., 1978b; Lindroth et al., 1981; Fisher et al., 1982; Mortola et al., 1982b; Radvanyi-Bouvet et al., 1982). This pattern results from closure of the glottis, and can be observed in all normal infants, not only those with respiratory distress syndrome, as thought earlier (Harrison et al., 1968). The frequency of breaths exhibiting these characteristics decreases with time (Yao et al., 1971), and they are an uncommon observation after a few days (Fig. 2.9). In the rat, as well, the frequency of breaths with complete ex-

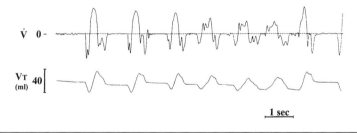

Figure 2.8. Recordings of respiratory airflow (*top*) and tidal volume (*bottom*) in a full-term normal human infant 10 min after birth. Note the frequent interruptions of the expiratory flow. Modified from Fisher et al., 1982.

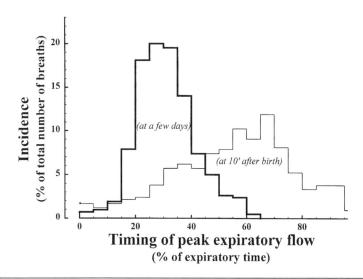

Figure 2.9. Histogram distribution of the occurrence of the peak flow in human infants, expressed in percent of total expiratory time. During the first minutes after birth (*thin line*), peak expiratory flow occurs most often during the last third of the expiratory phase of the breathing cycle, mostly because of the retarding action of the laryngeal braking. At a few days (*thick line*), peak flow occurs most frequently at about one-third of the expiratory phase, as in adults. Note also the greater variability at 10 min (*thin line*), compared to a few days (*thick line*). From data of Fisher et al., 1982.

piratory occlusions exceeds 50% at birth, and rapidly declines within the first postnatal days (Fig. 2.10). And in the opossum, a marsupial born in an extremely immature state, brief interruptions of the expiratory flow have been observed even at 4 weeks of postnatal life (Farber, 1978). During these short periods of upper airway occlusion in some newborns the diaphragm is relaxed (Fig. 2.11), whereas in others, such as in the pouch-young opossum (Farber and Marlow, 1978), the diaphragm may occasionally become active, suggesting incomplete coordination between inspiratory activity and the control of upper airway dilation. Probable mechanisms and mechanical implications of a pattern with increased expiratory resistance are discussed in Chapter 3, under "Expiration." It suffices here to emphasize that the function of this breathing pattern may be related not so much to the increase in mean lung volume as to the as-

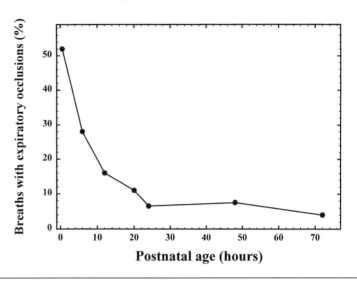

Figure 2.10. Percentage of breaths having periods of zero flow interrupting the expiratory phase (expiratory occlusions) in newborn rats during the first postnatal hours.

sociated increase in intra-airway pressure. An airway pressure maintained above the atmospheric value could represent an important backforce contributing to the clearing of the pulmonary fluid and to a uniform aeration of the lungs. Total closure of the airways in expiration is a frequent breathing pattern in amphibians and reptiles (Naifeh et al., 1970; Bartlett and Birchard, 1983; Seymour, 1989). In reptiles, which have very large lung volumes and compliance with respect to those of mammals of the same size (Bartlett et al., 1986; Milsom, 1989; Frappell and Mortola, 1998), filtration of fluid into the airspaces can occur during inspiration, owing to the characteristically elevated pulmonary artery pressure and low protein osmotic pressure, whereas reabsorption takes place during the expiratory phase (Burggren, 1982).

In some respects, the physiological lung edema of the newborn mammal in the first hours after birth represents a similar situation, and perhaps the positive airway pressure during expiration can be interpreted as a common strategy favoring fluid clearance. After a few hours this breathing pattern is retained only in the form of a partial narrowing of the larynx in infants (the *grunting* pattern), and probably in other medium-size

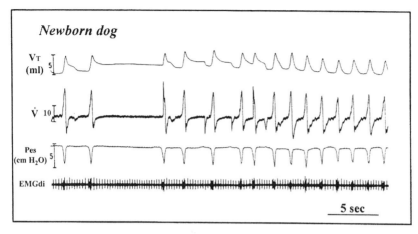

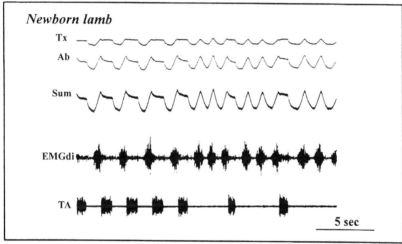

Figure 2.11. *Top:* Breathing in a newborn dog, under light barbiturate sedation. V_T, tidal volume, ml (inspiration upward); $\dot{V}$, respiratory airflow, ml/sec; Pes, esophageal pressure, cm H_2O; EMGdi, diaphragm electromyogram, with electrocardiogram superimposed. *Bottom:* Thoracic (Tx) and abdominal (Ab) motion, and the sum of the two (Sum), in a newborn lamb. EMGdi and TA are the electromyogram of, respectively, diaphragm and thyroarytenoid muscle. During the interruptions of expiratory flow, the diaphragm is not active; indeed, pleural pressure is not more negative than at end-expiration, indicating that these periods of maintained lung inflation are not contributed by activation of the inspiratory muscles. Rather, the TA, a vocal folds adductor, is active. From Mortola, 1984, and Andrews et al., 1985.

and large species, as a means of keeping mean lung volume elevated, whereas it is uncommon in the smallest mammalian species. The breathing pattern of adult seals, and probably of other diving mammals when resting ashore, is often punctuated by pauses at end-inspiration or in the middle of expiration. The newborns of these species present an extraordinarily high frequency of breaths having expiratory occlusions, even a few days after birth (Fig. 2.12).

In summary, in the first few hours after birth, breathing is at first rapid and irregular. Then, as the pulmonary fluid decreases, frequency also decreases. Complete closure of the larynx, thus interrupting expiration, is very common. Respiratory system and lung compliances gradually rise,

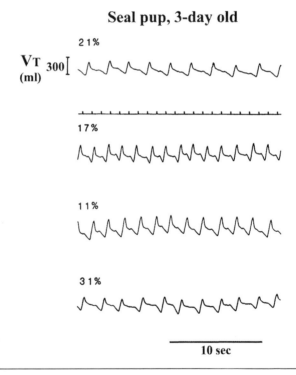

Figure 2.12. Spirometric record (inspiration upward) in a 3-day-old seal pup (gray seal, *Halichoerus grypus*), while breathing air (21% O_2), hypoxic gas (17%, 11%), and hyperoxic gas (31%). Note that breathing rate changes with the inspired oxygen concentration, but the tendency for occlusions in the middle of expiration remains unaltered. From Mortola and Lanthier, 1989.

while chest wall compliance decreases; these changes favor a progressive increase in the resting volume of the respiratory system. The time course of these changes (Fig. 2.13) is probably greatly dependent on the time course of fluid clearance; hence, the important determinants must be the duration of labor before delivery and species size, the latter because pulmonary fluid reabsorption is probably faster in the smaller species.

Pulmonary Ventilation and the Breathing Pattern

Even several hours after birth, the neonatal breathing pattern is irregular, with alternating periods of fast and slow rate, the latter often including apneas of variable durations (Fig. 2.14). These irregularities tend to be accentuated in a warm environment, whereas they are reduced under hypoxic or hypercapnic conditions. Changes in sleep state can also introduce variability, although in rats, cats, and guinea pigs during the first postnatal days sleep is mostly characterized by the desynchronized phase of rapid-eye-movement (REM) sleep (Jouvet-Mounier et al., 1970). In preterm infants, both irregular and regular breathing can be found in both REM and non-REM sleep (Rigatto et al., 1980).

Nose Breathing. Many newborns breathe through the nose, but whether they are obligatory or facultative nose-breathers, and to what extent the pattern differs from that of the adults, is unclear. Obstruction of the nasal passages, as with choanal atresia, can be a serious situation in human infants; and even some adult animals, such as the rat and the rabbit, have serious problems in switching to mouth breathing after closure of the nostrils. On closure of the nasal passages, asphyxia occurs in both lambs and ewes, more severely in the former (Wood and Harding, 1989). The airflow resistance of the nasopharyngeal pathway, on a per-kilogram basis, in kittens and puppies is lower than that in adult cats and dogs, but the resistance of the oral passages is not higher than that in the adults (Mortola and Fisher, 1981). Infants with nasal obstruction can breathe through the mouth while sucking pacifiers with an enlarged end-hole (Swift and Emery, 1973). Hence, difficulties in mouth breathing in newborns may not be due to anatomical or mechanical factors; rather, the process of switching from nose- to mouth-breathing may be a behavioral response gradually acquired through a learning process.

Breathing Pattern and Species Size. Allometric analysis indicates that neonatal pulmonary ventilation $\dot{V}E$ is proportional to body weight $W^{0.91}$

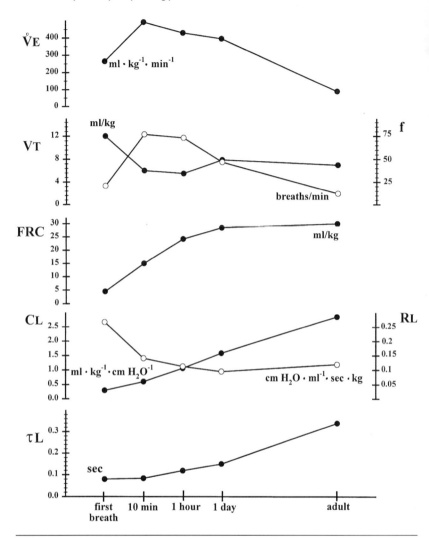

Figure 2.13. Average values of some respiratory parameters in infants during the first day of life, and in adults. V̇E, pulmonary ventilation; VT, tidal volume; f, breathing rate; FRC, functional residual capacity; CL, lung compliance (●); RL, total pulmonary resistance (○); τL, time constant of the lungs. Values are normalized by the average weight of the infant: 3340 grams at birth and 10 min, 3300 grams at 1 hour, 3160 grams at 1 day; adult body weight, 70 kg.

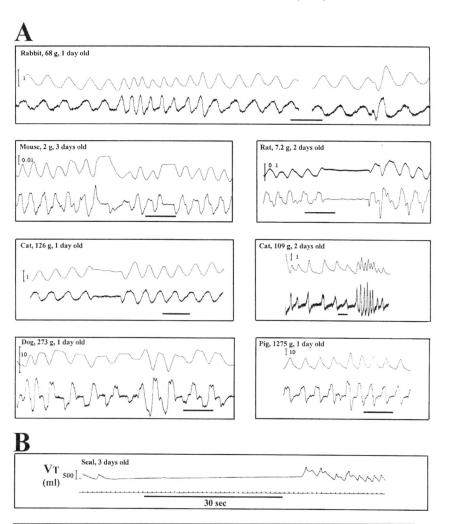

Figure 2.14. *A:* Breathing pattern in some newborn mammals, 1–3 days old. Record at top, in each case, is tidal volume (inspiration upward, calibrations indicated); record at bottom is airflow. Horizontal bars are 1 sec. Rapid bursts are usually associated with an increase in end-expiratory level. Note also the occasional breaths having interruptions of expiratory flow, a remnant of a phenomenon very common during the first hours. *B:* Spirometric record in a seal pup. The apnea, which would be extraordinarily long for most species, is not unusual for newborn seals. *A:* from Mortola, 1984; *B:* from Mortola and Lanthier, 1989.

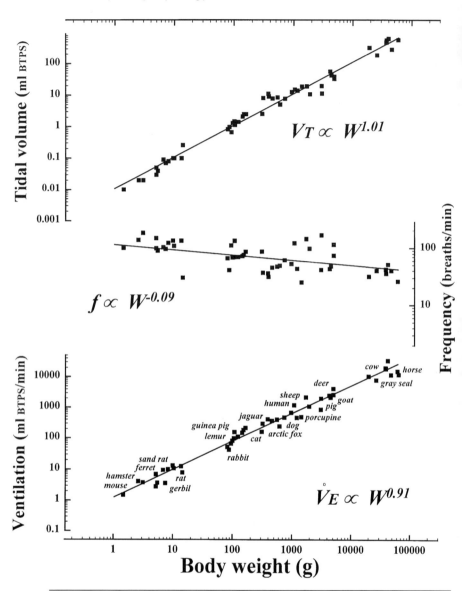

Figure 2.15. Allometric relationships of the parameters of the breathing pattern (V$_T$, tidal volume; f, breathing rate; V̇$_E$, ventilation in some newborn mammals. Although all data points are represented, the regression lines were calculated after averaging the values pertinent to any particular species. Data for rat are from[1,3,4,7]; sand rat[3]; rabbit[1,9]; mouse[1,3,9,20]; hamster[3,9]; cat[1–3,14,16]; guinea pig[1,3]; gerbil[3,9]; dog[1–3,7]; pig[1,3,9]; deer[3,9]; ferret[3]; lemur[3]; sheep[20]; jaguar[3]; por-

(Fig. 2.15); this exponent is slightly lower than unity and thus clearly higher than the 0.75 exponent of the adult allometric curve. Mean inspiratory flow (V_T/T_I, where V_T is tidal volume and T_I is inspiratory time) probably scales to body weight, as does $\dot{V}_E$, because the "duty cycle" T_I/T_{TOT},[3] where T_{TOT} is total cycle time, does not seem to vary with species size in a systematic manner (Mortola, 1984; Lekeux et al., 1984; Koterba and Kosch, 1987). Hence, two parameters independently measured, metabolism (see Figs. 2.4 and 2.5) and pulmonary ventilation, indicate a much greater uniformity among neonates than among adults.

As in adults, also in newborns the tidal volume is directly proportional to the species' body size ($V_T \propto W^{1.01}$). Among newborns of various species, however, the decrease in breathing frequency with the increase in animal size ($f \propto W^{-0.09}$) is much less obvious than it is among adults (-0.09 in newborns, -0.25 in adults). Hence, the greater interspecies uniformity in $\dot{V}_E$ among newborn mammals, as compared to adults, derives from the fact that breathing rate does not change with size among newborns as much as it does among adults. It could be questioned whether this pattern is determined by the fact that there is a mechanical limit to the maximal breathing rate that the smallest species can attain; in other words, whether the mechanical characteristics of their respiratory system may pose a limit to $\dot{V}_E$. For example, if the 2-gram newborn mouse were breathing as predicted by the allometric curves of adult mammals, its breathing rate should be ~250 breaths/min, instead of its normal 145 breaths/min. Because the inspiratory time is about 40% of the total breathing cycle, a rate of 250 breaths/min would imply an inspiratory time (T_I) of ~95 msec. A T_I this short would require of the respiratory system a very fast response time between muscle contraction

---

[3] Note that $\dot{V}_E = V_T/T_I \cdot T_I / T_{TOT} \cdot 60$.

---

cupine[3,9]; horse[8,10,19]; cow[11-13,18]; human[15,21-23]; gray seal[17]; goat[3,9]; arctic fox.[9] [1]Mortola, 1984; [2]Pedraz and Mortola, 1991; [3]Mortola et al., 1989; [4]Mortola, 1991; [5]Coté et al., 1988; [6]Moss et al., 1995; [7]Saiki and Mortola, 1996; [8]Gillespie, 1975; [9]Mortola and Lanthier, 1996; [10]Koterba and Kosch, 1987; [11]Lekeux et al., 1984; [12]Kiorpes et al., 1978; [13]Bisgard et al., 1974; [14]Mortola and Rezzonico, 1988; [15]Talbot, 1938; [16]Mortola and Matsuoka, 1993; [17]Mortola and Lanthier, 1989; [18]Reeves and Leathers, 1964; [19]Stewart et al., 1984; [20]Mortola and Tenney, 1986; [21]Mortola et al., 1992a; [22]Mortola et al., 1992b; [23]Polgar and Weng, 1979.

and pressure generation (i.e., a τrs of < 24 msec, where τrs is the time constant of the respiratory system).[4] Can the values of compliance and resistance of the respiratory system be so small as to permit such a short τrs? The data of passive respiratory mechanics in very small newborn mammals would suggest that it cannot (see Chapter 3, under "Being Very Small"). During spontaneous breathing, however, the effective inspiratory time constant can be substantially shorter than its passive value (see Chapter 3, under "Inspiration"). In fact, in many newborns the breathing rate can be much higher than that in resting conditions, indicating that the resting values are not close to the maximal attainable. For example, most newborns respond to hypoxia with a major increase in breathing rate (Mortola, 1996), and even animals as small as the newborn rat can sustain levels of $\dot{V}E$ almost double the resting value for several days with no signs of adaptation (Piazza et al., 1988; Rezzonico and Mortola, 1989). For the great majority of newborns, then, and with the possible exception of only extremely small marsupials, the hypothesis that during resting conditions the mechanical properties of the respiratory system may pose a constraint on $\dot{V}E$ seems very unlikely.

Dead Space and Alveolar Ventilation

Pulmonary ventilation ($\dot{V}E$) is the parameter of greatest interest in the analysis of air convection and its implications for the work, energetics, and control of breathing, but it is only an indirect indication of the air convection at the alveolar level, which is the important parameter for gas exchange. For the latter, the variable of interest is alveolar ventilation ($\dot{V}A$), defined as the product of breathing rate and the change in alveolar volume. Hence, $\dot{V}A$ is less than $\dot{V}E$, the difference being the ventilation of all those airways not participating in gas exchange, whether because of anatomical or functional reasons, and globally called *physiological dead space* (VD):

$$\dot{V}A = (VT - VD) \cdot f = \dot{V}E - \dot{V}D \qquad (Equation\ 2.2)$$

The relationship between $\dot{V}A$ and $\dot{V}E$ depends not only on VD but also on the breathing pattern. In fact, for the same $\dot{V}E$ and VD, the faster the

[4] In a first-order system, the time required for ~98% completion of an event is four times its time constant.

pattern the lower the $\dot{V}A$ (because $\dot{V}E$ is the same, a faster pattern must have a smaller VT; hence, the ventilation of the dead space represents a larger fraction of $\dot{V}E$). Among adult animals, several considerations, as well as measurements of dead space and blood gases (Lahiri, 1975; Tenney and Boggs, 1986), have led to the conclusion that during resting conditions VD/VT is an interspecies constant; hence, resting $\dot{V}A$ is also approximately a constant fraction of $\dot{V}E$ (Leith, 1983; Tenney and Boggs, 1986).

Among newborns of various species, does $\dot{V}A$ represent a constant fraction of $\dot{V}E$, and is VD/VT comparable to that in adults? Very few data on measurements of VD are available, and those that exist are almost exclusively taken from the human infant. Equally, $\dot{V}A$ has rarely been measured in species other than humans, and almost never in combination with $\dot{V}E$. Hence, no conclusive answers are at hand; rather, only some considerations based on indirect measurements can be addressed.

Tracheal volume, which represents an important fraction of the conductive airways, is directly proportional to body mass (Fig. 2.16). Because VT is also directly proportional to body mass (see Fig. 2.15), if tracheal volume is representative of anatomical dead space, then $\dot{V}A$ should be a fixed proportion of $\dot{V}E$. During breathing, as pressure is applied to intra- and extrapulmonary airways, changes in the volume of the conductive airways depend on airway compliance, which is high in the fetus and the newborn and decreases with age (Fig. 2.17). In newborn cats, dogs, and rabbits, tracheal compliance, normalized by body weight, averages between 0.04 and 0.06 (ml $\cdot$ cm $H_2O^{-1} \cdot kg^{-1}$),[5] or about twice the adult values (Olsen et al. 1967; Mortola and Sant'Ambrogio, 1978). Although the neonatal airways when studied in isolation show high compliance, changes in airway volume during breathing in vivo may not be great, because smooth muscle tone plays a stabilizing role in airway caliber and increases airway stiffness (Bhutani et al., 1986). Indeed, cineroentgenograms of the trachea in normal infants could barely detect changes in diameter during breathing (Wittenborg et al., 1967). In a pediatric pop-

[5] In one specimen from a stillborn dolphin, however, tracheal compliance averaged five to ten times less (J. P. Mortola, J. T. Fisher, G. Sant'Ambrogio, and F. B. Sant'Ambrogio, unpublished, 1978). In dolphins, as in other diving mammals, the airway cartilages form complete rings, unlike the incomplete, horseshoe-shaped cartilaginous structures of most land mammals.

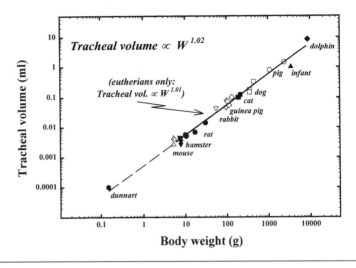

Figure 2.16. Allometric relationship of tracheal volume in newborns of various species. Tracheal volume is directly proportional to body weight, whether the marsupial species (*dashed line*) are included or not. Data from Mortola and Fisher (1980) and additional, unpublished data; although all data points are represented, the regression lines were calculated after averaging the values pertinent to any particular species.

ulation, VD was found to be essentially constant despite changes in breathing pattern (Rose and Froese, 1980), or despite changes in end-expiratory pressures between zero and 4–8 cm H_2O (Numa and Newth, 1996). The latter finding, which applied not only to children but also to a group of infants, is surprising, but it may indicate that volume- or pressure-dependent changes in VD are indeed very small.

Is the newborn's VD/VT higher than that in adults? The higher airway compliance in the newborn does not necessarily imply that, at end-expiration, the volume of the conductive airways is greater than that in adults, because the pressure distending the airways at end-expiration (which is the transpulmonary pressure at the resting volume) is lower in newborns (see Chapter 3, under "Static Coupling between Lungs and Chest Wall"). In studies of infants, several attempts have been made to quantify the intra- and extrathoracic anatomical VD, using single-breath wash-out techniques or by filling the extrathoracic airways with saline. Because all of the CO_2 expired originates from those alveolar regions of the lungs con-

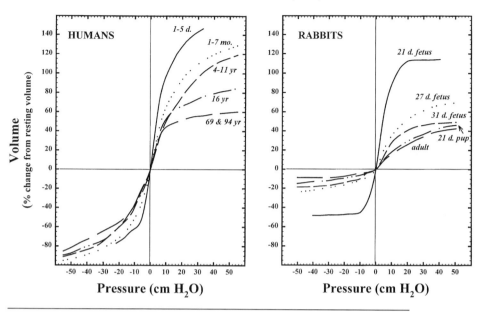

Figure 2.17. Pressure-volume relationship of tracheal segments in humans (*left*) and rabbits (*right*) of different ages. Volume is expressed as a percentage of the resting volume of the trachea (no pressure applied). Curves for humans are adapted from Croteau and Cook (1961) and represent one subject (16 yr old), or the mean of a few subjects of the age range indicated. Data for rabbits are from Bhutani et al. (1981) and are the mean curves of six or seven samples each. In the rabbit, fetuses at 21, 27, and 31 days correspond to, respectively, ~67%, ~87%, and 100% of gestation.

tributing to gas exchange, $V_T \cdot [CO_2]$expired $= V_A \cdot [CO_2]$alveolar. Substituting $V_T - V_D$ for V_A and rearranging, one obtains the Bohr equation:

$$V_D/V_T = ([CO_2]\text{alveolar} - [CO_2]\text{expired})/[CO_2]\text{alveolar},$$

(*Equation 2.3*)

where $[CO_2]$alveolar is measured from the expired gas at the end of expiration (*end-tidal* CO_2). The total physiological V_D has been measured using this equation. Results ranged between 1.1 and 2.8 ml/kg. The difficulty in measuring accurately *end-tidal* CO_2 in subjects breathing at high rates is the main reason for the large scatter in the results. A recent study (Numa and Newth, 1996) concluded that the intrathoracic V_D/kg is age-independent, thus similar between infants and adult men, whereas the extrathoracic VD/kg would be almost twice as great in infants. As the authors point out, however, measurements of the volume of the upper airways by filling the space with liquid could seriously overestimate the physiological dead space, for many reasons; for example, they would include the volume of the oropharyngeal cavity, normally not used by infants during breathing.

In summary, the physiological VD is probably not much higher in infants than it is in adults, whether normalized by body weight (VD/kg, ~2.2 ml/kg) or in relation to VT (VD/VT, ~0.3) (Strang, 1977; Polgar and Weng, 1979; Gaultier and Girard, 1980). To what extent this conclusion can be extrapolated to other species is impossible to say at present, because of the paucity of comparative data. In piglets, comparison of values of $\dot{V}A$/kg (Côté et al., 1992, 1996) with those of $\dot{V}E$/kg obtained from other studies would suggest that the piglet has a high VD/VT. In one study in foals during the first postnatal week, VD/VT averaged ~0.66, a value that seems extraordinarily large (Stewart et al., 1984).

In infants, the major phase of alveolar formation begins prenatally, whereas in the rat and possibly other small species it ensues postnatally (Brody and Thurlbeck, 1986; Burri, 1994). Therefore, it is possible that in the newborn of small species the intrathoracic VD/VT is higher than that in medium-size or large species, and also larger than in the adult.

Coupling of Ventilation and Metabolism

Both $\dot{V}O_2$ and $\dot{V}E$ scale to W with approximately the same exponent, 0.92 (see Fig. 2.4) and 0.91 (see Fig. 2.15), respectively. Hence, among newborn species at rest, $\dot{V}E$ is directly proportional to $\dot{V}O_2$ (Fig. 2.18). This proportionality is not surprising, since, as in adults, pulmonary ventilation is the only mechanism for gas exchange.[6] It implies, in fact, that the ventilatory equivalent (the quantity of $\dot{V}E$ for unitary $\dot{V}O_2$) is an interspecies constant. This value, computed from the respective allometric relationships, is 41 ml $\dot{V}E$ per ml $\dot{V}O_2$, which corresponds to 2.5% of $\dot{V}E$ extracted as O_2 (although both variables are measured in ml/min, $\dot{V}E/\dot{V}O_2$

---

[6] As mentioned earlier (Fig. 2.2), however, in the smallest newborn marsupials gas exchange through the skin can be an important fraction of the total.

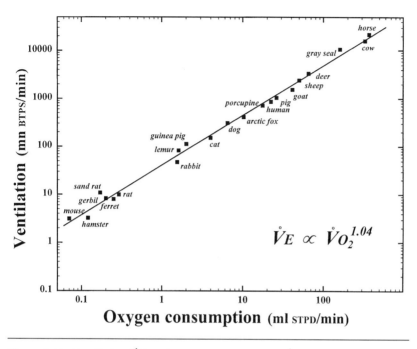

Figure 2.18. Ventilation (V̇E) and oxygen consumption (V̇O$_2$) in newborns of several species during resting conditions. Data are the average of the results of several studies. From sources indicated in legends of Figs. 2.4 and 2.15.

is not a dimensionless parameter, since V̇E is expressed under BTPS conditions, and V̇O$_2$ under STPD conditions). Similar derivations from allometric analysis in adult mammals have led to values of ventilatory equivalent either slightly higher (Drorbaugh, 1960) or slightly lower, between 32 and 37 (Stahl, 1967; Frappell et al., 1992; Frappell and Baudinette, 1995). Hence, in general, newborns probably fulfill their resting metabolic requirements with levels of V̇E not very different from those of adults, and, where differences are found, they are likely to be small. Simultaneous measurements of V̇E and V̇O$_2$ in rats of 50, 100, 250, and 400 grams showed that the ventilatory equivalent does not vary systematically with animal size (Mortola et al., 1994), and the average value found, 42, is comparable to that of newborn rats at thermoneutrality, and to that derived above from interspecies allometric analysis.

Blood gases, which are usually in equilibrium with alveolar gases, are an obvious indicator of gas exchange, and therefore of the relation between metabolic rate ($\dot{V}o_2$ and $\dot{V}co_2$) and ventilation.

Because the amount of carbon dioxide in the inspired air is minute (0.03%), the CO_2 concentration present in the alveoli represents the balance between what is produced ($\dot{V}co_2$) and what has been ventilated out of the alveolar region ($\dot{V}A$). Hence,

$$[CO_2] = \dot{V}co_2/\dot{V}A, \qquad (Equation\ 2.4)$$

which, if expressed in the form of alveolar partial pressure, yields the alveolar gas equation for CO_2

$$P_{ACO_2} = (\dot{V}co_2/\dot{V}A) \cdot Pb, \qquad (Equation\ 2.5)$$

where Pb represents the value of barometric pressure in dry conditions.

In the preceding section, I concluded that among newborn species in resting conditions, gaseous metabolism is proportional to $\dot{V}E$. Because V_T/kg is constant (see Fig. 2.15), if V_D/V_T is also an interspecies constant, then $\dot{V}A$ should be directly proportional to $\dot{V}E$ and to metabolic rate. This conclusion could be verified by measurements of blood gases, and in particular of arterial Pco_2, according to Equation 2.5.[7] The results are highly variable both within and between species (Table 2.1). This variability is not unexpected, considering the influence of postnatal age on blood gas parameters (see Fig. 1.9). Nevertheless, from what is available it seems that the values of arterial Pco_2 of newborns at rest do not systematically vary with the size of the species, and are within the range observed in adult mammals (Tenney and Boggs, 1986).

Interspecies Comparisons

With the possible exception of the smallest pouch-young marsupials, newborn mammals rely on the lungs for gas exchange. Per unit W, in many species, oxygen consumption ($\dot{V}o_2$) is higher in newborns than in adults, and also higher in newborns compared to adults of their own size from different species. An exception is represented by the very small

---

[7] In adult men, the alveolar and arterial values of Pco_2 are highly similar. This, however, may not be the case in the newborns of other species, especially in the first hours after birth.

Table 2.1. Arterial blood gases in newborns of various species

Species	Age	PaO$_2$	PaCO$_2$	pH	Reference[1]
Guinea pig	2 d	80	28	—	Clark and Fewell, 1996
Cat	5 d	85	36	7.3	Matsuoka and Mortola, 1993
	12 d	72	35	7.37	From data of Rohlicek et al., 1996
Dog	11 d	95	39	—	Rohlicek et al., 1998
Monkey	2 d	63	34	—	Woodrum et al., 1981
Pig	15 d	98	37	7.43	Waters et al., 1996
	7 d	—	37	—	Côté et al., 1992
	18 d	—	40	—	Côté et al., 1996
	6 d	102	37	7.42	Rosen et al., 1993
Sheep	6–8 d	73	38	7.45	Robinson et al., 1985
	1 wk	99	40	7.37	Moss et al., 1995
	1–6 d	79	29	7.39	Moss et al., 1987
	8 wk	86	44	7.40	Fahey and Lister, 1989
	1 d	82	43	7.41	Delacourt et al., 1996
	1–4 wk	99	38	7.41	Moss et al., 1996
Human	1 d	73	33	7.37	Polgar and Weng, 1979
Cow	1 d	71	44	7.40	Adams et al., 1993
	1–3 d	—	49	7.37	Reeves and Leathers, 1964
	17 d	93	44	—	Lekeux et al., 1984
	4–6 wk	73	41	7.38	Kiorpes et al., 1978
	6–8 wk	86	39	7.34	Bisgard et al., 1974
	1 d	65	44	7.44	Reeves and Leathers, 1967
	3 wk	94	43	7.37	Donawick and Baue, 1968
	2 d	62	42	7.43	Stenmark et al., 1987
Horse	1 d	68	45	7.39	Stewart et al., 1984
	1 d	72	43	7.38	Gillespie, 1975
	1 d	90	45	7.37	Rossdale, 1968
	1 d	100	43	7.41	Lavoie et al., 1990

Note: Age in days (d) or weeks (wk).
[1]When more than one age group was studied, the values reported here refer to the group closer to 1 day old.

species, where $\dot{V}o_2/W$ is similar to or smaller than that in adults. In the latter cases, in the days after birth, $\dot{V}o_2/W$ increases in eutherians, whereas it remains constant in marsupials. $\dot{V}o_2/W$ is larger in newborns of smaller species than in those of larger species, a pattern qualitatively similar to that well known for adults. However, the allometric exponent of the interspecies neonatal $\dot{V}o_2$ is 0.92 (i.e., higher than the classic 0.75 of the adult relationship), indicating that $\dot{V}o_2/W$ among newborns is much more uniform than it is among adults. One factor contributing to this interspecies homogeneity could be a systematic difference in the degree of development at birth, the precocial species being preferentially large and the altricial species preferentially small.

A proportionality with W very similar to that observed for $\dot{V}o_2$ applies also to pulmonary ventilation ($\dot{V}E \propto W^{0.91}$), and probably to alveolar ventilation. These relationships result from the tendency for breathing rate to decline with size ($f \propto W^{-0.09}$), as VT/W remains constant ($VT \propto W^{1.01}$). It follows that among newborn species the ventilatory equivalent ($\dot{V}E/\dot{V}o_2$) is very similar. From these relationships one could also anticipate that the values of alveolar and blood gases should be within a narrow range for all newborns; indeed, some measurements in a few species support this prediction.

Clinical Implications

In the early hours after birth, the breathing pattern is very irregular, commonly characterized by breaths of short duration and brief interruptions of the expiratory phase. In infants this pattern usually subsides within a day, but it can persist in infants with low lung compliance, because of a delay in the clearance of the pulmonary fluid (transient tachypnea syndrome). The brief interruptions of the expiratory flow reflect the narrowing of the glottis during expiration, a normal mechanism probably contributing to the process of fluid clearance. Maternal anesthesia and the mode of delivery employed seem to have minor, if any, effects on the early breathing pattern of the infant.

Obstruction of the nasal passages, as in choanal atresia, can have serious consequences, because the human infant is preferentially a nose-breather. Presumably, the nose-breathing attitude is not dictated by mechanical or anatomical constraints; rather, mouth-breathing may be a pattern behaviorally acquired through a learning process.

The bronchomotor tone is likely to be the major factor compensating for the high compliance of the infant's airways, which in vivo are little affected by changes in lung volume and airway pressure. But alterations in the control of the airway smooth muscle or in the airway mechanical characteristics, as occurs in airway spasm or tracheomalacia, can lead to an abnormal increase in airway resistance.

Summary

1. At birth, $\dot{V}O_2$/kg increases. During the following hours or days, in some species (small eutherians) it continues to increase; in others (marsupials) it remains constant.

2. In the first hours after birth, the breathing pattern is very irregular, and short periods of apneas can alternate with bursts of high rate. Often the expiratory flow is briefly interrupted, either partially or totally, by adduction of the laryngeal folds. This pattern keeps mean lung volume elevated, and probably contributes to the clearance of the pulmonary fluid. Most newborns are nose-breathers, and the ability to switch from nose- to mouth-breathing may be a learned behavioral response.

3. $\dot{V}O_2$/kg of newborns is usually larger than that in the adults of the same species, and also larger than that in adults of other species with their same body weight. Only in the smallest species, and in marsupials, is $\dot{V}O_2$/kg less than would be expected for adults of their size.

4. By comparing newborns a few days old from different species, we find that $\dot{V}O_2$/kg varies much less than it does among adults. In fact, the allometric exponent is 0.92, rather than the adult 0.75.

5. In newborns, pulmonary ventilation ($\dot{V}E$) scales to $W^{0.91}$; hence, just as $\dot{V}O_2$/kg, $\dot{V}E$/kg is also much more uniform among newborns than it is among adults ($W^{0.75}$). Interspecies differences are introduced by differences in breathing rate, which decreases with body mass ($f \propto W^{-0.09}$), although less than it does among adults. Tidal volume/kg is an interspecies constant, its average value (~ 11 ml/kg) slightly higher among newborns than in adults.

6. Few data are available on dead space, alveolar ventilation ($\dot{V}A$), or blood gases of newborn mammals. Nevertheless, arterial PCO_2 does not seem to change systematically with the size of the species, and values for newborns are comparable to those of adults. Hence, it is possible that, as is $\dot{V}E$, $\dot{V}A$ is directly proportional to metabolic rate.

Chapter 3

Mechanical Behavior
of the Respiratory Pump

Numerous factors are involved in the translation of the neural output into ventilation, the process that determines the dynamic performance of the respiratory system. In particular, following neural activation, the muscle force produced depends on the physical properties of the respiratory muscles; its translation into inspiratory pressure depends on the configuration and the geometrical and mechanical characteristics of the structure to which the force is applied. Finally, of the total muscle pressure generated, part is required to overcome the elastic properties of the respiratory system in order to change lung volume (V), and part is dissipated to overcome the frictional resistance of the respiratory system in order to move air, that is, to generate flow ($\dot{V}$) (Fig. 3.1).

This chapter reviews the main characteristics of the respiratory muscles and the basic features of the mechanical properties of the respiratory system of the newborn mammal, including their contribution to the neonatal breathing pattern. Some additional concepts and their applications specifically to measurements in the human infant are given in Appendix A.

The Respiratory Muscles

The mass and structure of the respiratory muscles, their intrinsic mechanical characteristics, resistance to fatigue, and blood perfusion are among the main factors contributing to the generation and maintenance of force during muscle activation.

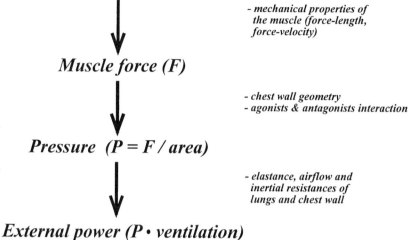

Activation of Respiratory Muscles

- *mechanical properties of the muscle (force-length, force-velocity)*

Muscle force (F)

- *chest wall geometry*
- *agonists & antagonists interaction*

Pressure (P = F / area)

- *elastance, airflow and inertial resistances of lungs and chest wall*

External power (P • ventilation)

Figure 3.1. Schematic summary of the processes involved in the translation of muscle activation into ventilation.

Mass and Fiber Type

Interspecies analysis suggests that the diaphragm mass is a constant proportion of body mass ($\sim$0.4%) in both newborn and adult mammals. This is probably true also for other respiratory muscles (Fig. 3.2). Diaphragm mass is proportional to body mass in the growing animal (Davidson, 1968; Lieberman et al., 1972; Powers et al., 1991) (Fig. 3.3), as well as among normal adult men of different body weight (Arora and Rochester, 1982). As is true of the other skeletal muscles (Aherne et al., 1971), increase in the diaphragm's muscle mass during postnatal development is almost uniquely due to muscle fiber hypertrophy (Bowden and Goyer, 1960; Mayock et al., 1987), and growth among different diaphragmatic regions is homogeneous (Maxwell et al., 1989). Similarly, the axons of the phrenic nerve increase in diameter and myelination (Schwieler, 1968; Wozniak et al., 1982), almost reaching the adult value within the first postnatal year in infants, and at about 2 months in kittens (Marlot and Duron, 1979b; Teixeira et al., 1992). The mechanical characteristics of the central tendon organ may play a pivotal role in the early development of the muscle component of the diaphragm (Griffiths and Berger, 1996).

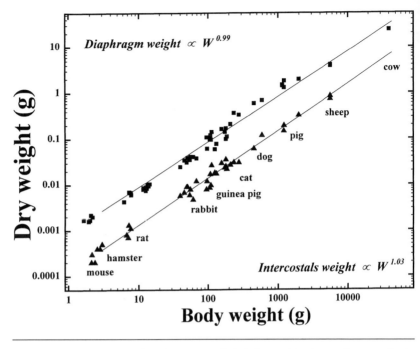

Figure 3.2. Diaphragm (■) and intercostals (▲) dry weight (grams) versus body weight (W, grams) in several newborn mammals. Intercostal weights represent the total weight of both internal and external muscles from two adjacent intercostal spaces. Although all data points are represented, the regression lines have been calculated from the average values of each species. The slopes of the log-transformed equations did not differ significantly from unity.

Morphologic, immunohistochemical, and biochemical criteria have indicated a change in the profile of the diaphragm fiber type during growth in numerous species, including the rat, hamster, rabbit, baboon, sheep, and horse. The slow-oxidative fibers are underrepresented in the early postnatal period (Sieck and Fournier, 1991). In addition, electrophoretic analysis of the native myosin heavy chain of the newborn diaphragm has revealed an embryonic-neonatal component not expressed at later postnatal stages, at the expense of other myosin isoforms (Fig. 3.4). The developmental pattern in diaphragm fiber typing appears to be consistent, at least qualitatively, between small immature (altricial) and large mature (precocial) newborns (D'Albis et al., 1991; Finkelstein et al., 1992; Watchko and Sieck, 1993; Brozanski et al., 1993; Johnson et al.,

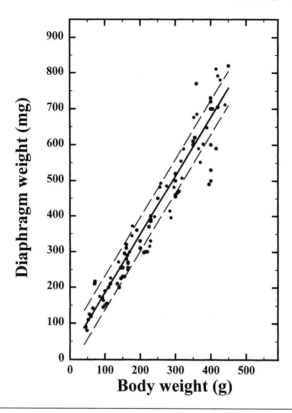

Figure 3.3. Diaphragm wet weight (mg) from rats of different body weight (grams): a linear proportionality between the two is maintained during the growth of the rat. *Dashed lines* represent 95% confidence intervals. Modified from Davidson, 1968.

1994; Cobb et al., 1994; Mayock et al., 1994). Qualitatively similar transformations have been reported in other respiratory muscles, including the intercostals and an expiratory muscle of the abdomen (Watchko et al., 1992b; Vazquez et al., 1993).

Fatigue and Fatigue Resistance

The issue of fatigue of the neonatal respiratory muscles and its potential contribution to chest distortion and breathing strategies have attracted much attention. In the premature infant and newborn rabbit, the diaphragmatic concentration of type I slow-twitch, fast-oxidative ("fatigue-

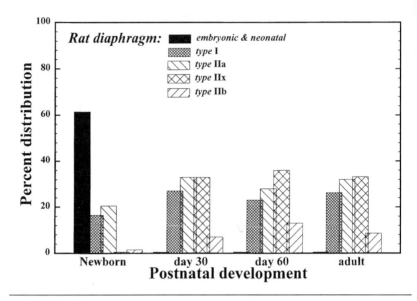

Figure 3.4. Percentage distribution of the myosin heavy chain isoforms in rats of different ages. In the newborn, the embryonic-neonatal isoform variety largely substitutes for the I and II complex. Averages of data from Brozanski et al., 1993; Watchko et al., 1987, 1992b; Vazquez et al., 1993; Sugiura et al., 1992; Talmadge and Roy, 1993; Gosselin et al., 1994; Mortola and Naso, 1995.

resistant") fibers was found to be low (Keens et al., 1978; Le Souëf et al., 1988). A decrease in the tonic activity of the inspiratory muscles was found to coincide with the occurrence of periodic breathing (O'Brien, 1985); the diaphragm aerobic capacity, evaluated from metabolic parameters including myoglobin concentration, was reported to be lower in the newborn lamb than in the adult (Griffiths et al., 1994). These data corroborated earlier suggestions that muscle fatigue was more likely in the newborn than in the adult; some changes in diaphragm electromyogram (EMG) were also interpreted as supporting this view (Lopes et al., 1981). The mechanical performance of a muscle, however, does not depend only on its fiber-type composition. Fiber size and capillarization are also important, and are known to change drastically during the development of many skeletal muscles, including the diaphragm (Tamaki, 1985; Smith et al., 1989; Sieck and Fournier, 1991). Indeed, contrary to the original proposition of a great propensity for fatigue in the newborn's respiratory muscles, and in agreement with the frequency distribution of slow- and fast-oxidative fibers, many morphological, biochemical, and functional studies on several species have indicated that the diaphragm is well prepared for high respiratory workloads even at young ages (Lieberman et al., 1972; Maxwell et al., 1983; Powers et al., 1991). In fact, postnatal changes in diaphragm fatigue resistance seem to correlate poorly with fiber oxidative capacity (Sieck and Fournier, 1991).

The diaphragm of the newborn baboon is as fatigue-resistant as that of the adult, and recovers more quickly (Maxwell et al., 1983). In rats and cats the diaphragm's resistance to fatigue is actually greater in newborns than in adults (Sieck et al., 1991; Watchko et al., 1992a; Trang et al., 1992; Watchko and Sieck, 1993; Prakash et al., 1993), presumably because of a more favorable balance between the energetic requirements of its fibers and their oxidative capacity (Watchko and Sieck, 1993). Changes in the power spectra of the diaphragmatic EMG, considered a sign of muscle fatigue, did not correlate with the infant's episodes of periodic breathing and apneas (Nugent and Finley, 1985). It is also known that ventilation in human infants and other newborn mammals can be maintained well above the resting value for long periods of time. For example, newborn rats in chronic hypoxic or hypercapnic conditions can sustain a level of ventilation almost double the resting value for several days with no signs of adaptation (Piazza et al., 1988; Rezzonico and Mortola, 1989). Thus it seems that, in the past, the occurrence of fatigue in the neonatal respiratory muscles has been overemphasized; although it may happen in some pathological conditions (Nichols, 1991), the information currently available does not convincingly indicate that respiratory muscle fatigue represents a greater potential problem in the newborn than it does in the adult.

Blood Perfusion

The perfusion of blood to the diaphragm and intercostals is greater (per unit of body, or of muscle, weight) in the newborn lamb than in either the fetus, during breathing movements, or the adult sheep (Soust et al., 1989a; Berger et al., 1994). This finding was interpreted as reflecting the higher metabolic and ventilatory requirements of the newborn, as compared to the adult, and the greater need to stabilize the rib cage in the newborn. With the rise in ventilation, diaphragmatic EMG and blood flow increased approximately in proportion to the metabolic demands of the muscle (Soust et al., 1989a, 1989b). In the lamb, only at very fast stimulation rates (100/min), probably exceeding its highest breathing rates, did the increase in O_2 delivery not fulfill the O_2 demands of the diaphragm (Nichols et al., 1989). Diaphragmatic blood flow was also shown to increase in newborn piglets when breathing with inspiratory resistive loads (Watchko et al., 1985). With growth, in the diaphragm as in other skeletal muscles, fiber size increases; for this reason, despite an increase in the total number of capillaries and in the number of capillaries per fiber, the muscle capillary density decreases (Tamaki, 1985; Smith et al., 1989).

Mechanical Properties

Only scattered information is available concerning the intrinsic mechanical properties (force-length and force-velocity relationships) of the respiratory muscles during the early postnatal period. In the rat diaphragm during postnatal growth (Table 3.1) the increase in fiber size and Lo (op-

Table 3.1. Diaphragm properties: newborn versus adult

Muscle mass/body weight	newborn = adult
Capillary density	newborn > adult
Fiber length	newborn < adult
Fiber cross-sectional area	newborn < adult
Fiber type: I/IIa,b,c	newborn < adult
Lo0 (length at which maximal twitch force occurs)	newborn < adult
Twitch force/muscle cross-sectional area	newborn < adult
Fusion frequency (minimum stimulus frequency necessary to evoke maximal tetanic force)	newborn < adult
Maximal tetanic force/muscle cross-sectional area	newborn < adult
Twitch contraction time (time to peak force)	newborn > adult
Half relaxation time (time for force to decline to ½ of peak twitch force)	newborn > adult
Fatigue resistance (by various criteria)	newborn > adult
Oxidative capacity (SDH, Mb, and other criteria[1])	newborn < adult
Myosin heavy chain phenotype	
embryonic/neonatal	in newborn only
I (slow)	newborn < adult
complex II (a,x,b)	little in newborn

[1]SDH, succinic dehydrogenase activity; Mb, myoglobin levels.

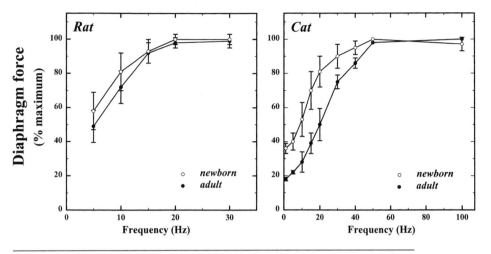

Figure 3.5. Relationship between frequency of muscle stimulation and diaphragmatic force (as a percentage of maximum) in newborn and adult rats (*left*) and cats (*right*). In newborns, the curve is shifted to the left, because of the longer twitch time. Symbols are mean values; bars are 1 SD. Right: from data of Sieck et al., 1991; left: from data of Watchko et al., 1992a.

timal muscle fiber length for maximum isometric force generation) is accompanied by increases in maximum tetanic force and shortening velocity, both probably contributed by the developmental changes in myosin heavy-chain composition mentioned above (Prakash et al., 1993; Johnson et al., 1994; Ianuzzo and Hood, 1995). In kittens, as well, the speed of diaphragm shortening increased with age, although the newborn vs. adult difference was small (Glebovskii, 1961). Some of the neonatal diaphragm responses to direct muscle stimulation, including the slower twitch time and leftward shift of the stimulation frequency-muscle force relationship (Fig. 3.5), indicate developmental differences in the excitation-contraction coupling, the mechanisms of which are not known, but may be common to those of other skeletal muscles in the process of aging (McCarter, 1990; Sarnat, 1992). Changes similar to those observed between newborns and adults have also been observed between the late fetus and the newborn rat, by use of a phrenic nerve-diaphragm in vitro preparation (Martin-Caraballo et al., 2000).

If one considers that the force of a muscle is proportional to its cross-sectional area, and that the pressure generated is the ratio between the

force and the surface area on which it is applied, the proportionality between respiratory muscle mass and body mass (see Figs. 3.2 and 3.3) offers the potential for similar values of inspiratory and expiratory pressures, irrespective of animal size or age. In other words, the smaller
respiratory muscles of the newborn generate less force than those of the
adult; this is true even after normalization for muscle cross-sectional area
(*specific* force) in the rat (Watchko et al., 1992a), but perhaps not in the
rabbit (Moore et al., 1993). But because this force is applied on a pro-
portionately smaller chest, the resulting inflating pressure can be similar
to that of the adult. And according to the Young-Laplace relationship, for
the same muscle force the pressure is inversely proportional to the radius
of curvature; hence, in infants, the small radius of curvature of the diaphragm dome favors the development of high inflating pressures, in comparison with those developed by the stronger muscles of older or larger
individuals. The result is that during breathing at rest both newborns and
adults inflate their lungs with similar swings in transpulmonary pressure
P_L (5–7 cm H_2O). During the first inspiration at birth the muscles generate very high inspiratory P, between 30 and 100 cm H_2O (see Fig. 1.13).
The infant's maximal inspiratory pressures during crying (Shardonofsky
et al., 1989) and the values of maximal expiratory P developed by children (Gaultier and Zinman, 1983) are close to the maximal static pressures of adults. In apparent contrast would seem to be the report that, in
piglets, the transdiaphragmatic pressure generated during supramaximal
stimulation of the phrenic nerves against occluded airways is progressively greater with the age of the animal (Watchko et al., 1986a). It should
be noted, however, that even when the airways are occluded and lung volume remains constant, diaphragm contraction is not isometric (Newman
et al., 1984b); hence, the force and pressure produced by the diaphragm
during phrenic stimulation are reduced, depending on the degree of diaphragmatic shortening, which is greater in the younger animals because
of the greater chest distortion.

Experiments with newborn piglets would seem to indicate that, as is
known to be the case in the adult, the mechanical efficiency of the diaphragm decreases during hypoxia (Watchko et al., 1986b) or severe hypercapnia (Watchko et al., 1987). The same has been observed in newborn rats, by the use of a phrenic nerve-hemidiaphragm preparation
(Bazzy, 1994). In the neonatal rat the force generated by the diaphragm

decreases at high frequencies of phrenic nerve stimulation, which suggests neuromuscular transmission failure (Bazzy and Donnelly, 1993; Feldman et al., 1991; Fournier et al., 1991). Hypoxia, however, seems to impair the diaphragm's neuromuscular transmission less in the newborn rat than in the adult (Bazzy, 1994).

The Newborn Thorax

The change most readily apparent in the infant thorax during growth is its progressive modification in shape. From an almost circular profile at birth, the cross section of the thorax becomes more elliptical, its lateral diameter exceeding the anterior-posterior dimension. In addition, the angle of attachment of the ribs to the vertebral column, initially close to 90°, gradually increases with age (Takahashi and Atsumi, 1955; Howett and Demuth, 1965; Openshaw et al., 1984) (Fig. 3.6). The pull due to the hydrostatic force generated by the abdominal content in the vertical posture is a likely explanation for these changes, probably combined with the increased muscle mass and the progressive mineralization of the soft thoracic structures. Functionally, this results in a progressive stiffening of the chest wall, which contributes to the increase in the resting volume of the respiratory system (Vr) and to a greater chest stability against changes in pleural pressure.

During growth, the end-tidal configuration of the diaphragm follows the changes in thorax shape and the development of the visceral organs. With the progressively more oblique orientation of the ribs, the costal attachments of the diaphragm, with respect to the central tendon organ, descend along the body's axial direction, enhancing the domelike shape of the muscle and increasing the area of apposition[1] (Devlieger et al., 1991). This development has functional implications for the translation of muscle force into rib cage expansion and lung inflation. In fact, as the diaphragm contracts, the transmission of the abdominal pressure to the lower ribs is larger the greater the apposition area. In the newborn, the small apposition area implies that the diaphragm exerts a lesser inflatory action on the rib cage. In addition, the greater compliance of the ab-

---

[1] The apposition area corresponds to the lower region of the rib cage; there, the diaphragm is apposed to the rib cage without interposed lung.

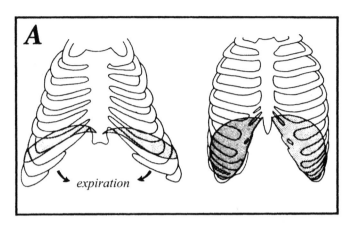

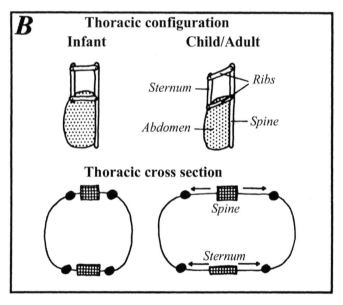

Figure 3.6. Schematic representation of the main differences in the shape of the human thorax and diaphragm dome (*A*) and of the thoracic configuration (*B*) between newborn infants (*left*) and adults (*right*). In most mammalian species that retain the quadrupedal posture into adulthood, the differences may be less apparent than they are in a bipedal species like humans. (*A:* as illustrated by Devlieger et al., 1991; *B:* as illustrated by Openshaw et al., 1984.

dominal wall does not favor the rise in abdominal pressure. Diaphragmatic efficiency is improved by abdominal loading, such as binding of the infant's abdomen (Fleming et al., 1979), or by breathing in the prone position. Both conditions decrease abdominal compliance and widen the apposition area (Wolfson et al., 1992), with the effect of improving both tidal volume and, in many cases, arterial oxygenation in comparison to breathing in the supine posture (Hutchison et al., 1979; Wagaman et al., 1979; Numa et al., 1997), without measurable changes in functional residual capacity (FRC) (Aiton et al., 1996; Numa et al., 1997).

In infants, the frequency of sudden infant death syndrome (SIDS) is usually reported to be higher among infants placed in the prone position, indicating that in the pathophysiology of SIDS other factors are much more important than the mechanical advantage of this posture. A synergistic activation of the abdominal muscles would also improve the mechanical efficiency of diaphragmatic contraction; but in infants and probably newborns of many other newborn species, during unassisted spontaneous breathing the abdominal muscles are usually little or not active (O'Brien et al., 1980; Kosch and Stark, 1984; South et al., 1987), and therefore do not contribute to maintaining a low abdominal compliance and optimal diaphragmatic length. Hence, neonatal breathing in the supine posture requires large diaphragmatic excursions (Laing et al., 1988; Wolfson et al., 1992) and, possibly, greater diaphragmatic work than is needed in the prone posture (Guslits et al., 1987). During CO_2-induced hyperventilation, not only the intercostal muscles (Hershenson et al., 1989) but also the abdominal muscles (South et al., 1987; Watchko et al., 1990; Praud et al., 1991) are recruited, and probably contribute to the hyperventilation by protecting the length of the diaphragm and the mechanical efficiency of its contraction.

Mechanical Interaction between Lungs and Chest Wall

The mechanical characteristics of the lungs and the chest wall are of fundamental importance for the translation of neural output into pulmonary ventilation. Lung volumes, and particularly the end-expiratory level, depend on the mechanical interaction between lungs and chest wall, and the response time of the respiratory system to the contraction of the inspiratory muscles depends on the compliance and resistance of the respiratory components. Hence, the relationship between pulmonary ventilation and metabolic requirements discussed in Chapter 2 must be largely dependent on the interspecies relationships of the mechanical parameters. In this section we will examine the passive mechanical properties of the newborn's lungs and chest wall; later (under "The *Active* Mechanical Behavior," below), we will consider the same properties under conditions of spontaneous breathing.

Terminology[2]

The pressure across a structure (respiratory system: Prs; lungs: PL; chest wall: Pw) is the pressure difference between inside and outside; positive is inflating, negative is deflating. At the resting volume (Vr), by definition, the respiratory system (rs) tends neither to inflate nor to deflate, and the transrespiratory system pressure (Prs = Palveolar − Pbody surface) is nil. At Vr, however, the lungs tend to collapse (PL > 0) and the chest wall tends to expand (Pw < 0), with equal and opposite pressures (PL = −Pw).

As lung volume increases above Vr, Prs increases, because of the increase in PL. When lung volume drops below Vr, by contrast, Prs falls below zero, because the outward recoil of the chest more than compensates for the inward recoil of the lungs. Hence, any change in lung volume (V) from Vr requires muscle contraction to offset Prs; its magnitude depends on the elastic properties of the respiratory system (Ers), plus a component dependent on flow ($\dot{V}$) determined by the airflow resistance of the respiratory system (Rrs). These concepts are summarized by a simplified version of the equation of motion of the respiratory system

$$Prs = P_{elastic} + P_{flow\text{-}resistive} = Ers \cdot V + Rrs \cdot \dot{V}.$$

(Equation 3.1)

When lung volume does not change ($\dot{V} = 0$), the flow-resistive component is nil, Prs = Ers · V, and the system is in a *static* mode. Conversely, when $\dot{V} \neq 0$, the system is in a *dynamic* mode. Each of these two modes can be *active* or *passive*, depending on, respectively, the presence or absence of activity of the respiratory muscles. For example, in many adult

---

[2] Additional terminology and concepts pertinent to some measurements of respiratory mechanics are given in Appendix A.

mammals, during spontaneous breathing in resting conditions, a large part of expiration is a *passive-dynamic* event, since the respiratory muscles are usually inactive; on the other hand, a forced expiration is an *active-dynamic* process. Breath-holding is an *active-static* maneuver, whereas during relaxation against closed glottis at end-inspiration (a phenomenon very common in the early hours after birth; see Chapter 2, under "Mechanical Events in the First Hours after Birth") the respiratory system is in a *passive-static* mode.

It should be noted that an additional component of Prs is that required to accelerate the gas, its magnitude depending on the inertia of the gas and of the respiratory structures. During resting breathing, this component is usually small and, for simplicity, neglected. But during rapid changes in $\dot{V}$, as during high-frequency ventilation, the pressure losses due to inertia can contribute substantially to Prs.

Mechanical Properties of the Lungs

Unlike that of other internal organs, a large part of the development of the lungs occurs postnatally. Only a small percentage of the adult quantity of alveoli is present in the lungs of newborn infants (Dunnil, 1962), or in the lungs of newborns of other species of similar or greater size (Engel, 1953). In the lungs of newborn rodents and other altricial species are alveolar sacs rather than true alveoli (Reid, 1967; Amy et al., 1977; Thurlbeck, 1977).

In newborns of many species the mass of the lung is approximately a fixed fraction of body mass (about 0.4% for the dry lung mass, or 2% for the wet mass) (Fig. 3.7). As is also true for the heart and other internal organs of the newborn (Webster et al., 1947; Webster and Liljegren, 1949), lung mass is a greater proportion of body mass than it is in adults (Coppoletta and Wolbach, 1933; Webster and Liljegren, 1949; Avery and Cook, 1961; Amy et al., 1977; Thurlbeck, 1977; Fisher and Mortola, 1980a; Fisher and Mortola, 1981; Yokoyama and Nambu, 1981; Nardell and Brody, 1982; Gaultier et al., 1984), and decreases rapidly with postnatal growth (Fig. 3.8). In the fetus the lung is an even greater proportion of body mass than it is in the newborn (Suen et al., 1994).

Differences in the time of appearance of surfactant during gestation and in the ability to retain air during deflation have been observed between the upper and lower lobes of lamb and monkey fetuses (Howatt et

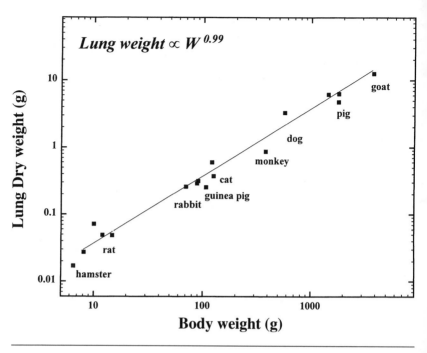

Figure 3.7. Relationship between body weight (grams) and lung dry weight (grams) of neonates of several species. Each symbol represents the mean value of a particular study. Although all data points are represented, the regression line was calculated after averaging the values pertinent to any particular species. The slope of the log-transformed equation did not differ significantly from unity. Data for hamster are from [1,4]; rat[1-3]; rabbit[2]; cat[2,5]; guinea pig[2,6,7]; dog[2]; pig[2,8,9]; goat[10]; monkey[11]. [1]Mortola, 1991; [2]Fisher and Mortola, 1980a; [3]Nardell and Brody, 1982; [4]Schumacher et al., 1965; [5]Mortola et al., 1984b; [6]Gaultier et al., 1984; [7]Lechner et al., 1986; [8]Sullivan and Mortola, 1985; [9]Standaert et al., 1991; [10]Avery and Cook, 1961 (assuming dry/wet ratio = 0.2); [11]Hodson et al., 1977.

al., 1965; Kotas et al., 1977). Studies of lamb and rabbit fetuses, moreover, have shown a time lag in the morphological aspects of lung maturation (Brumley et al., 1967; Kikkawa et al., 1971), the upper lobes leading the lower lobes. These differences decrease with gestational age, and toward the end of gestation and in full-term newborns no interlobal differences seem to exist, either in morphological appearance, dry-to-wet weight ratio, phospholipid content, or elastic behavior (Howatt et al., 1965; Brumley et al., 1967; Mortola et al., 1984b).

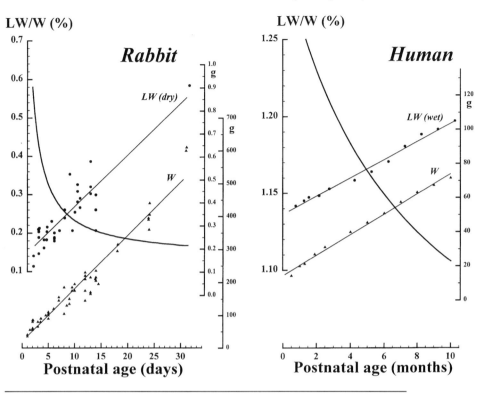

Figure 3.8. Lung weight (LW) and body weight (W) during the early phases of postnatal development in rabbits (*left*) and humans (*right*). From the LW and W regression lines, the values of the corresponding lung weight-body weight ratios have been calculated, and are represented by the *thick continuous lines*. Lung mass is a higher proportion of body mass during the early postnatal stages. Data on rabbits from data of Fisher and Mortola, 1980a; lung weights of infants from data of Coppoletta and Wolbach, 1933; body weights of infants from standard curves for Caucasian populations.

Estimates of the lungs' elastic properties are commonly obtained from the changes in V per unitary change in P_L, in static conditions. In vivo this necessitates the measurement of esophageal pressure, a technically difficult measurement in small species, and one not necessarily reflecting mean pleural pressure in conditions of chest-wall distortion. Whether in vivo or in vitro, several methodological factors can introduce substantial variability; among these, the most important are the control of absolute lung volume and the effect on it of gas compression, the lung-volume *his-*

tory (which is the pattern of lung inflation before the measurements), and the magnitude and rate of V (or PL) changes. Despite these uncertainties, some features of the newborn lung PL-V curve seem consistent, at least in qualitative terms, among several species.

The inflation limb of the PL-V curve often tends to be S-shaped (Agostoni, 1959; Fisher and Mortola, 1980a), even when care is taken to avoid lung collapse and lung-volume history has been carefully controlled. For a unitary change in PL, the change in V is at first rather small, then increases at higher pressures to decrease again above 15–20 cm H_2O. This probably indicates a process of opening and recruitment of alveoli or airways that had been closed at Vr, or complex changes in the shape of the surface tension area (Wilson, 1979). Indeed, only some of the alveoli are aerated in the first few days of life (Gruenwald, 1947). Agostoni pointed out that the peculiar shape of the PL-V inflation limb of the newborn lung resembles that of the edematous lung (Cook et al., 1959), suggesting the presence of some fluid in the alveoli or in the interstitial spaces.

The deflation curve is more concave toward the P axis in the newborn than in the adult; in this respect, the exponential fitting of the deflation PL-V curve from total lung capacity (TLC) to 50 percent TLC (Colebatch et al., 1979; Gibson et al., 1979), which has been found adequate in many adult mammals (Schroter, 1980), is not appropriate in most newborns (Fagan, 1976, 1977; Saetta and Mortola, 1985a). The only known exception would appear to be the newborn guinea pig (Gaultier et al., 1984), a precocial species at birth with a high degree of lung development (Lechner and Banchero, 1980, 1982). At end deflation, in all species, air is retained in the airways, and the importance of surfactant foam production, and of lung maturity for this phenomenon, is addressed in Chapter 2, under "Coupling of Ventilation and Metabolism."

The Liquid-Filled Lung. The mechanical properties of the liquid-filled lung are thought to depend on connective-tissue elements of the lung parenchyma and pleura (Setnikar, 1955; Mead, 1961). The liquid-filled lung and strips of subpleural lung tissue have therefore been used to assess the mechanical behavior of the lung tissue alone, excluding contributions of surface forces, and their changes with growth (Martin et al., 1977; Fedullo et al., 1980; Nardell and Brody, 1982). Of course, this assumes that the mechanical behavior of the pulmonary tissue in the liquid-filled condition is the same as when surface forces are operating, an assumption that may not necessarily be correct (Forrest, 1976; Wilson, 1979). The collagen and elastin constituents are low in the newborn lung (Bradley et al., 1974; Keeley et al., 1977; Powell and Whitney, 1980; Nardell and Brody, 1982); their progressive increase seems to parallel, respectively, the increase in the lung's resistance to rupture and the formation of new alveoli (Stanley et al., 1975; Kida and Thurlbeck, 1980; Nardell and Brody, 1982). Other aspects of lung mechanics do not correlate well with the changes in elastin and collagen contents (Nardell and Brody, 1982), probably indicating that maturational changes in other lung-tissue constituents, including the ground substance, may also be involved.

Stress relaxation after a step volume change (i.e., the drop in P_L at $\dot{V} = 0$), an index of the viscous properties of lung tissue, is more pronounced in newborns than in older rats (Nardell and Brody, 1982), a finding similar to what is observed in air-filled lungs of rat pups and piglets (Sullivan and Mortola, 1987; Peslin et al., 1991; Pérez Fontán et al., 1992). These stress-relaxation phenomena are the major factors underlying the frequency dependence of lung compliance at low flow rates (Sullivan and Mortola, 1986; Bigos and Pérez Fontán, 1994), and are therefore an important contributor to the active stiffening of the respiratory system during breathing (see "Inspiration," below).

Just as in air, in the liquid-filled condition the deflation P_L-V curve is more concave toward the P axis and, with respect to the total volume inflated, more liquid volume is retained at end-deflation in newborns than in adults (Nardell and Brody, 1982). The P_L-V hysteresis, which in liquid-filled conditions reflects the viscous and elastic properties of the lung tissue, decreases with growth (Fedullo et al., 1980; Nardell and Brody, 1982). But when these properties are studied in lung strips, in the form of tension-length hysteresis, the developmental aspects are less clear; tension-length hysteresis was found to increase in humans as they age, to diminish in rats and rabbits, and to change little with age in horses and monkeys (Martin et al., 1977). Thus, the mechanical properties of newborn lung tissue differ from those of adults, as is to be expected, considering the differences in structural components, and the viscous properties of the newborn lung tissue are evidently more important, with respect to the elastic properties, than those of adult lung tissue. On the

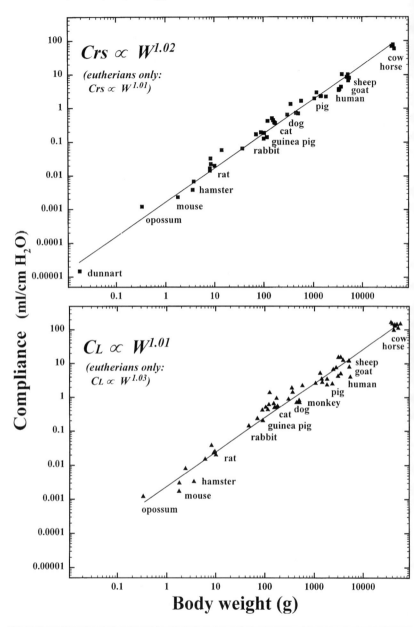

Figure 3.9. Allometric curves of (*top*) respiratory system compliance (Crs) and (*bottom*) lung compliance (CL) in newborn mammals; neither exponent differs significantly from unity. Each symbol is the mean value of a particular study. Although all data points are represented, the regression lines were calculated af-

one hand, this viscoelastic behavior decreases the efficiency of the system, since part of the elastic pressure stored during inspiration is lost as heat, therefore reducing the P_L available during deflation (Mortola et al., 1985a); on the other hand, the less pronounced lung recoil favors the maintenance of air in the pulmonary airways.

Allometry of Lung Compliance. The comparison of values of lung compliance (C_L) among newborn mammals of a rather limited range of body size has indicated that C_L scales approximately in proportion to body mass (Fisher and Mortola, 1980a). The same finding ($C_L \propto W^{1.01}$) emerges when data obtained from the use of different techniques are pooled together over a wide range of species (Fig. 3.9). Some departures from the proportionality constant (~ 2.6 ml $\cdot$ cm $H_2O^{-1} \cdot kg^{-1}$) are significant, but, as is often the case in this type of interspecies relations, it is difficult to determine whether differences are simply an expression of the variability introduced by combining data obtained with different methodologies or in fact have a physiological meaning (as is the case with the newborn opossum; see "Being Very Small," below). The interspecies constancy of C_L is also manifest when lung weight, rather than body weight, is used as the normalizing factor (Fisher and Mortola, 1980a), owing to

---

ter averaging the values pertinent to any particular species. Data for hamster are from [1,2]; rat[2–4,6,32–34,45,46]; rabbit[2,3,6,35,36]; mouse[1,2]; opossum[1]; cat[2,3,6–10,45]; guinea pig[2,6,11–13,37]; dog[2,3,6,10,14,15,38]; pig[2,3,6,15–17,39,47–49]; monkey[18]; sheep[2,19–22,45,50]; horse[23,24]; cow[2,25–29]; infant[5,30,40–44]; goat[31]; dunnart[51]. [1]Frappell and Mortola, 1989; [2]Bartlett and Areson, 1977; [3]Fisher and Mortola, 1980a; [4]Newman et al., 1984a; [5]Mortola et al., 1993; [6]Mortola, 1983; [7]Mortola et al., 1984b; [8]Rossi and Mortola, 1987; [9]Fisher et al., 1987; [10]Fisher et al., 1990; [11]Gaultier et al., 1984; [12]Lechner and Banchero, 1980; [13]Lechner et al., 1986; [14]Agostoni, 1959; [15]Mortola et al., 1987a; [16]Sullivan and Mortola, 1985; [17]Standaert et al., 1991; [18]LaFramboise et al., 1983; [19]Shaffer et al., 1976; [20]Shaffer et al., 1978; [21]Davis, 1988b; [22]Davis, 1988c; [23] Koterba et al., 1994; [24] Gillespie, 1975; [25]Kiorpes et al., 1978; [26]Slocombe et al., 1982; [27]Lekeux et al., 1984; [28]Inscore et al., 1990; [29]Leblanc et al., 1991; [30]Polgar and Weng, 1979; [31]Avery and Cook, 1961; [32]Adolph and Hoy, 1960; [33]Marlot and Mortola, 1984; [34]Saetta and Mortola, 1985b; [35] Mortola et al., 1984c; [36]Thach et al., 1976; [37]Clerici et al., 1989; [38]Griffiths et al., 1983; [39]Clement et al., 1986; [40]Mortola et al., 1990a; [41]Mortola et al., 1982b; [42]Mortola et al., 1984a; [43]Taeusch et al., 1974; [44]Tepper et al., 1984; [45]Suen et al., 1994; [46]unpublished data; [47]Mundie et al., 1994; [48]Potter et al., 1997; [49]Dreshaj et al., 1994; [50]Davey et al., 1998; [51]Frappell and Mortola, 2000.

the above-mentioned interspecies constancy of the ratio between lung and body weight (see Fig. 3.7). That C_L/kg is similar among newborns of different sizes should indicate that the major source of lung recoil (i.e., the collapsing pressure generated at the alveolar-liquid interface) is also similar among species. This component of lung recoil is proportional to surface tension and inversely related to alveolar diameter. Since surface tension is likely to be an interspecies constant, then it should follow that, at any given P_L, alveolar dimensions are similar among different-sized newborns. This prediction was supported by experimental measurements (Bartlett and Areson, 1977). It should also follow that the major structural difference among newborn lungs of different sizes lies mostly in the total *number* of alveolar units rather than in their dimensions.

A comparison of C_L between newborns and adults of the same species on the basis of the animals' body mass yields results that do not necessarily agree with those obtained when lung weight is used as the normalizing parameter, because of the substantial decrease in the contribution of the lungs to total body mass during growth (see Fig. 3.8). In newborns, C_L is small and tends to increase progressively with age (Avery and Cook, 1961; Fisher and Mortola, 1980a, 1981; Yokoyama and Nambu, 1981; Gaultier et al., 1984). This pattern is consistent for many of the species investigated, including human infants (Cook et al., 1958), and is probably due to the smaller volume of the newborn air spaces per unit of lung mass, rather than to a lower tissue distensibility (Saetta and Mortola, 1985a). When rats have reached a body weight of about 350 grams, C_L (per unit of lung weight) begins to decrease, a phenomenon related to the modification of lung tissue that accompanies the process of aging (Sahebjami and Vasallo, 1979).

Mechanical Properties of the Chest Wall

As defined by respiratory physiologists, and in the broadest functional sense, the chest wall comprises any set of structures *outside the lung* that move during passive changes in lung volume. More specifically, once the functional arrangements of the main muscles involved with breathing have been taken into account, it is convenient to consider the *chest wall* as being characterized by two compartments acting as aspiration pumps on the lung, the abdomen (including the diaphragm) and the rib cage (or thorax). The dynamic mechanical performance of these two pumps (examined in "Chest Wall Motion," below) depends not only on the intrinsic properties of the muscles, but also on their configuration, their elastic and resistive characteristics, and the timing and intensity of the muscle action involved.

The P-V curve of the chest is commonly obtained by determining the difference between the P-V curve of the respiratory system (lungs and chest combined) and that of the lungs alone. In the tidal volume range, particularly in the smallest mammalian species, these two curves are almost superimposed. The P-V curve of the chest is therefore almost parallel to the V axis and little offset from it, that is, the compliance of the chest wall (Cw) is extremely high relative to that of the lung, and its recoil pressure is low (Agostoni, 1959; Avery and Cook, 1961; Nightingale and Richards, 1965; Fisher and Mortola, 1980a). Below Vr, Cw may decrease (Avery and Cook, 1961), although the occurrence of airway closure complicates the precise definition of this portion of the curve.

In infants, the partitioning of Crs in its lung (CL) and chest wall (Cw) components requires measurements of mean pleural pressure, which are often much more problematic than they are in the adult and need to be carefully interpreted (LeSouëf et al., 1983; Coates and Stocks, 1991). Richards and co-workers (Richards and Bachman, 1961; Nightingale and Richards, 1965) found that the Crs value was similar to known values of CL, and therefore concluded that the human infant Cw is very high, and that the chest wall contributes little to the recoil of the respiratory system. Their conclusion agrees with later measurements of Cw obtained as the ratio of the change in lung volume to the corresponding change in pleural pressure during artificial inflation of the lungs (Reynolds and Etsten, 1966; Gerhardt and Bancalari, 1980). Hence, whereas in man Cw and CL have similar values, in infants Cw is about 4–5 times higher than CL (Polgar and Weng, 1979); this difference may become even larger at higher lung volumes, because with increased lung inflation the compliance of the lung decreases more than that of the chest wall. Some indirect computations would indicate that in the 5-year-old child the Cw-CL ratio is about 1.5–2, and remains approximately steady around this value throughout adolescence (Sharp et al., 1970).

Allometry of Crs. Figure 3.9 (above) shows the allometric relationship of Crs, which is obtained by pooling together data from various studies. The curve has an allometric exponent similar to that of the lung; in fact, not only the slope but also the proportionality constants are rather similar ($\sim$1.8 ml $\cdot$ cm $H_2O^{-1} \cdot kg^{-1}$ for the respiratory system and $\sim$2.6 ml $\cdot$ cm $H_2O^{-1} \cdot kg^{-1}$ for the lungs). This reaffirms what was mentioned earlier, that on average the newborn Crs is mostly contributed by CL. Hence, CL/Cw is lower in newborns than in adults (Fig. 3.10). The similarity between the exponents of the allometric functions scaling CL to body mass (1.01) and Crs to body mass (1.02) indicates that the relative contribution of the neonatal chest wall to the elastic recoil of the respiratory system is not only small but also almost the same fraction for all species. The latter conclusion is surprising, since one would expect a relatively stiffer chest wall in the newborns of the larger species (lambs, goats, foals, etc.), which are able to maintain the standing position almost immediately after birth. It should be emphasized, however, that these measurements can be highly sensitive to postnatal age. For example, in rats, Crs increased twelvefold from fetal day 19 to postnatal day 4, and about half of this change occurred postnatally (see Fig. 1.18).

With growth, Crs normalized by body mass decreases, especially in the smallest species, indicating a progressive stiffening of the chest wall (Fig. 3.11). A decrease in volume-specific Crs (e.g., Crs/FRC) has also been observed in humans, by pooling together data from different subjects between 22 months and 18 years of age (Sharp et al., 1970).

From the viewpoint of pulmonary function testing, one practical implication of the very high ratio between Cw and CL in the infant is that measurements of Crs can be interpreted as reflecting CL, which is convenient because Crs is much more easily measured than CL (see Appendix A). In newborns, the high value of Cw/CL contributes to peculiar patterns of chest wall motion, with distortion during inspiration.

Chest Wall Motion. A striking characteristic of the dynamic behavior of the chest wall in infants is the paradoxical inward movement of the upper rib cage (rc) during inspiration, which is particularly evident during rapid-eye-movement (REM) sleep (Finkel, 1972; Knill et al., 1976; Henderson-Smart and Read, 1978; Curzi-Dascalova, 1982) (Fig. 3.12). This pattern is easily observed in many newborns, especially in the smallest species, and it has also been demonstrated by an ultrasound technique during fetal breathing (Patrick et al., 1978; Poore and Walker, 1980). The paradoxical motion of the rib cage occurs even more commonly in preterm infants than in term infants, regardless of sleep state (Davi et al.,

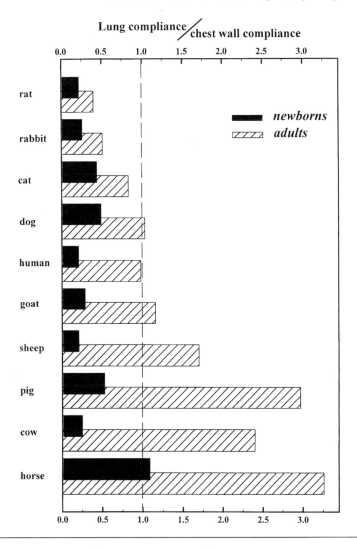

Figure 3.10. Ratio between lung compliance and chest wall compliance in newborns (*filled bars*) and adults (*hatched bars*) of several species. Values for newborns from Agostoni, 1959; Avery and Cook, 1961; Fisher and Mortola, 1980a; Polgar and Weng, 1979; Davis et al., 1988b; Slocombe et al., 1982; Standaert et al., 1991; Koterba et al., 1994. Data for adults from Agostoni, 1959; Avery and Cook, 1961; Fisher and Mortola, 1980a; Polgar and Weng, 1979; Crosfill and Widdicombe, 1961; Gillespie, 1983; Gallivan et al., 1989; Mutoh et al., 1991, 1992; Koterba et al., 1994; Leith, 1971; Chelucci et al., 1993; Spells, 1969/70.

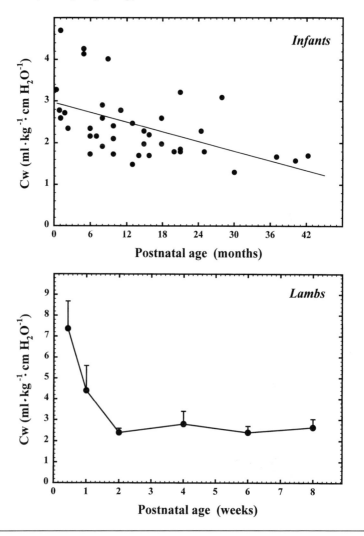

Figure 3.11. *Top:* Compliance of the chest wall (Cw), normalized by body weight, in infants of different postnatal age. The line is the regression line (slope −0.037). *Bottom:* Compliance of the chest wall in lambs between 3 days and 8 weeks after birth (means ± SEM). *Top:* modified from Papastamelos et al., 1995; *bottom:* modified from Davey et al., 1998.

1979). With postnatal growth, the frequency of paradoxical rib cage motion decreases (Gaultier et al., 1987), and in adolescents the rib cage expands in inspiration during REM sleep as well, albeit less than during non-REM sleep or wakefulness (Tabachnik et al., 1981).

Inflation of a paralyzed animal by external means is accompanied by outward movement of both rc and abdomen (ab) (Fig. 3.13). When the diaphragm is relaxed and free passive tension, transdiaphragmatic pressure is nil and the pressure across the chest wall (Pw) is the same as that across its individual components (Prc and Pab); hence, Pab can be seen as the pressure moving the rc (Pab = Prc). Goldman and Mead (1973) found, in adult humans, that this equation holds true even when the diaphragm is under passive tension by abdominal compression. And in the adult man, particularly during resting breathing in the upright posture, the chest moves close to the passive curve (Konno and Mead, 1967; Sharp et al., 1975; De Troyer and Estenne, 1984). Goldman and Mead (1973) thought that during diaphragmatic contraction the negative pleural pressure is offset both by the direct action of the diaphragm on rc at the level of the costal insertions and by the expanding effects of Pab on the apposition area; they proposed therefore that "contraction of the diaphragm alone will move the chest wall along the relaxation characteristic."

An alternative possibility is that the coupling between rc and ab is not as efficient as was proposed by Goldman and Mead, and that the diaphragm contraction does distort the chest wall from its passive configuration, determining the inward movement of rc; this effect, however, would be minimized through the contraction of extradiaphragmatic muscles. Such a possibility finds support in observations that in some subjects, during inspiration, ab and rc expand, respectively, more and less than in the passive condition (Sharp et al., 1975; Mortola et al., 1985b), and in tetraplegic patients rc moves paradoxically inward (Mortola and Sant'Ambrogio, 1978; Danon et al., 1979), a finding not compatible with Goldman and Mead's proposition. Similarly, in infants with tetraplegia or under spinal anesthesia, in whom the diaphragm is presumably the only functioning inspiratory muscle, paradoxical motion of the rib cage is the consistent finding (Thach et al., 1980; Pascucci et al., 1990). It would follow, then, that during active breathing the movement of the chest wall close to, or along, the relaxation line represents *compensated* distortion, and not absence of it. The two concepts are quite different when exam-

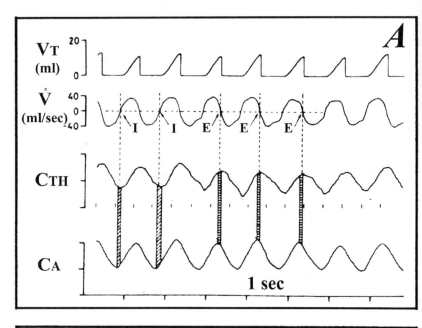

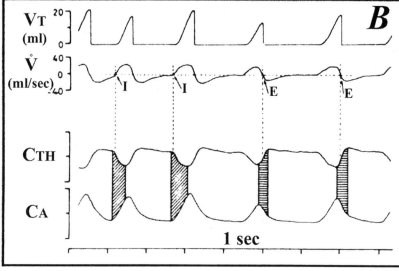

Figure 3.12. Records of tidal volume (VT, inspired volume only), airflow (V̇), chest circumference (CTH), and abdominal circumference (CA) in an infant during quiet sleep (A), and during active (REM) sleep (B). The phase difference between the minimum values (in A) of CA and CTH and that between the maximum

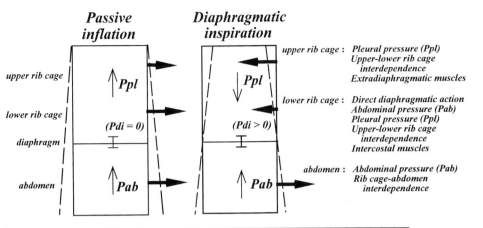

Figure 3.13. Schematic representation of the pressures applied to the chest wall and its components (upper and lower rib cage) and abdomen during passive inflation (*left*) and spontaneous inspiration with diaphragm only (*right*). During passive inflation, all three compartments expand (*heavy arrows*). During diaphragmatic contraction, Ppl drops, causing an inward motion of the upper rib cage. Dashed lines indicate the expected directional change in the silhouette of the chest wall in the two conditions. At extreme right, a summary of the pressures and forces determining motion of the upper rib cage, lower rib cage, and abdomen during spontaneous (diaphragmatic) inspiration. Ppl, pleural pressure; Pdi, transdiaphragmatic pressure; Pab, abdominal pressure.

ined in light of the energetics of breathing, since compensated distortion implies recruitment of intercostal muscles for the sole purpose of stabilizing the ribs and improving the diaphragm's ventilatory efficiency.

A functional evaluation of the net effect of chest distortion on lung volume can be done by comparing lung volumes (or PL) under both active and passive conditions for the same level of Pab. Equally informative is the relationship between lung volume and the displacement of the abdominal wall, since, in the absence of abdominal expiratory activity, changes in Pab and abdominal motion are closely related (Fig. 3.14). Normally, infants during tidal breathing present a "volume loss," which

values (in *B*) between CA and CTH are indicated by the striped areas. Note that the phase differences increase in active sleep. Vertical dotted lines indicate the onset of inspiratory flow (I) or expiratory flow (E) Slightly modified from Andersson et al., 1983.

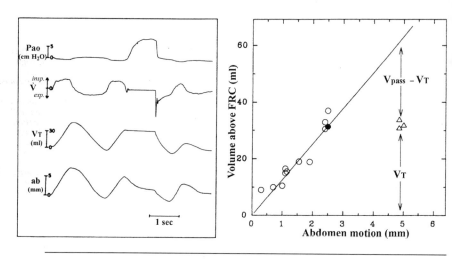

Figure 3.14. *Left panel:* From top to bottom, records of the pressure at the airway opening (Pao), airflow ($\dot{V}$), tidal volume (VT), and anterior-posterior motion of the abdomen (ab) in an infant during spontaneous breathing. At the end of the second inspiration, the airways have been occluded to trigger the Hering-Breuer reflex. Once P reaches a plateau, the respiratory system is relaxed (static-passive mode). From the passive V and ab values of this and other occlusions, the passive ab-V relation is constructed (*right panel*); ● represents the occlusion shown in the record at left. The slope of the line represents the passive ab-V curve. △ represent the ab-V end-inspiratory values of three breaths (static-active mode). At the same ab, the V difference between the passive curve and end-inspiration (Vpass − VT) represents the V loss due to distortion.

means that tidal volume is less than the passive volume at the same ab or Pab (Mortola et al., 1985b); in fact, the volume loss can be as large as tidal volume itself.

The volume loss is expected to be greater during REM sleep, when inward movement of the rc in inspiration is common (see Fig. 3.12), and in the supine posture more than in the prone position, owing to the greater abdominal compliance in the former case. In piglets, in the absence of pulmonary vagal feedback, the deeper breathing increases the volume of distortion (Clement et al., 1986). A large volume loss and increased diaphragmatic work are also to be expected under conditions of abnormally low CL, as in infants with bronchopulmonary dysplasia (Allen et al., 1991), in prematures (Heldt, 1988), or when airway resistance is increased. By contrast, a very small or absent volume loss should indicate increased activity of the extradiaphragmatic muscles, as under conditions of increased chemical drive (Hershenson et al., 1989). Seen from this perspective, a major difference in chest-wall dynamics between infants and adult humans would lie in the compensatory activity of the extradiaphragmatic inspiratory muscles. In adults, the intercostal muscles and scaleni muscles are active particularly in the upright posture (Koepke et al., 1955; Taylor, 1960a; De Troyer and Estenne, 1984); in infants, especially during REM sleep, the intercostals are minimally, if at all, active (Muller et al., 1979; Lopes et al., 1981). Proper contraction of the intercostal muscles to counteract the diaphragmatic distortion of the rc requires functional development of the γ-loop of the muscle proprioceptive organs (i.e., spindles), whereby the activity of the α-extrafusal fibers is continuously modulated by the state of contraction of the γ-intrafusal fibers. Experiments on anesthetized kittens and rabbits indicated that the spindle-control mechanism of the intercostal muscles is not fully operative in the newborn (Schwieler, 1968; Trippenbach and Kelly, 1983), which accords with similar observations in other skeletal muscles (Schulte, 1981).

Static Coupling between Lungs and Chest Wall

The higher levels of Cw/C_L in newborns than in adults should result in a lower static relaxation volume of the respiratory system, Vr, because at any lung volume the outward pull of the chest on the lung is reduced (Fig. 3.15). Indeed, measurements of Vr (per unit of lung mass) or of P_L at Vr support this expectation. In newborns, from the very small opossum to the very large foal, P_L at Vr was found to range between 0.5 and 2.5 cm H_2O (Agostoni, 1959; Avery and Cook, 1961; Fisher and Mortola, 1980a; Mortola et al., 1984b; Frappell and Mortola, 1989; Koterba et al., 1994), with a small trend toward an increase with species size. In most newborn species, the value of P_L at Vr is about half than in adults, but in the foal the values for newborns and adults are similar. Small values of P_L have been measured in the human infant at end-expiration by the use of the esophageal balloon technique (Helms et al., 1981).

The lung volume at relaxation, Vr, has been measured in newborns of several species, and found to be directly proportional to species body weight (the proportionality constant being about 30 ml/kg; Fisher and

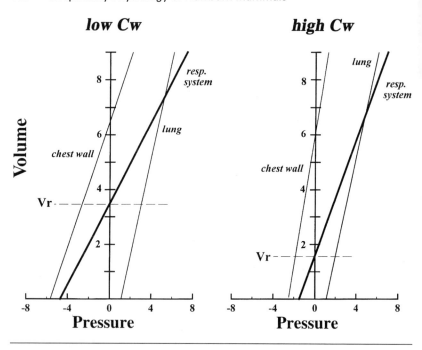

Figure 3.15. Schematic pressure-volume relations of lung, chest wall, and respiratory system in conditions of low (*left panel*) or high (*right panel*) chest wall compliance (Cw). Linear relations have been drawn for simplicity. With higher Cw, the relaxation volume of the respiratory system (Vr) and the mean recoil pressure of the lungs at Vr both decrease.

Mortola, 1980a, 1981) or to lung weight (6 ml/kg; Fisher and Mortola, 1980a). Comparisons with adults have been made on the basis of lung capacity, animal size, body weight, or lung weight. The type of normalization selected is crucial for the conclusion. Normalized by lung weight, Vr in the newborn is in general less than that in the corresponding adult (Fisher and Mortola, 1980a). Similarly, in infants, despite the fact that FRC is dynamically maintained above Vr (see "Expiration," below), FRC per lung weight is smaller than that in the adult man (Cook et al., 1958). Like infants, other newborn mammals minimize the potential problem of the low Vr by maintaining a dynamically elevated FRC (see "Expiration").

The narrow range of PL at Vr among newborns of different body and lung size indicates, as it does among adults, that the weight of the lung does not determine PL at Vr. It also implies that the difference in PL between dependent and nondependent lung regions is less than if PL at Vr were size-dependent. This finding accords with the view that the respiratory system is designed to maintain PL within close limits in all animals (Agostoni and D'Angelo, 1970/71).

The value of pleural pressure generated by the balance of recoil forces of lungs and chest wall is also referred to as pleural *surface* pressure (Ppl); its difference from airway pressure determines the value of transpulmonary pressure. The actual pressure of the liquid in the pleural space (Ppl(liq)), which varies among regions, is determined by the various factors that reabsorb the liquid and maintain it in a dynamic state. Ppl(liq) can be more subatmospheric than Ppl, although the exact mechanisms underlying this difference are still debated (Agostoni and D'Angelo, 1991). In the puppy, Ppl(liq) is about -3 cm H_2O (Agostoni, 1959; Miserocchi et al., 1984), both on the costal side and the mediastinal side (Miserocchi et al., 1984). This value is only slightly more negative than Ppl. In contrast, in the supine adult dog at Vr, Ppl(liq) is $4-6$ cm H_2O more negative than the surface pressure on the costal side (Setnikar et al., 1957; Miserocchi et al., 1981) and up to $8-10$ cm H_2O more negative than the surface pressure on the mediastinal side (Miserocchi et al., 1984). Because the negativity of Ppl(liq) is thought to be largely a reflection of the various factors involved in the absorption of fluid from the thoracic cavity, the less-negative values in the newborn should indicate that at this age the mechanisms for avoiding fluid accumulation in the pleural cavity are less effective than they are in adults.

Resistance to Airflow

By contrast with efforts to determine the elastic properties of the newborn's respiratory system, measurements of the resistance R to airflow have been performed more rarely and in fewer species. Because of the alinearity of the pressure-flow relationship, the value of R depends on the flow rate chosen for its computation. Measurements of R are highly sensitive to experimental conditions, including absolute lung volume and lung volume history. In addition, seemingly minor technical details, such as the size of the endotracheal tube, body posture, or head position, can have major effects on the numerical results. Notwithstanding the numerous potential sources of variability, the interspecies comparisons of

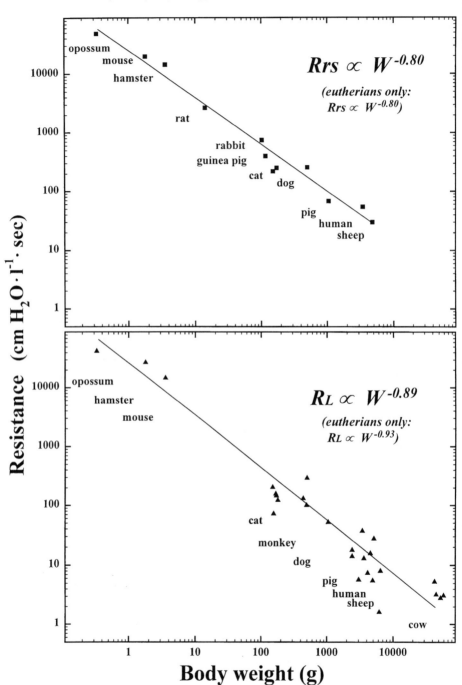

the available data show a clear trend for both the total resistance of the respiratory system (Rrs) and that of the lungs (RL) to decrease with increasing body weight, as expected; this decrease, however, is not linearly proportional to body mass. In fact, $RL \propto W^{-0.89}$ and $Rrs \propto W^{-0.80}$ (Fig. 3.16), both exponents being significantly different from -1. In other words, in the smallest newborns Rrs and RL, relative to mass, are lower than in newborns of larger species. The structural reasons for the deviation from strict proportionality are unclear.

The Time Constant. One physiological implication of the allometric relationship of R is that the passive response time of the system, conveniently expressed by the time constant τ, changes with the newborn's size. In fact, if for simplicity we equate the respiratory system to a first-order mechanical system with constant Crs and Rrs, then $\tau = Crs \cdot Rrs$. Because of the almost direct proportionality of C to body mass ($Crs \propto W^{1.02}$, Fig. 3.9) and the nondirect proportionality of R ($Rrs \propto W^{-0.80}$, Fig. 3.16), τrs is *not* an interspecies constant, scaling $\propto W^{1.02} \cdot W^{-0.80} = W^{0.22}$. Similarly, an allometric exponent significantly greater than 0 can be calculated for the time constant of the lungs ($\tau L \propto W^{1.01} \cdot W^{-0.89} = W^{0.12}$). In other words, from the data of C and R currently available for newborn species, one would expect the response time of the passive respiratory system to be shorter in the smallest species, which is not unlike what was observed, and better documented, among adult mammals (Bennett and Tenney, 1982; Leiter et al., 1986). A few direct measurements support

---

Figure 3.16. (*opposite*) Allometric relationship of the total resistance of the respiratory system (Rrs, *top panel*) and the total pulmonary resistance (RL, *bottom panel*) of newborn mammals. Each symbol is the mean value of a particular study. Although all data points are represented, the regression lines were calculated after averaging the values pertinent to any particular species. Data for hamster are from[1]; rat[2]; rabbit[2]; mouse[1]; opossum[1]; cat[2-5,18]; guinea pig[2]; dog[2,5,19]; pig[2,20-23]; monkey[6]; sheep[7,8,14,16,17]; cow[9-12]; infant[13,15]. [1]Frappell and Mortola, 1989; [2]Mortola, 1983; [3]Rossi and Mortola, 1987; [4]Fisher et al., 1987; [5]Fisher et al., 1990; [6]LaFramboise et al., 1983; [7]Shaffer et al., 1976; [8]Shaffer et al., 1978; [9]Kiorpes et al., 1978; [10]Slocombe et al., 1982; [11]Lekeux et al., 1984; [12]Inscore et al., 1990; [13]Polgar and Weng, 1979; [14]Davey et al., 1998; [15]Mortola et al., 1982b; [16]Shaffer et al., 1985; [17]Robinson et al., 1985; [18]unpublished data; [19]Sly and Lanteri, 1990; [20]Dreshaj et al., 1994; [21]Dreshaj et al., 1996; [22]Potter et al., 1997; [23]Martin et al., 1995.

this expectation. For example, in infants τrs is ∼0.2 sec, in piglets ∼0.16 sec, whereas in the much smaller newborn hamsters, mice, and opossums τrs is about half these values (Frappell and Mortola, 1989). Hence, the passive properties of the respiratory system vary among newborns in a way that appears to meet, at least qualitatively, their ventilatory requirements, the shortest response times occurring in the smallest species with the highest metabolic and ventilatory rates (see Chapter 2, "Coupling of Ventilation and Metabolism"). In addition, numerous dynamic factors can substantially modify the passive mechanical behavior, in both inspiration and expiration, thus fine-tuning the coupling between the mechanical properties of the respiratory system and the ventilatory necessities (see "The *Active* Mechanical Behavior," below).

The two allometric relationships (RL and Rrs) are very similar, suggesting that RL is the major component of Rrs, and that the resistance of the chest wall offers only a small contribution. To verify this possibility experimentally it is necessary to measure RL and Rrs in the same newborns simultaneously. This has been done only on rare occasions. In tracheostomized and anesthetized kittens and piglets, RL was ∼76% of Rrs (Mortola, 1983; Rossi and Mortola, 1987), and 74% is the value that can be calculated from published data on human infants (Polgar and Weng, 1979).

Factors Contributing to R. Interspecies comparisons of the components of Rrs are not available. A few data, mostly on infants (Polgar and Weng, 1979; Mortola, 1987), would suggest that the peripheral airways contribute a somewhat larger fraction of Rrs in newborns than in adults. In an open-chest ventilated animal, after careful puncture of the visceral pleural and parenchyma, it is possible, with miniature pressure transducers, to measure a pressure presumably very similar to the alveolar pressure, thus permitting the partitioning of RL in its airways and tissue components. These measurements would indicate that a very large component of RL is represented by the pulmonary tissue, more than 70% in young dogs (Sly and Lanteri, 1990), and between 40 and 75% in piglets (Dreshaj et al., 1994; Martin et al., 1995; Dreshaj et al., 1996; Potter et al., 1997). In the adult dog, pulmonary tissue resistance is 65–85% of RL (Ludwig et al., 1987). The variability in these values is probably introduced by differences in the absolute lung volume at which the measurement is performed, and by differences in lung volume history. In fact, pulmonary tissue resistance is recognized to be essentially a reflection of the tissue's viscoelastic properties. Because these properties are substantially more marked in the newborn than in the adult, it is likely that the tissue resistance in the newborns does not represent the same fraction of R_L as does that in the adult.

Being Very Small: The Case of Newborn Marsupials

The allometric relationships are a useful tool in appreciating trends over a wide range of body sizes. Nevertheless, the logarithmic representation of the data compresses individual variations from the overall mean; some of these variations could represent signals with physiological meanings. An interesting deviation from the interspecies relations discussed above is offered by the newborn opossum, a 300-mg animal born after only 13 days of gestation (Fig. 3.17). In this species C_L/kg, and C_{rs}/kg, are two to three times higher than they are in mice, hamsters, and probably most of the other newborns of this order of body size. Conversely, $R_L \cdot kg$ and $R_{rs} \cdot kg$ in the opossum are lower than what would be expected for a new-

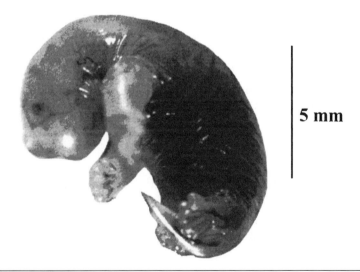

5 mm

Figure 3.17. A one-day-old newborn opossum (*Didelphis virginiana*), the only North American marsupial. Born after a gestation of only ~13 days, this embryolike newborn, less than 1 cm long and about 300 mg in weight, continues its development in the mother's pouch, firmly attached to a nipple. Courtesy of Jay P. Farber, University of Oklahoma, Oklahoma City.

born animal of its size (Fig. 3.18, left panels). The opossum, like many other marsupials (Gemmell, 1986), is born at an extremely early stage of development; in the lungs, the terminal sacs are not fully formed and true alveoli begin to appear only at about 45 days postpartum. This primordial lung is nevertheless capable of efficient gas exchange, and the surfactant-secreting type II cells are already present at birth. In the opossum, not only Crs but also the resting volume Vr is, with respect to body mass, larger than in newborns of larger species. Hence, if normalization were performed on the basis of Vr, rather than body weight, the volume-specific compliances and resistances would not differ markedly from those of other species (Fig. 3.18, right panels). The newborn opossum thus appears to have a lung disproportionately large for its size, perhaps a compensation for the limits imposed by the incomplete pulmonary architecture on the gas-exchange surface area. It is also possible that the design of the mammalian lung has a lower limit in size, beyond which its function would be impaired or excessively costly. This would imply the necessity for a disproportionately large organ in those very small newborns that depend on the lungs for gas exchange (Frappell and Mortola, 1989).

By contrast, in the dunnart pouch young, which at birth weighs about 15 mg and does not depend on the lungs for gas exchange (see Figs. 2.1 and 2.2), values of Crs/kg are similar to those of larger eutherian newborns. Its lungs, characterized by large sacs, constitute a large portion of the visceral content, and the linear dimensions of the trachea are large relative to the animal's size (Frappell and Mortola, 2000).

The *Active* Mechanical Behavior

An important question is whether or not the mechanical properties of the respiratory system are the same during spontaneous breathing as they are during passive conditions. The answer is that neither during inspiration nor during expiration does the system behave, mechanically, as would be expected on the basis of the passive measurements.

Inspiration

The main reason for the active-passive mechanical difference has to do with the uneven distribution of pressures on the chest wall during muscle contraction, leading to distortion. The fact that during inspiration the

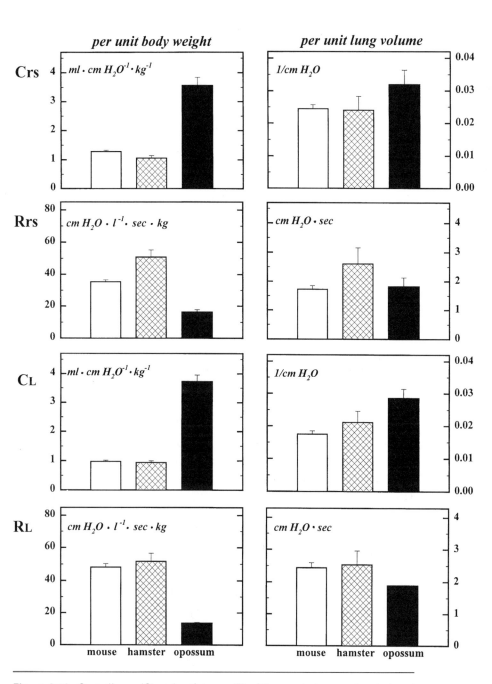

Figure 3.18. Compliance (C) and resistance (R) of the respiratory system (rs) and of the lung (L) in the newborn mouse, hamster, and opossum. *Left:* Data are normalized by body weight (kg). *Right:* Data are normalized by the passive resting volume of the respiratory system (Vr, ml). Bars are 1 SD. (From data of Frappell and Mortola, 1989.

respiratory muscles need to generate a P higher than the P measured in passive conditions means that *active* Crs is less than the corresponding *passive* value. This fact has various functional implications, including that of a response time shorter than would be expected on the basis of passive measurements.

An active shortening in τrs is also contributed by the decrease in lung compliance CL during breathing. In fact, at resting breathing rates, in human infants as in other newborn species, dynamic CL[3] (CLdyn) is commonly less than the static value (CL), in contrast to what is usually found in the adult. Such a difference also affects Crs, which, just as CL, is lower in dynamic conditions than in static conditions in many newborns, including the human infant (Kugelman et al., 1996). In ventilated children, the frequency dependence of Crs has been measured both before and after thoracotomy (Nicolai et al., 1993). The difference between dynamic and static CL seems to be related to the higher viscous properties of the newborn lung, more than to inequalities in the time constants of the peripheral airways (Nardell and Brody, 1982; Sullivan and Mortola, 1986, 1987). In addition, the lung deformation that accompanies chest wall distortion contributes to some of the drop in CLdyn (Sullivan and Mortola, 1985). Like CL, resistances, too, can decrease with breathing frequency, when tissue stress relaxation is marked and tissue viscosity represents a large component of Rrs (Kochi et al., 1988). The latter mechanisms have been estimated to decrease τrs by about 30% in newborn opossums, mice, and hamsters (Frappell and Mortola, 1989).

Expiration

If the entire expiration were a passive process strictly governed by the recoil pressure of the respiratory system at end-inspiration and by airway resistance, the air volume above Vr at any time during expiration (VTE) could be calculated according to the exponential function

$$\text{VTE} = \text{V}_0 \cdot e^{-\text{TE}/\text{τrs}} \qquad (Equation\ 3.2)$$

[3] The term *dynamic* CL can create confusion, because, by definition, compliance is a static, not a dynamic, measurement. CLdyn is computed during continuous ventilation (from which the use of the term *dynamic* arises), from the P and V changes between two points of zero flow, the onset of inflation and the end of inflation (see also Appendix A).

where V_0 is the end-inspiratory volume above Vr, at the onset of the expiratory time ($T_E = 0$), and τ_{rs} is the passive mechanical time constant of the respiratory system. Volumes can be converted into pressures according to the compliance of the respiratory system (Crs), and Equation 3.2 becomes

$$\text{PTE} = P_0 \cdot e^{-T_E/\tau_{rs}} \qquad (Equation\ 3.3)$$

where P_0 is the recoil pressure of the respiratory system at end-inspiration, and PTE represents the recoil pressure at any time during expiration. This means that at end-expiration, P and V depend on the corresponding values at end-inspiration (P_0 and V_0) and on the ratio between the time available (T_E) and the time required to complete expiration (τ_{rs}).

A numerical example is presented in Fig. 3.19, where VT/kg is 9 ml · kg^{-1} and Crs/kg is 1.8 ml · cm H_2O · kg^{-1}. These parameters are interspecies constants; hence, P_0 is approximately constant among species, at ~5 cm H_2O (Fig. 3.19, *top panel*). In a human infant during resting breathing with a TE of 0.5 sec and a passive respiratory time constant (τ_{rs}) of ~0.22 sec (Appendix A), the value of TE/τ_{rs} is ~2.3. From the values of P_0 and TE/τ_{rs} we can then calculate the expected PTE at about 0.5 cm H_2O (Fig. 3.19, *bottom panel*). Hence, from simple calculations based on the data of passive respiratory mechanics one would conclude that the internal recoil pressure of the system at end-expiration is very small, or, stated differently, that TE is sufficiently long for full expiration. A similar conclusion could be obtained from data of passive mechanics in the newborns of other species. Hence, if the respiratory system during expiration were to behave according to its passive mechanical characteristics, one would be left with the conclusion that during resting breathing in newborn mammals the end-expiratory level (FRC) equals the passive resting volume (Vr), with no positive end-expiratory pressure, or "internal PEEP." Experimental data indicate that this conclusion is not correct. In fact, notwithstanding the validity of these theoretical calculations, FRC can be substantially greater than Vr. The reason for the difference between what is observed in the spontaneously breathing animal and the predictions based on passive mechanical data has to do with the fact that the mechanical behavior of the respiratory system in expiration deviates substantially from the system's passive properties.

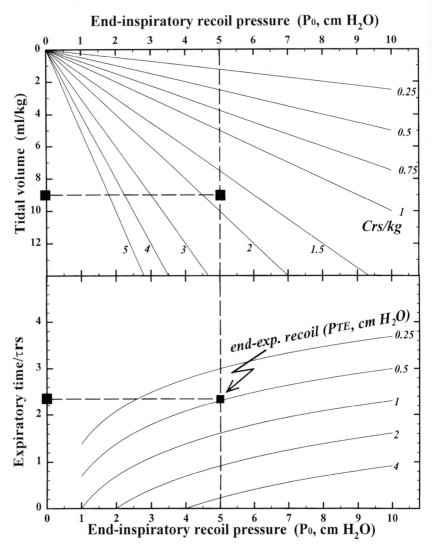

Figure 3.19. Correlation of the main variables determining the end-expiratory volume in the theoretical case of a purely passive expiration. *Top:* The end-inspiratory recoil pressure (P0) is the pressure driving the expiratory flow, and its magnitude depends on tidal volume and respiratory system compliance (Crs). *Bottom:* The recoil pressure at end-expiration (PTE) depends on P0, the total duration of expiration (TE), and the passive time constant of the respiratory system (τrs). The *broken lines* join average interspecies values, from the proportionality constants of the newborn allometric relationships.

Two dynamic mechanisms are the primary factors modifying the expiratory mechanical behavior of the respiratory system: the postinspiratory activity of the inspiratory muscles, which produces an inspiratory pressure, Pmus(I), and the laryngeal control of expiratory flow. The braking action of the inspiratory muscles slows the expiratory emptying and, coupled with the narrowing of the glottis in expiration, prolongs the expiratory time constant [τrs(E)] above the passive τrs value. In human infants this results in a dynamic elevation of FRC above Vr (Mortola et al., 1982b, 1984a; Kosch and Stark, 1984).

Graphically, the retarding action of Pmus(I) and laryngeal braking on expiratory flow can be appreciated by comparing the expiratory flow-volume loop during tidal breathing with the flow-volume curve obtained after release of an end-inspiratory airway occlusion, which, because of the Hering-Breuer reflex, determines muscle relaxation (Mortola et al., 1982b; Mortola et al., 1984a) (Fig. 3.20). The two expiratory curves differ, the tidal expiratory flow-volume curving at the left of the passive curve, indicating that the expiratory time constant τrs(E) is longer than τrs. From analysis of these curves, the postinspiratory action of the inspiratory muscles was found to persist through much of the expiration, in some infants up to 80% of it (Mortola et al., 1984a). The complex interplay of the factors controlling expiratory flow explains why attempts to find a correlation between the pattern of tidal expiration (e.g., the expiratory V̇-V loop) and the passive compliance and resistance of the respiratory system have not been satisfactory (Lodrup Carlsen et al., 1994; Seddon et al., 1996). In fact, some relationship between the tidal expiratory flow and passive mechanical properties should emerge only if passive respiratory mechanics were the dominant factor in determining expiratory flow, which, as we have seen, is often not the case in the neonatal period. Differently, in some disease states (e.g., airways obstruction) (Banovcin et al., 1995), the respiratory mechanical properties can become so greatly altered as to constitute the dominant factor in determining the expiratory flow profile.

Does the dynamic prolongation of the expiratory time constant, and the consequent elevation of FRC, offer a major advantage to neonatal breathing? Undoubtedly, it represents an important protective mechanism against the low Vr caused by the passive characteristics of lungs and chest wall (see Fig. 3.15). Indeed, the common clinical observation that

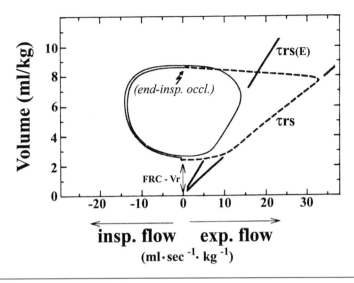

Figure 3.20. Schematic representation of the airflow-volume loop of two breathing cycles in a spontaneously breathing infant. Inspiration is at left (negative flows). The *continuous loop* represents one resting breathing cycle. At end-inspiration of the second breathing cycle, the airways are briefly occluded by the investigator (end-insp. occl.); following release of the occlusion, the slope of the deflation flow-volume curve (*dashed line*) represents the passive time constant of the respiratory system (τrs). Most often, the tidal expiratory flow-volume curve is at the left of the passive curve, indicating that during tidal breathing the expiratory time constant [τrs(ε)] exceeds τrs. By extrapolating the linear portion of either the passive or the expiratory flow-volume curves to zero flow, one can estimate the dynamic elevation of the end-expiratory level above the resting volume of the respiratory system (FRC − Vr).

in artificially ventilated and intubated babies (i.e., with neither postinspiratory muscle activity nor laryngeal braking) blood gases deteriorate unless an end-expiratory pressure is added to the ventilatory circuit (Gregory et al., 1971; Berman et al., 1976) supports the view that a large FRC is a crucial necessity for the human infant.

It should be considered, however, that whenever lung volume is elevated above Vr, some muscle pressure must be generated to overcome the recoil of the respiratory system before any inspiratory flow can be produced; this can be expected to lower the mechanical efficiency, not dissimilarly to the effects of continuous positive airway pressure (Lorino et al., 1996). In addition, the higher the FRC the more pronounced the input from the airway stretch receptors inhibiting inspiratory activity (see Chapter 4, under "Airway Receptors"). Both factors tend to decrease $\dot{V}_E$, and therefore may not be desirable in the smallest species, which have high metabolic requirements and the necessity for high ventilatory rates. In fact, one may think that the smallest species are at risk of a serious dynamic hyperinflation, because of the exponential effect of T_E on P_{TE} (Equation 3.3, above). In reality, this does not occur. In the smallest newborns the dynamic hyperinflation, or FRC-Vr difference, is not larger than in species of medium or large size; in fact, in some cases it can even be very small (Mortola et al., 1985a). For example, in newborn rats, which breathe at a rate three to four times faster than infants, the FRC-Vr difference is almost nil. The reason for this is that τexp is shorter in the smaller species, not only because the passive time constant is shorter, but also because of the differences in the magnitude of the dynamic phenomena controlling expiratory flow mentioned above. In addition, at the very high breathing rates of the smallest newborn species, C_Ldyn can be substantially less than the passive C_L, further shortening the $\tau rs(E)$ (Fig. 3.21). The dynamic elevation of FRC in newborns, then, is an appropriate solution to the mechanical characteristic of a low Vr, but it could pose some problems in the smallest species, which require very high ventilatory rates. Perhaps this is the reason why the dynamic elevation of FRC is pronounced only in species of large or medium size, including the human infant, in which the metabolic and ventilatory demands are not as great as those of the smallest species.

Energetics

As a first approximation, the cost of breathing can be computed from the values of respiratory work and mechanical efficiency (cost = work/efficiency). The passive respiratory work can be calculated by planimetry from the pressure-volume diagram during assisted ventilation (Cook et al., 1957), or by assuming a sinusoidal flow pattern and a constant compliance and resistance, as proposed by Otis et al. (1950). With the latter approach, from the average published data of Crs, Rrs, VT, and f of various neonatal species, and arbitrarily defining the efficiency of the respiratory muscles at 10%, during resting conditions the respiratory cost/

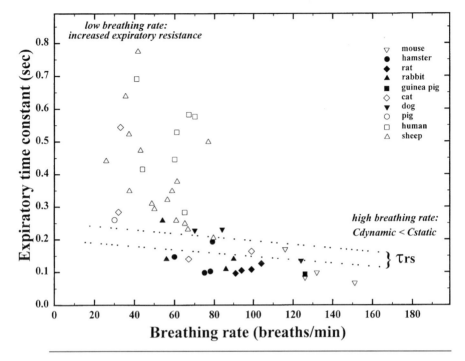

Figure 3.21. Expiratory time constant τrs(E) measured during spontaneous breathing in some newborn mammals, plotted at the breathing rate of the animals. The dotted lines represent the approximate range of the passive time constant, τrs, for these species. In the slowest-breathing animals, τrs(E) is usually longer than τrs, whereas the opposite can occur in the newborn of the fastest-breathing species, because for them dynamic compliance is less than passive compliance. From data of Mortola et al., 1985a, and Côté et al., 1988.

minute is a very small fraction of oxygen consumption ($\dot{V}o_2$), between 1 and 3% (Fig. 3.22). For the same pulmonary ventilation $\dot{V}E$, a small breathing rate disproportionately increases the elastic component of the work of breathing; conversely, high breathing rates disproportionately increase the resistive component. As in adults, in newborn mammals including infants the resting breathing frequencies fall within the range of minimal work.

It should be emphasized that the above figures are based on values of passive mechanics, and therefore reflect only the external component of the total respiratory work. In active conditions the energetic losses are

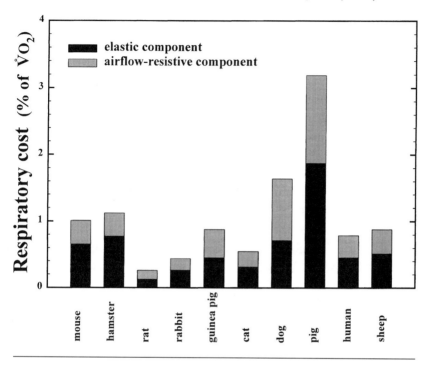

Figure 3.22. Total cost of breathing during resting conditions, derived from computations based on passive values of compliance and resistance, and expressed as a percentage of resting oxygen consumption. The total cost is separated into its elastic and airflow-resistive components; the latter was computed by assuming that the upper airway provides about 20% of the total airflow resistance. As in adults, also in newborns at rest the cost of breathing represents a minimal component of the total oxygen consumption, usually less than 2%.

invariably higher, because of factors such as the distortion of the system and the antagonistic or isometric muscle contractions, components that are not included in the above computation. Estimates of the total muscle force required to breathe have been attempted from the airway pressure developed during respiratory efforts at constant lung volume. This approach, however, is based on numerous unverified assumptions (Mortola et al., 1982b; Mortola, 1987). Nevertheless, even if the respiratory work during active breathing were double the passive value, it would still be correct to conclude that the cost of breathing in the newborn mammal represents a minute component of its total $\dot{V}o_2$.

Interspecies Comparisons

The total mass of the diaphragm, intercostals, and probably other respiratory muscles is directly proportional to body weight among newborns of the various species, as well as within a species during growth. Lung mass/body weight is also an interspecies constant, but within any given species it decreases with growth. Alveolar size changes little among species, and the direct proportionality between lung volume and body size is due primarily to differences in the number of alveoli. Both lung compliance and respiratory system compliance scale in direct proportion to body weight, whether or not data of the altricial marsupials are included in the comparisons. In all newborns the chest wall compliance is characteristically high, such that the lung-chest wall compliance ratio is consistently lower than that in adults. As a consequence, the resting volume of the respiratory system is low. The potential risks of a low resting volume are usually compensated for by combining three mechanisms: high breathing rate, postinspiratory activity of the inspiratory muscles, and increased upper airway resistance. The latter two mechanisms are adopted particularly by newborns of medium and large size. Values of pleural pressure at end-expiration and during breathing are similar among species, as they are between newborns and adults. Resistances decrease with body size, but not in direct proportion to it. Hence, the time constant of the respiratory system increases the larger the newborn, scaling to $W^{0.22}$. The work of breathing, even if one includes the internal losses due to distortion, is a small fraction of total oxygen consumption for newborns of all species studied.

Clinical Implications

From the information available, and contrary to what was believed earlier, the respiratory muscles of the infant are at least as resistant to fatigue as are those of adults. This does not exclude the possibility of muscle fatigue in cases of a severe and prolonged increase in respiratory work, or in prematures. Body posture has implications for the mechanical efficiency of the respiratory muscles and gas exchange; the prone position would be more advantageous than the supine position, but in the prone posture the incidence of sudden infant death syndrome (SIDS) is higher.

Chest distortion, with paradoxical inward movement of the rib cage during inspiration, is a normal attribute of the infant's breathing, especially during REM sleep, because of the high ratio of chest wall compliance to lung compliance. Distortion can increase in conditions of low lung compliance, as in prematures with bronchopulmonary dysplasia, or of high airway resistance. The opposite situation, uniform expansion of rib cage and abdomen during inspiration, could be a sign of increased ventilatory stimuli, as in hypoxia. Because of the important difference between active and passive respiratory mechanics, the flow-volume function during resting breathing is not closely related to the passive mechanical properties; but in disease states with low lung compliance or high resistance the abnormal mechanical characteristics can become the major determinants of the flow-volume profile. In infants, the glottis is instrumental in keeping the end-expiratory level (FRC) above the passive volume of the respiratory system. Hence, after intubation, blood gases deteriorate unless a positive end-expiratory pressure is added to the ventilatory circuit.

Summary

1. The mass of the respiratory structures is proportional to the body size of the newborns, across species; lung mass is ~2%, and diaphragm mass ~0.4%, of body mass; some deviations may occur in extremely small newborns, as those of marsupials. With growth, the lung weight-body weight ratio decreases.

2. The neonatal diaphragm differs from that of the adult in numerous morphological, biochemical, and mechanical ways; several data suggest that it is at least as resistant to fatigue as that of the adult.

3. Among newborns, across species, the compliance of the respiratory system (Crs) and that of the lungs (CL) are almost directly proportional to the animal's body size, the proportionality constants being, respectively, ~1.8 and ~2.6 $\text{ml} \cdot \text{kg}^{-1} \cdot \text{cm H}_2\text{O}^{-1}$.

4. The similarity of Crs to CL is due to the high compliance of the chest wall (Cw). In fact, one of the most important mechanical characteristics of the neonatal respiratory system is the high ratio of Cw to CL.

5. Two of the major implications of the high Cw/CL are the likelihood of chest distortion during inspiration and a low resting volume of the respiratory system (Vr).

6. In some newborns, usually the smallest, with their high metabolic and ventilatory requirements, the end-expiratory level (FRC) almost coincides with Vr; in others, such as the human infant, FRC is dynamically maintained at a level above Vr by mechanisms that combine a short expiratory time with muscle control of expiratory flow and increased laryngeal resistance.

7. In newborns, the rib cage moves out of phase with airflow. Several structural and functional characteristics contribute to this behavior. The shape of the thorax and the high compliance of the chest wall are not as favorable in expanding the rib cage during diaphragmatic contraction as they are in adults. In addition, the inspiratory extradiaphragmatic muscles are probably less effective in stabilizing the rib cage than they are in the adult.

8. Pulmonary resistance and total respiratory system resistance are lower, relatively to body mass, in the newborns of the smaller species. Hence, the passive time constant is shorter in the newborns of the smallest species, a property that favors their needs for high ventilatory rates.

9. In active conditions, the mechanical behavior of the respiratory system deviates from its passive properties because of the many additional mechanisms operative in inspiration and expiration. As a result, the cost of breathing is higher than would be estimated on the basis of the passive mechanical characteristics, but is still a very small fraction of the total oxygen consumption.

Chapter 4

Reflex Control of the Breathing Pattern

In first approximation, breathing can be seen as a mechanism of gas convection serving the metabolic requirements of the organism. Within this scope, the neural output of a central pattern generator intermittently activates the inspiratory muscles, producing a tidal volume at a defined frequency; the adequacy of the resulting pulmonary ventilation is then monitored by chemosensors for levels of oxygenation and carbon dioxide. Indeed, chemoreceptors and ventilatory chemosensitivity, functions to be discussed in the next chapter, are central aspects of the regulation of breathing for the proper matching between ventilation and metabolism. But even if the neuromechanical unit were reduced to its most essential aspects, it would be unrealistic to think that it could function properly with only feedback information from the oxygen and carbon dioxide chemoreceptors. In fact, because the respiratory muscles also have postural tasks, and because of the continuous interference of nonrespiratory functions, the monitoring of lung expansion as it progresses during inspiration, with immediate adjustments if necessary, is an important aspect of the proper operation of the respiratory system. Monitoring of this sort is achieved through information originating from the airways and from anatomical structures within the chest, both forming the basis of the reflexes to be discussed in this chapter.

Reflex Control of Breath Amplitude and Duration

As a first approximation, the breathing pattern of a mammal may look like an oscillatory event resembling a sinusoidal function, of which the amplitude is tidal volume (V_T) and the period is the total breath duration

(TTOT). Of course, closer analysis reveals major deviations from the sinusoidal function. Especially in newborns, irregularities in the breathing pattern are common events (some of their characteristics are discussed above, in Chapter 2). Nevertheless, one cannot avoid being amazed by the consistent similarity of the spirogram among species, with their enormous variations in size and mechanical loads, and by its constancy within a subject under various circumstances. For example, VT changes little with posture, including changes in neck or trunk position, whether during mouth breathing or nose breathing, or even during modest exercise such as slow walking. In fact, the stability of VT can be further demonstrated experimentally by applying a load to the respiratory system. Breathing through a narrow tube increases the total resistance ("resistive load"), and breathing from a container reduces the total compliance ("elastic load"). Yet even in these cases the reduction in VT following the introduction of the load is substantially less than one would expect from a consideration of the changes in the mechanical properties of the respiratory system. In other words, there is a compensation against the mechanical load, and a large fraction of it takes place within the first loaded breath.

Load compensation is a fundamental aspect of the breathing act. The respiratory pump, lying by design in the middle of the body, shares muscles and structures with multiple nonrespiratory functions, including locomotion, and is therefore subjected to many continual changing loads. If there were no mechanism to control the mechanical result of the neural output, breathing would probably be a chaotic and energetically demanding act, with catastrophic effects on both the distribution of ventilation and blood-gas homeostasis. Humoral stimuli, including blood gases, reach, and reflexively influence, neurons controlling the respiratory motor output. These inputs act on a rather long time scale, in some cases over periods of minutes or hours, and are most effective in the control of ventilation in relation to metabolic requirements. The control of tidal volume, by contrast, is based on neural feedback loops from specific receptors placed in the lungs and the chest, which operate within the duration of the breath itself.

The response of the breathing pattern to a load that would limit VT and pulmonary ventilation V̇E is based therefore on fast-acting neuromechanical mechanisms and on slower-acting humoral stimuli, including

changes in blood gases. The latter assume progressively greater importance as the load is sustained over time. Several ingenious experiments
have been performed on adult humans and animals to understand the
mechanisms of the immediate load compensation, and numerous reviews are available (e.g., Milic-Emili and Zin, 1986; Daubenspeck, 1995).
The afferent information received from the chest wall muscles, reflexively projected at the spinal and supraspinal levels, and that received
from the pulmonary airways via the vagus nerves are recognized as the
primary mechanisms ensuring the immediate protection of tidal volume.

Load Compensation in Newborns

Relative to the abundant literature on adults, the information on the effectiveness of load compensation during the early phases of postnatal development is limited. Because of the developmental changes in respiratory system compliance (Crs) and resistance (Rrs), comparison among
subjects of different size is usually done by the use of loads that represent fixed percentages of the subject's Crs and Rrs. Most studies have
been concerned with the immediate load compensation; only a few have
investigated the breathing pattern during sustained respiratory loads, in
which changes in the chemical stimuli can trigger responses not elicited
within the first loaded breaths.

Immediate Compensation. The immediate capabilities for load compensation are studied by measuring the change in mechanical or neural
output when the mechanical impedance of the respiratory system is suddenly changed, as would occur with a reduction in Crs or an increase in
Rrs. The mechanical output is measured in the form of VT, airway or
pleural pressure, and the neural output is quantified from the electromyogram (EMG) of the diaphragm or other respiratory muscles, or
the activity of the phrenic nerve. Mechanical loads are multiples of the
subject's Crs or Rrs. Often, however, a brief total occlusion of the airways
at end-expiration is the preferred test. This represents an infinite load
easily reproducible among animals and across conditions, although it can
be performed only during anesthesia or sleep, to eliminate the behavioral
responses that would confound the automatic reflexes.

In infants, the duration of an inspiratory effort after occlusion of the
airways at end-expiration is prolonged in comparison to the preceding
nonoccluded breaths. Furthermore, the peak airway pressure produced

during the effort is larger than one would expect from the subject's V_T and Crs (e.g., Olinsky et al., 1974a, 1974b; Frantz III and Milic-Emili, 1975; Adler et al., 1976; Gerhardt and Bancalari, 1981; Fisher et al., 1982; Mortola et al., 1995a) (Fig. 4.1). These findings indicate that in the newborn infant, mechanisms of compensation occur within the first loaded breath. Similar conclusions were drawn from experiments on newborn rabbits (Thach et al., 1976; Wyszogrodski et al., 1978; Fisher and Mor-

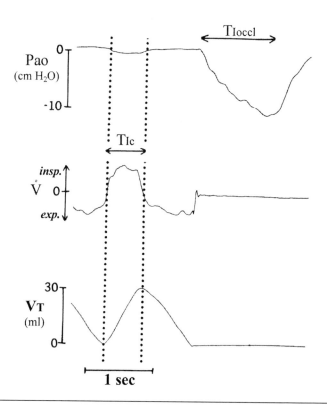

Figure 4.1. A human infant, one day old, is breathing through a tiny face mask connected to a pneumotachograph for measurements of airflow ($\dot{V}$) and tidal volume (V_T), and equipped with a side port for measurements of pressure at the airway opening (Pao). The inspiratory time of a control breath is indicated (T_{IC}). On closure of the mask outlet at end-expiration, the following inspiratory effort results in a lowering of Pao, with no volume change. The inspiratory duration of the effort (T_{Ioccl}) is substantially longer than T_{IC}. Slightly modified from Mortola et al., 1995a.

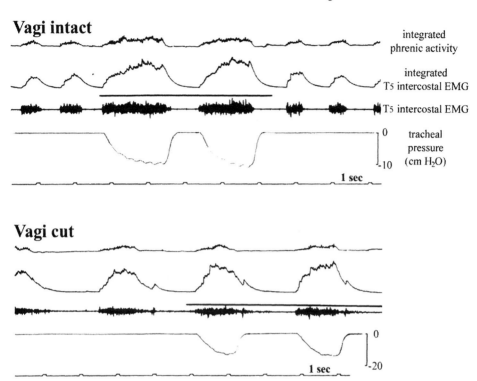

Figure 4.2. Response to occlusion of the airways at end-expiration before and after vagotomy in one-day-old anesthetized kitten. *Top panel,* vagi intact; *bottom panel,* postvagotomy. In each panel, from *top to bottom,* are the integrated activity of the phrenic nerve, the integrated and raw electromyogram of the 5th intercostal nerve, and tracheal pressure. The *horizontal bars* midway in each panel indicate the duration of the end-expiratory occlusion. During the occluded efforts, before vagotomy, all the muscles substantially prolonged and increased their inspiratory activity, neither of which happened after vagotomy. Slightly modified from Trippenbach and Kelly, 1983.

tola, 1980b), kittens (Trippenbach and Kelly, 1983), and lambs (Webb et al., 1994) (Fig. 4.2).

Experiments in humans and anesthetized rabbits have shown that the reduction in VT and V̇E in response to elastic loads of various magnitudes does not differ appreciably between newborns and adults (Taeusch et al., 1976; Fisher and Mortola, 1980b). With respect to the responses to the end-expiratory airway occlusion, some studies indicated an increase

(Taeusch et al., 1976; Thach et al., 1978; Gerhardt and Bancalari, 1981), and others a decrease, in load compensation with advancing gestational or postnatal age (Olinsky et al., 1974a, 1974b; Rabbette and Stocks, 1998). Kosch, Fox, Stark, and co-workers found no differences in load compensation between full-term and premature infants (Kosch et al., 1986; Fox et al., 1988). These discrepant conclusions could reflect methodological aspects. First, the inspiratory time TI computed from a mechanical signal ($\dot{V}$, VT, or pressure) may not correspond to the neural TI, such as that computed from an EMG record, especially during resistive loading (Miserocchi and Milic-Emili, 1976; Kosch et al., 1986). Second, in newborns, functional residual capacity FRC is often above the passive volume of the respiratory system (see Chapter 3, under "Expiration"); this implies that part of the inspiratory muscle output must overcome the internal recoil pressure before airway pressure can be lowered below the atmospheric value. Hence, in infants the negative pressure generated during the effort against the occlusion at end-expiration is only an approximation of the corresponding neural output (Mortola, 1987). In anesthetized pigs breathing through a tracheal cannula, the prolongation of TI during the end-expiratory effort was less in newborns than in adults (Clement et al., 1986), as was the case with younger compared to older kittens (Trippenbach et al., 1981). Tracheostomy, which is common practice in anesthetized animal preparations, has been shown to modify the magnitude of load compensation in lambs (Webb et al., 1994).

Maintained Respiratory Loads. Newborn monkeys confronted by resistive loads for 10–20 min did not maintain VT and $\dot{V}E$ as well as 3-week-old monkeys; their blood gases, however, were unaffected (LaFramboise et al., 1987). This intriguing result raises the possibility that a mechanism of adaptation to a maintained load in newborns may be a decrease in metabolic rate, as it occurs during hypoxia (see Chapter 5, under "Acute Neonatal Hypoxia"). Indeed, in infants and piglets, oxygen consumption $\dot{V}O_2$ dropped during breathing with added resistances (Duara et al., 1991, 1993). The mechanisms for the $\dot{V}O_2$ response to a sustained load remain speculative.

Postural changes alter the configuration of the chest wall, and modify FRC. In adult animals, especially if anesthetized, tilting from the head-down or horizontal posture to the head-up posture affects breathing rate, because of the lung's volume-sensitive vagal reflexes. But in newborn rabbits, rats, and dogs, similar maneuvers were of little consequence for breathing rate (Mortola, 1980), presumably because postural changes in small animals have less impact on the mechanical characteristics of the system. An increase or decrease in lung volume also occurs with, respectively, positive-pressure (PEEP) and negative-pressure (NEEP) breathing. In newborn rats, either condition determined only minor changes in breathing pattern when the animals were conscious, whereas a clear decrease in $\dot{V}E$ with PEEP, and an increase with NEEP, were observed under anesthesia (Marlot and Mortola, 1984). In the unanesthetized immature opossum, breathing rate was not consistently modified by PEEP (Farber, 1983), whereas it dropped in anesthetized newborn rabbits (Trippenbach et al., 1985b). In infants, continuous lung distention by PEEP lowered breathing rate (Martin et al., 1978; Stark and Frantz, 1979).

Thus, the quantitative aspects of the postnatal development of load compensation remain an open issue. Nevertheless, it is clear that the conscious newborn mammal has the ability to protect V_T and $\dot{V}E$ against mechanical perturbations, with immediate compensatory mechanisms and additional strategies when the load is sustained. As it has in the adult, it has been shown in newborns of many species that the responses depend largely on vagal inputs, because bilateral vagotomy greatly reduces or totally abolishes the immediate load compensation (e.g., Schwieler, 1968; Thach et al., 1976; Trippenbach et al., 1981; Trippenbach and Kelly, 1983; Clement et al., 1986; Webb et al., 1994).

Further analysis of the contribution of the vagus to the stability of the breathing pattern in the newborn has been done following conceptual and methodological approaches similar to those adopted on adults. These include the analysis of the effects of vagotomy on breathing, recording from pulmonary vagal afferents, and the effects of manipulations of lung volumes and pressures on both vagal afferent activity and ventilatory reflex responses.

Vagal Reflexes

The notion that bilateral section of the vagi has profound effects on the breathing pattern dates back at least 140 years, since the investigations by Hering and Breuer. Later, Coombs and Pike (1930), working on cats and dogs, suggested that the consequences of bilateral vagotomy were more dramatic in the newborn, possibly leading to death within a few hours. These results were essentially confirmed on both anesthetized and decerebrated kittens by Schwieler (1968), who also recognized that the postvagotomy respiratory failure did not necessarily have to be attributed only to the elimination of the pulmonary afferent traffic. Indeed, vagal section in the neck eliminates afferent and efferent control to many intrathoracic organs, including the airways and the heart. In animals with intact upper airways, cervical vagotomy eliminates a large component of the efferent innervation to the larynx, via the inferior (or "recurrent") laryngeal nerve, which can cause narrowing of the glottis and increased airway resistance.

Bilateral vagotomy has been performed in the lamb fetus, with minimal effects on the incidence, frequency, or amplitude of fetal breathing movements (Dawes et al., 1972; Adamson et al., 1991; Hasan and Rigaux, 1992b). But after bilateral section of the vagi nerves in the newborn the breathing pattern changes in a manner qualitatively similar to that of adults, with a prolongation of the inspiratory time (TI) and deeper VT. In addition, in newborn cats (Marlot and Duron, 1979a), rats (Fedorko et al., 1988), and monkeys (LaFramboise et al., 1985), vagotomy resulted in a very marked lengthening of the expiratory phase of the breathing cycle (TE); this effect, by greatly reducing breathing rate and V̇E, could be the first phase in the development of respiratory insufficiency. These dramatic effects are not a consistent finding (Rossi and Mortola, 1987), however, nor do they apply to all species. For example, in the newborn pig (Clement et al., 1986) and rabbit (Mortola et al., 1984c) the prolongation of TE after vagotomy was modest. Conscious newborn lambs after bilateral vagotomy experienced a mild acidemia and slight hypercapnia, but no changes in oxygenation, and were still capable of responding to hypercapnia with an increase in VT (Marsland et al., 1975). In the pouch-young opossum, as opposed to the 3-month-old, bilateral vagotomy had almost no effects on the breathing pattern (Fig. 4.3), despite the fact that reflexes attributable to vagal afferents can be demonstrated. Since almost invariably these observations were performed on anesthetized animals, apparent interspecies differences could result from the type and depth of anesthesia.

Vagal Innervation and Lung Mechanics. Vagal innervation can affect respiratory mechanics by altering the bronchomotor tone and the tonic

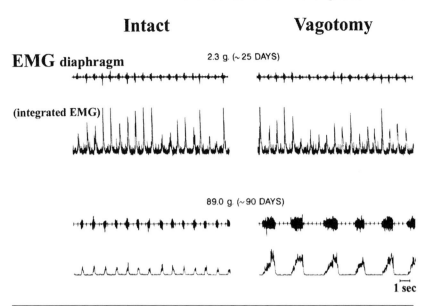

Figure 4.3. Effects of bilateral vagotomy on the breathing pattern of anesthetized suckling opossums at 25 days (*top*) and 90 days (*bottom*). In each panel the two records are, respectively, the raw and integrated electromyograms of the diaphragm. In the younger animal, vagotomy had almost no effect. Slightly modified from Farber, 1991.

activity of the respiratory muscles, as well as indirectly, because changes in the breathing pattern influence lung compliance. After vagotomy in newborn rabbits, Crs increased, a result attributed to the deeper VT produced (Mortola et al., 1984c). In piglets and kittens, which had only a small increase in VT after vagotomy, Crs changed very little (Clement et al., 1986; Rossi and Mortola, 1987).

Vagotomy was found to decrease total pulmonary resistance in spontaneously breathing kittens (Rossi and Mortola, 1987) and neonatal calves (Slocombe et al., 1982), but not in newborn dogs artificially ventilated with a constant pattern (Waldron and Fisher, 1988), nor in lambs unless ventilated with small tidal volumes (Pérez Fontán and Kinloch, 1993). These findings could suggest that the changes in lung mechanics observed after vagotomy in spontaneously breathing animals may be due to the changes in their breathing pattern. The newborn's airway smooth muscle has the ability to contract when the vagi nerves are electrically stimulated (Fisher et al., 1990). Hence, the lack of effects of vagotomy on airway resistance would indicate that in the newborn the vagal innervation provides little basal tone to the airway smooth muscle.

This interpretation, however, may be too simplistic. In fact, it is now established that vagal efferent control of the airway smooth muscle is characterized not only by the contracting effect of the cholinergic component but also by a relaxing component ("nonadrenergic noncholinergic innervation") of as yet ill-defined mediators. In newborns, both aspects of vagal motor controls are present, although quantitative assessments of their postnatal development are still inconclusive (Fisher et al., 1998). Stimulation of pulmonary C-fibers by chemicals injected in the blood stream promotes reflex bronchoconstriction in newborn pigs and dogs (Anderson and Fisher, 1993; Nault et al., 1999), but in earlier studies in piglets and newborn rabbits, no such effects were found (Haxhiu-Poskurica et al., 1991; Ducros and Trippenbach, 1991). In newborn guinea pigs, the bronchoconstrictor response to vagal C-fiber stimulation was present and as marked as that in older animals (Murphy et al., 1994). But in the same species the bronchoconstrictor response to infusion of histamine, which presumably acts both directly on the airway smooth muscle and indirectly by vagal activation of the cholinergic system, required a larger dosage in newborns than in adults (Clerici et al., 1989).

Vagal innervation could play a role in the secretion of surfactant (Klaus et al., 1962b; Goldenberg et al., 1967; Alcorn et al., 1980), and therefore in the maintenance of an adequate C_L, by mechanisms presumably related to the periodic expansion of the lung during breathing (Massaro and Massaro, 1983). This role of the vagal pulmonary innervation could be responsible for the difficulties encountered by prenatally vagotomized newborn lambs in maintaining an adequate gas exchange after birth (Wong et al., 1998). By contrast, in rats and rabbits raised after monolateral vagotomy, morphological and functional measurements indicated that the function of the lung of the denervated side did not differ significantly from that of the controlateral innervated lung, nor from that of control animals (Mortola et al., 1987b; Dotta and Mortola, 1992b). Hence, vagal innervation does not seem essential for the normal postnatal growth of the lungs. In addition, because the breathing pattern of the monolaterally vagotomized animals was deeper and slower than that of the controls, sustained differences in breathing pattern do not appreciably modify postnatal lung growth.

Changes in Lung Volume. Much understanding of the role of the vagi on the reflex regulation of breathing has been gathered from experiments designed to change lung volume in a controlled fashion while observing the corresponding effects on the breathing pattern, before and after vagotomy. This approach, initiated by Hering and Breuer and later combined with electrophysiological recordings of respiratory muscles or neural activity (Coleridge and Coleridge, 1986; Euler, 1986), has also been applied to studies of vagal control in the newborn.

A sustained increase in lung volume, as can be obtained by occluding the airways at end-inspiration or by suddenly increasing FRC with PEEP or a negative body-surface pressure applied to the thorax, invariably delays the onset of the next inspiratory effort (Fig. 4.4). This has been observed in newborns of all species investigated, from preterm infants (Sankaran et al., 1981) to the most immature pouch-young marsupials (Farber, 1972; Frappell and Mortola, 2000). Even in fetuses, lung expansion decreases the breathing movements (Ponte and Purves, 1973). The inhibition of breathing activity is more pronounced the larger the lung volume (Fig. 4.5), a response known as the Hering-Breuer inflation (expiratory-promoting) reflex. The opposite response (i.e., a facilitation of breathing activity when lung volume is lowered below FRC, the Hering-Breuer deflation reflex) has also been observed in many newborn species, although in this case unusual response patterns, such as the end-inspiratory pauses of the pouch-young opossum (Farber, 1972), have also been noticed. Vagotomy either abolishes or greatly reduces these responses.

The duration of the apnea during maintained lung inflation can be quite variable among species, even after the response is normalized by the control expiratory time (inhibitory ratio = $T_E$inflation/T_E). The fact that most observations have been made on animals under anesthesia is probably a major source of variability. In fact, sedation or anesthesia usually magnifies the strength of reflexes inhibitory to breathing, with different effects depending on type and dosage. Furthermore, anesthesia can lower body temperature, and both ambient and body temperature have major effects on the neonatal Hering-Breuer inflation reflex (Merazzi and Mortola, 1999a). It should also be noted that the reflex apnea during maintained lung inflation is not simply a measure of the strength of the vagal inhibition; rather, the duration of the apnea also depends on the ventilatory sensitivity to hypoxia and hypercapnia. Hence, a lower chemo-

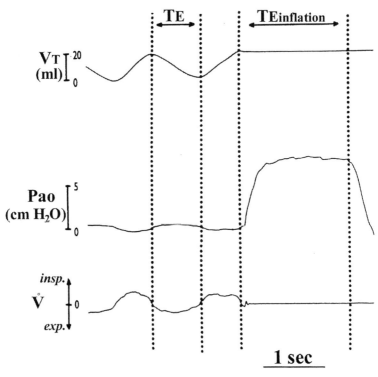

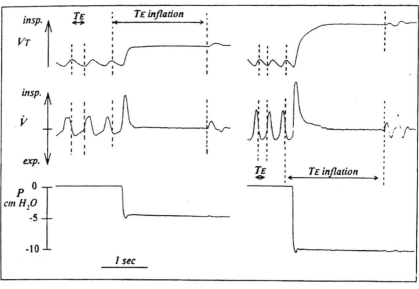

sensitivity can contribute to a stronger reflex response. Indeed, the interaction with the metabolically dependent hypoxia and hypercapnia has been considered the major reason for the T-sensitivity of the Hering-Breuer reflex in newborn rats (Merazzi and Mortola, 1999a, 1999b). Finally, because Crs and CL are not necessarily the same among species, or between newborns and older subjects, comparison of the reflexes on the basis of changes in lung volume may not yield the same results as when the comparison is based on changes in airway pressure or in transpulmonary pressure (Fig. 4.5).

The reflex is more pronounced in conscious newborn rats than in human infants, especially if results are compared on the basis of similar changes in airway pressure rather than lung volume, therefore taking into account their differences in respiratory system compliance (Fig. 4.6). This is reminiscent of what is seen among adults of various species, among which the human is the least responsive to lung inflation (Widdicombe, 1961). Comparisons between newborns and adults have been done rarely and only during anesthesia, with mixed conclusions (Gaultier and Mortola, 1981; Smejkal et al., 1985). In the newborn infant the reflex is readily evoked, contrary to the response of the conscious adult human, and this is the basis for the commonly held view that vagal control is stronger in the newborn. But the newborn-adult comparison in humans may be complicated by the behavioral and cortical responses of the conscious adult human (Gautier et al., 1981). Many studies have attempted an analysis of the developmental progression of the reflex in infants at different postnatal ages, since the early work by Cross et al.

Figure 4.4. (*opposite*) Hering-Breuer inflation (expiratory-promoting) reflex in a newborn infant (*top*) and a newborn rat (*bottom*). In the infant, lung volume was maintained at an elevated level by occlusion of the face-mask outlet at end-inspiration. As the occlusion is performed, the lungs remain inflated, and the pressure in the airways (Pao) rises to the passive recoil value of the respiratory system, until the onset of the first inspiratory effort. In the rat, lung volume was suddenly increased by the application of negative body surface pressure (P) on the thorax, -5 cm H_2O in the example at left, -10 cm H_2O in the example at right. (VT, tidal volume; $\dot{V}$, airflow.) The inflated expiratory time (TEinflation) is longer than the usual control expiratory time (TE). The larger the inflation, the greater the inspiratory inhibition (see Fig. 4.5). Recordings on infant from Mortola et al., 1995b; on rat from Merazzi and Mortola, 1999; both slightly modified.

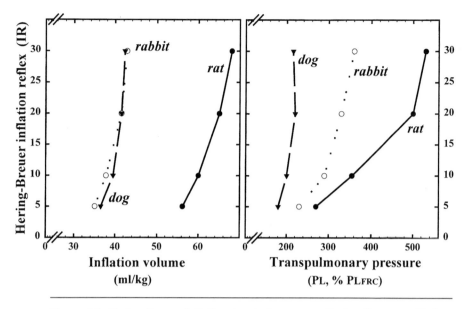

Figure 4.5. Hering-Breuer inflation (expiratory-promoting) reflex quantified as the ratio between the expiratory time during the inflation, TEinflation, and the control expiratory time, TE (inhibitory ratio, IR = TEinflation/TE), in newborn rats, rabbits, and dogs anesthetized with barbiturate. Data are represented as a function of the change in lung volume/kg body weight (*left*) and as a function of the change in transpulmonary pressure (PL) relative to the value at functional residual capacity (PLFRC) (*right*). From data of Gaultier and Mortola, 1981.

(1960), but the results have yielded discordant information (Chan and Greenough, 1992; Rabbette et al., 1991, 1994; Rabbette and Stocks, 1998; Trippenbach, 1981, also for earlier references).

The Paradoxical Inflation Reflex. In addition to the inhibition of inspiration and the facilitation of expiration, a rapid rise in lung volume can at first elicit a very brief inspiratory effort. Originally described by Head (1889), the reflex has been demonstrated in many newborns and adults of several species to be of pulmonary origin, since it is abolished by vagotomy, and has also been observed in the fetus (Hughes et al., 1967). Cross, who observed it in human infants, suggested that it may be relevant to the mechanisms of lung expansion during the onset of breathing at birth (Cross et al., 1960; Cross, 1961). It is possible that the neural ba-

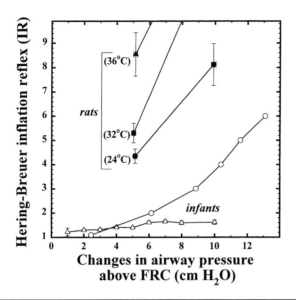

Figure 4.6. Hering-Breuer inflation (expiratory-promoting) reflex quantified as the ratio between the expiratory time during the inflation and the control expiratory time (inhibitory ratio, IR = TEinflation/TE), in conscious newborn rats (three to four days old, at three ambient temperatures) and human infants (one day old). Changes in airway pressure were computed from the changes in lung volume and the respiratory system compliance. The reflex inhibition is less in infants than in rats. Bars are SEM. From data of Mortola et al., 1995a (△) Cross et al., 1960 (○), and Merazzi and Mortola, 1999 (■, ●, ▲).

sis for the reflex is the same as that triggering the spontaneous deep breath, or sigh (Thach and Taeusch, 1976) (Fig. 4.7). Augmented inspiration can also be triggered by artificial ventilation, especially in cases of low compliance (Greenough et al., 1984). In newborns, deep breaths become particularly frequent during hypoxia, especially in those species that respond to hypoxia with limited or no hyperpnea and a rapid and shallow pattern. The periodic deep breath expands airways that tend to collapse during shallow breathing. Its occurrence in newborns of all species, from the premature human infant (Hoch et al., 1998) to the immature opossum (Farber, 1972), suggests that the periodic sigh is an essential component of the normal breathing pattern.

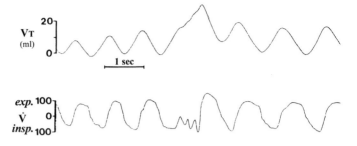

Figure 4.7. Records of tidal volume (*top*) and airflow (*bottom*) in a 4-day-old infant breathing through a face mask. Approximately in the middle of the segment is a spontaneous deep breath, with the characteristic biphasic pattern and multiple peaks in the inspiratory flow profile. Slightly modified from Thach and Taeusch, 1976.

Airway Receptors

One of the aims in the study of airway receptors in newborns is to identify developmental characteristics that could explain, or contribute insights into, the reflex responses of pulmonary origin. The most common general approach is to record the afferent activity from the peripheral cut end of the vagus as the airways are subjected to various mechanical or chemical challenges. Once the behavior of the receptors and their response to a given maneuver are established, an attempt is made to link the receptor response to the reflex effect provoked by that maneuver. There is much evidence from studies in adult animals that the lung's volume-dependent inhibition of inspiratory activity is due to the activation of slowly adapting airway receptors (SARs) (Euler, 1986). Far fewer studies have been performed with newborns, since some of the experimental tools used in adult studies do not find easy application to the developing animal. For example, because sulfur dioxide (SO_2) selectively inhibits SARs in adult rabbits, it has been used to study the reflex effects of this group of afferents with the remaining vagal innervation intact (Davies et al., 1978). SO_2, however, is ineffective in newborns (Mortola et al., 1984c). In addition, techniques for selective blocking of large-diameter myelinated fibers by local application of direct current or cold are less likely to be effective on the vagi nerves of newborn animals, because fewer fibers are myelinated (Fig. 4.8) and most are of small di-

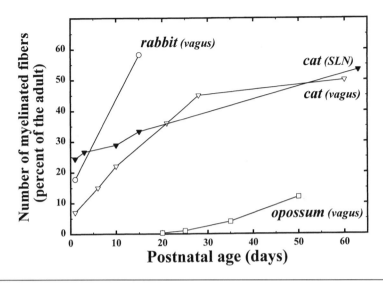

Figure 4.8. Number of myelinated fibers in the vagus nerve and in the internal branch of the superior laryngeal nerve (SLN) at various postnatal ages, in three mammals, expressed as a percentage of the total number of myelinated fibers in the adult. Vagus: modified from Fisher and Sant'Ambrogio, 1985, based on data of Marlot and Duron, 1979b (cat), DeNeef et al., 1982 (rabbit), Krous et al., 1985 (opossum); SLN: based on data of Miller and Dunmire, 1976, and Miller and Loizzi, 1974.

ameter (Schwieler, 1968; Marlot and Duron, 1979b; Sachis et al., 1982; Pereyra et al., 1992; Hasan et al., 1993). Although the travel distance is shorter, the narrow diameter and minimal myelinization are such that in newborns the conduction times can be longer than those in adults (Gregory and Proske, 1985).

Slowly Adapting Receptors. In newborn dogs, the SARs were found to be preferentially located in the large airways, the trachea providing about one-third of the total population (Fisher and Sant'Ambrogio, 1982). This is a distribution not notably different from that of the adult (Fig. 4.9). In kittens, almost half of all the SARs were located in the trachea and the lobar bronchi (Marlot et al., 1982). In adult cats, as well as in newborns, SARs have a respiratory modulation, with a progressive increase in firing rate the greater the transpulmonary pressure (PL), and a slow adaptation when the lungs are kept inflated (Fig. 4.10 and Fig. 4.11,

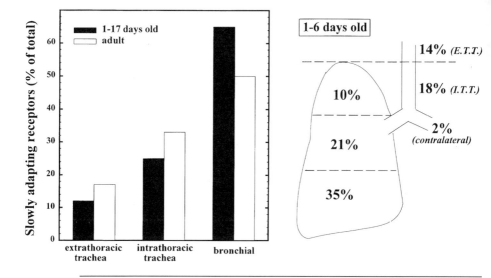

Figure 4.9. *Left panel:* Percentage distribution of the pulmonary slowly adapting stretch receptors in neonate and adult dogs. *Right panel:* Percentage distribution of the pulmonary slowly adapting stretch receptors in dogs at a few days of age. E.T.T., extrathoracic trachea. I.T.T, intrathoracic trachea. About 2% of the receptors are located contralaterally. In newborns, as in adults, a high percentage of receptors is located in the large airways. Modified from Fisher and Sant'Ambrogio, 1982.

left panel). But recordings in several species (cat, dog, opossum, sheep) have consistently indicated that the firing rate in the newborn during breathing, and at any given P_L, is less than that typically observed in the adult (Schwieler, 1968; Marlot and Duron, 1979a; Fisher and Sant'Ambrogio, 1982; Marlot et al., 1982; Farber et al., 1984) (Fig. 4.11, right panel). Both the passive resting volume of the respiratory system (Vr) and P_L at Vr are low in newborns (see Chapter 3, under "Static Coupling between Lungs and Chest Wall"). Hence, despite the fact that FRC is often maintained above Vr (see Chapter 3, under "Expiration"), the end-expiratory activity of the SARs ("tonic activity") in spontaneously breathing newborns is less than that in adults (Fisher et al., 1991).

In the sheep fetus, SARs can be activated by lung expansion, and some activity persists even at end-expiration (Ponte and Purves, 1973); whether or not these SAR inputs contribute to the inhibition of breath-

Opossum, 20 days old, 1.9 g

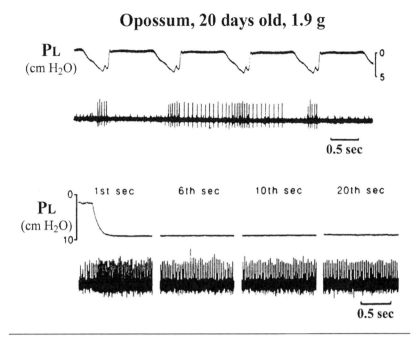

Figure 4.10. Transpulmonary pressure (PL) and action potentials from a pulmonary slowly adapting stretch receptor in a 20-day-old opossum. *Top:* Activity related to the breathing cycle. During the first and last of the breaths represented, the activity occurs only during inspiration (phasic activity). In the two intermediate breathing cycles, in addition to the phasic component, some activity is continuously present at end-expiration (tonic activity). *Bottom:* Discharge pattern during maintained lung inflation, showing small adaptation even after 20 sec. From Farber et al., 1984.

ing during the fetal life remains conjectural. As mentioned earlier, in the lamb fetus bilateral vagotomy had minimal effects on fetal breathing movements.

Rapidly Adapting "Irritant" Receptors. In addition to the rapid adaptation to a maintained stimulus, these receptors differ from the SARs in having an erratic pattern during tidal breathing. In newborns of several species the proportion of rapidly adapting "irritant" receptors (RARs) has been consistently found to be less than that in adults (Fig. 4.12), and in studies of vagal afferents in kittens (Schwieler, 1968; Marlot and Duron, 1979a; Marlot et al., 1982) there is no mention of any receptor having these characteristics. It should be kept in mind, however, that if RARs

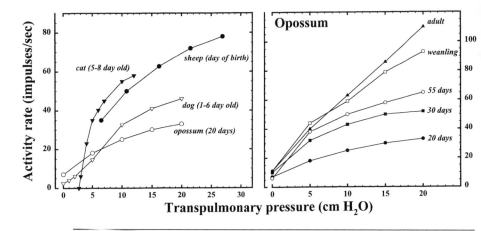

Figure 4.11. *Left:* Average discharge rate of the airway slowly adapting stretch receptors at various transpulmonary pressures (PL) in newborns of four species. Data were collected in static conditions (i.e., maintaining PL constant). *Right:* Discharge rate of slowly adapting stretch receptors in the opossum at various postnatal ages. Data of sheep from Ponte and Purves, 1973, after correction of tracheal pressures for PL according to average data of compliance; dog from Fisher and Sant'Ambrogio, 1982, combining "tracheal" and "bronchial" receptors; opossum from Farber et al., 1984; cat from Marlot et al., 1982; right panel slightly modified from Farber et al., 1984.

were mostly nonmyelinated afferents, their low incidence could in part be an artifact of the greater difficulty in recording from these fibers.

Cough and sneeze are complex motor defense mechanisms largely dependent on the activation of RARs. Although it is now recognized that cough probably involves the coordinated action of more than one group of receptors (Coleridge and Coleridge, 1986), it is of interest that the paucity of RARs in newborns parallels the difficulty in provoking cough as a response to tracheolaryngeal stimulation (Fig. 4.13). In intubated premature infants, stimulation of the tracheobronchial mucosa often failed to elicit cough, and could inhibit breathing (Fleming et al., 1978). Sneeze, a reflex of trigeminal origin, is characterized in kittens by expiratory efforts without the inspiratory preparatory phase observed in adults (Wallois et al., 1993).

Other Ventilatory Reflexes of Pulmonary Origin. In the airways of the adults of several species, a widespread distribution of C-fibers has been

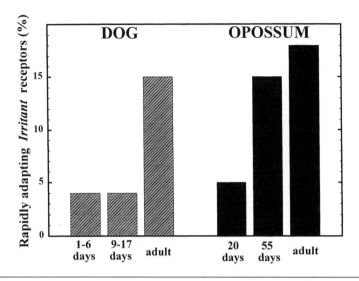

Figure 4.12. Frequency of rapidly adapting "irritant" receptors, expressed as a percentage of all the airway receptors having respiratory modulation (slowly adapting and rapidly adapting). From data of Fisher and Sant'Ambrogio, 1982, and Farber et al., 1984.

identified. They are activated by conditions such as embolism and lung congestion, and they respond selectively to chemicals injected into the right side of the heart, chemicals such as phenyl diguanide and capsaicin, resulting in bronchoconstriction, rapid and shallow breathing, increased airways secretion, and often bradycardia and hypotension (Coleridge and Coleridge, 1986). Many of these nonmyelinated vagal afferents are located close to the pulmonary capillaries, from which their frequent designation as J-receptors (from "juxtapulmonary capillary receptors") arises, but several other locations along the bronchial tree have also been found (Paintal, 1973; Coleridge and Coleridge, 1986).

As mentioned previously, chemicals injected in the blood stream and believed to have stimulated the pulmonary C-fibers promoted reflex bronchoconstriction in newborn pigs, dogs, and guinea pigs (Anderson and Fisher, 1993; Murphy et al., 1994; Nault et al., 1999). In kittens, none of the cardiorespiratory responses were observed until they were at least 3 weeks old, unless extraordinarily high dosages were injected (Kalia, 1976), whereas the full battery of responses was observed in 2-day-old

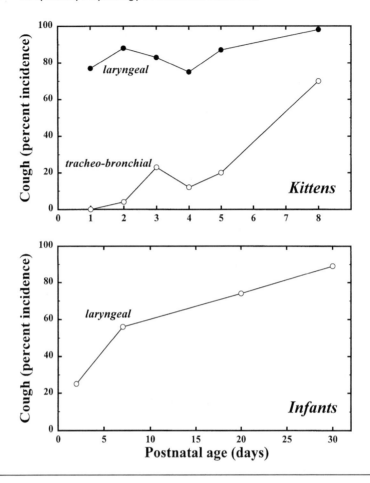

Figure 4.13. Incidence of cough when stimuli are applied at the laryngeal or tra-cheobronchial mucosa in infants (*bottom*) and anesthetized kittens (*top*). Data on infants from Miller et al., 1952; on kittens from Korpas and Kalocsayova, 1973.

piglets (Schleman et al., 1979). Lactic acid injection induced rapid and shallow breathing in 2-day-old rabbits, but with dosages double those used in adults (Ducros and Trippenbach, 1991). Hence, despite the small number of studies and the usual limitations introduced by the use of anesthesia, it could be tentatively concluded that in newborns the vagal reflexes originated by C-fibers are qualitatively similar to those of the adult, but their threshold is substantially higher in the less precocial species. In the newborn, the presence of pulmonary interstitial fluid and the high pulmonary artery pressure should represent continuous stimuli to the C-fibers. Hence, the high threshold in the early postnatal phases perhaps reduced the occurrence of unnecessary cardiorespiratory reflexes.

Vagal Control during Hypoxia

The immediate $\dot{V}E$ response to hypoxia in newborns is a transient hyperpnea, followed by a decline of the $\dot{V}E$ level, with a rapid and shallow pattern (see Fig. 5.15, below). Given the importance of the vagus nerves in regulating the depth and frequency of breathing, the possibility has been raised that vagal control may be involved in the poor hyperpneic response to hypoxia characteristic of the newborn. Because the slowly adapting airway receptors can be stimulated by hypoxia (Fisher et al., 1983b), the vagal inhibition during inspiration could be enhanced, favoring a shortening of T_I and the hypoxic tachypnea. In addition, if hypoxia induces airways constriction, the increased airway resistance could represent a mechanical limitation to the hypoxic hyperpnea. These hypotheses, however, did not find much support from experimental evidence; vagotomy did not appreciably alter the pattern of the immediate hypoxic response of the phrenic output or $\dot{V}E$ in rabbit pups (Trippenbach et al., 1985a), newborn piglets (Lawson and Long, 1983), or newborn monkeys (LaFramboise et al., 1985). In lambs, some differences in the response to hypoxia before and after vagotomy were mostly attributable to the different motor output to the larynx, rather than to the afferent information (Delacourt et al., 1995).

During hypoxia, because of the occurrence of hypometabolism and the consequent reduction in inspiratory "drive" (see Chapter 5, under "Acute Neonatal Hypoxia"), one may expect that stimuli normally inhibitory to breathing could have greater effects than during normoxia. One attempt to investigate this possibility was made in conscious neonatal rats by measuring the strength of the vagal Hering-Breuer inflation reflex. Hypoxia clearly reduced the strength of the reflex in 8-day-old pups, as it does in adults, but in younger pups the reflex was unaffected or even magnified by hypoxia. It may be that the low chemosensitivity characteristic of these young ages, coupled with the reduction in ventilatory drive during hypoxic hypometabolism, enhanced the relative efficacy of the vagal inhibition (Matsuoka and Mortola, 1995). In addition, hyperthermia in rat pups increased the strength of the Hering-Breuer reflex (Merazzi and Mortola, 1999a), which is of interest considering that in hypoxia even a normal body temperature could represent a condition of relative hyperthermia (see Chapter 5, under "Acute Neonatal Hypoxia"). Finally, when hypoxia and hyperthermia occur in combination the Hering-Breuer reflex can be stronger than that during normoxia (Merazzi and Mortola, 1999b). Hence, the data available seem sufficient at least to entertain the possibility that in the acutely hypoxic newborn the hypometabolic response decreases the $\dot{V}E$ drive, creating the basis for a greater efficacy of inputs inhibiting breathing. Associated circumstances, such as a reduction in chemosensitivity or hyperthermia, could further amplify these inhibitions (Fig. 4.14). This possibility is of interest also for the interpretation of breathing irregularities during hypoxia in newborns and human infants. Indeed, the association of hypoxia and hyperthermia has often been suspected in the pathophysiology of neonatal apneas and of sudden infant death (Stanton, 1984; Sawczenko and Fleming, 1996).

Ventilatory Reflexes of Extrapulmonary Origin

Probably all organs provide afferent inputs which, whether directly or indirectly, affect breathing. Distention of the abdominal and thoracic viscera, including the bladder and the esophagus, stimulation of the skin, activation of pain receptors, or increases or decreases in blood pressure are some of the numerous examples that reflexively modify the breathing pattern. Whether the resulting effect is a stimulation or an inhibition of tidal volume, frequency, or both can vary, depending on the intensity of the stimulus and its association with other inputs. Here, we will discuss those extrapulmonary reflexes best known for their probable implications for the regulation of the breathing pattern in newborn mammals.

Upper Airway Reflexes

The upper airway, defined as the air passages from the mouth and nose to the larynx, is innervated by branches of the trigeminal (V), glossopharyngeal (IX), and vagal (X) nerves, respectively the fifth, ninth, and tenth cranial nerves. This is thus a busy area, involved in many tasks often concurrently performed. Various stimuli coexist; for example, with inspira-

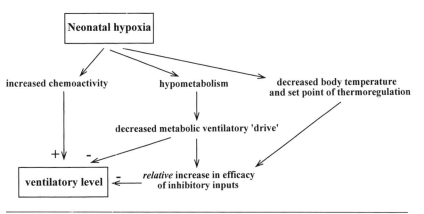

Figure 4.14. Possible events accompanying a hypoxic episode in the newborn mammal. Hypoxia stimulates pulmonary ventilation (V̇ᴇ) via the peripheral chemoreceptors, even while it reduces it because of the resulting hypometabolism. Because in newborns the former is not very pronounced, whereas the latter is very common, their combination can result in a reduced V̇ᴇ level. The decreased metabolic ventilatory "drive" favors a *relative* increase in the efficacy of inhibitory inputs, such as those originated by the pulmonary afferents or the upper airways, especially in conditions of hyperthermia. The latter is more likely to occur in hypoxia than normoxia, because in hypoxia the set point of thermoregulation is lowered. The greater *relative* inhibition can reduce V̇ᴇ to the point of entering a vicious cycle that, by further reducing V̇ᴇ and aggravating the hypoxia, could lead to irreversible apnea.

tion, as air passes through the upper airway, inputs related to air flow, pressure, and temperature, as well as olfactory information, are generated. It is even possible that in some animals there are nasal chemosensors capable of detecting the air gas composition (Arieli, 1990). The effect on the breathing pattern of these simultaneously occurring inputs represents the net effect of the individual reflex responses, and therefore can vary not only quantitatively but also qualitatively with seemingly small changes in experimental conditions. The approach taken for the experimental study of the reflex effects of upper-airway stimuli is to consider one or a few inputs at a time, as if they were generated by anatomically and functionally separated entities, and then attempt to reunify the individual results into an integrated view.

The bulk of information on upper-airways function, and specifically on their involvement with the breathing act, has been obtained from studies of adult animals and humans (Bartlett, 1986, 1989; Widdicombe, 1986; Iscoe, 1988). Data regarding the physiology of the upper airways in newborns and during development is mostly confined to the laryngeal area. The role of the glottis in controlling upper-airways resistance, expiratory flow, and FRC was mentioned in Chapters 2 and 3; here, the reflex effects on the breathing pattern will be discussed.

Laryngeal Afferents. In newborns the superior laryngeal nerve (SLN), like the vagus, contains only a few myelinated fibers of small diameter (see Fig. 4.8). This is probably also the case with the inferior laryngeal "recurrent" nerve and with other nerves supplying the upper airways (Fisher et al., 1991). No information is available on the C-fiber population of the upper-airway nerves in the newborn. Single-fiber recording from the peripheral cut end of the SLN has revealed receptors responding primarily to mechanical events, such as the tension developed within the laryngeal structure during inspiration, and others sensing the drop in temperature that can accompany air inflow. The percentage distribution of laryngeal receptors responding primarily to inspiratory or expiratory mechanical event and those sensitive to cold temperature differs little between newborn and adult dogs, with a large proportion of receptors sensing mechanical events, probably through changes in tension in their microenvironment. For the SLN, however, as was mentioned above for the vagus, the firing rates are generally lower in newborns than in adults (Fig. 4.15).

Many of these receptors probably sense more than one stimulus. It is known that laryngeal afferents can respond to mechanical but also chemical stimuli, such as those introduced by water, foreign milk, or other liquids on the mucosa (Johnson et al., 1975; Harding et al., 1978b). This chemical sensitivity is specific for molecules lacking Cl^- or other small anions that could functionally replace it (Boggs and Bartlett, 1982). At the same time, stimuli known to activate laryngeal endings in the adult, such as CO_2 or smoke, would not be effective in the newborn (Fisher et al., 1991).

Reflex Effects on Breathing Pattern. Several experimental approaches have indicated that stimulation of the upper airways has important effects on breathing. Almost invariably, these effects are inhibitory to $\dot{V}E$ and are more marked in newborns than in adults. As is so for the neural afferents, the studies on the reflex responses have concentrated mostly on the la-

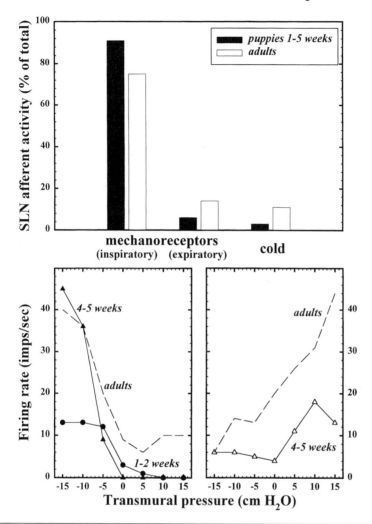

Figure 4.15. *Top:* Percentage distribution of receptors recorded from the superior laryngeal nerve of few-week-old puppies and adult dogs. *Bottom:* Average firing rates of laryngeal mechanoreceptors sensing predominantly negative pressures (*left panel*) or positive pressures (*right panel*). Values obtained from data provided by Fisher et al., 1985, 1991, combining groups originally classified as "pressure" or "drive" receptors.

ryngeal area, which seems to be the region most responsible for the ventilatory responses to mechanical or thermal stimuli applied to the upper airway. Even the inhibition of breathing determined by head immersion (the "diving reflex"), which in conscious animals is of trigeminal origin, seems in anesthetized lambs to be caused predominantly by stimulation of the larynx (Tchobroutsky et al., 1969). Electrical stimuli applied to the whole SLN, therefore combining inputs from many different types of receptors, provoke respiratory arrest (Sutton et al., 1978).

In all the newborns tested (cat, dog, sheep, pig, and opossum), water and other liquids, including foreign milk and distilled water but not normal saline, when instilled in the laryngeal region can produce prolonged apneas (Downing and Lee, 1975; Johnson et al., 1975; Farber, 1978; Harned et al., 1978; Kovar et al., 1979; Lucier et al., 1979; Fagenholz et al., 1979; Boggs and Bartlett, 1982; Lanier et al., 1983) (Fig. 4.16). Apnea, a reflex response that depends on the integrity of the SLN, is likely to be functionally similar to what is observed in some infants during feeding (Plaxico and Loughlin, 1981) or during experimental application of water boluses in the pharyngeal area (Davies et al., 1989; Wennergren et al., 1989). The sensitivity observed in these cases does not vary with

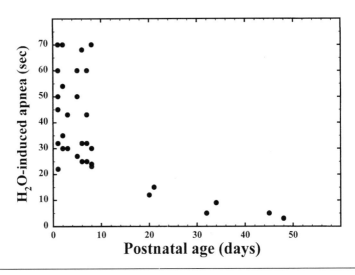

Figure 4.16. Duration of apnea induced by water instilled in the larynx of dogs of various ages under barbiturate anesthesia. Modified from Boggs and Bartlett, 1982.

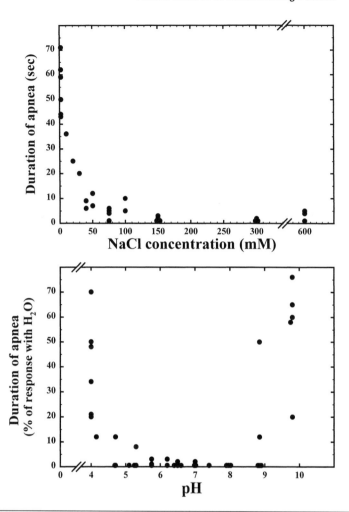

Figure 4.17. Apnea in newborn dogs under barbiturate anesthesia. *Top:* Duration of apnea induced by instillation of liquid boluses of various NaCl concentrations in the larynx. *Bottom:* Apnea induced by instillation of saline solutions of various pH in the larynx, expressed as a percentage of the apneic duration provoked by water. Modified from Boggs and Bartlett, 1982.

changes in the osmolarity of the liquid. Rather, the important variables are the ion composition and pH level (Kovar et al., 1079; Boggs and Bartlett, 1982); small $[Cl^-]$, high $[K^+]$, and very low or very high pH, especially if in combinations, elicit the strongest apneas (Fig. 4.17).

Long apneas, often persisting beyond the removal of the stimulus, can be provoked in newborns by unidirectional airflow through the isolated in situ upper airways, or by sustained negative pressures (Al-Shway and Mortola, 1982; Mortola et al., 1983; Fisher et al., 1985). Also in these cases the response depends on laryngeal afferents, and is not seen in the adult. The cooling of the mucosa produced by the airflow has been shown to be the main stimulus in determining the apnea response (Mathew et al., 1990), presumably through the activation of the laryngeal cold receptors.

In infants breathing through a tracheostomy, sustained negative pressures applied by suction at the nostrils decreased inspiratory flow and tidal volume, but did not provoke apnea (Thach et al., 1989). These effects, more modest than those observed in newborn animals, could be due to the fact that the negative pressures provoked closure of the oropharyngeal areas, hence did not stimulate the laryngeal region. It is also possible that the experimental results in animals were magnified by the anesthesia, which is often the case for reflexes inhibitory to breathing. In conscious kittens with chronic bilateral section of the SLN, the breathing pattern did not differ from that of intact kittens, whether in normoxia or hypoxia (Mortola and Rezzonico, 1989).

To conclude, there is abundant evidence that in newborns the upper airway, and specifically the laryngeal region, can be the source of reflexes inhibitory to breathing, but the inhibition is probably small under normal breathing in conscious conditions. It may be that these inhibitory reflexes can be triggered by special circumstances, for example when foreign material accidentally reaches the larynx, as during gastroesophageal reflux. In these cases, following the apnea, coughing or swallowing would be the "appropriate" response to laryngeal stimulation, but neither one is common in the newborn. In cases of upper-airway obstruction the reflex inhibition of the diaphragm could help to break a vicious cycle that otherwise would only aggravate the occlusion. Finally, one cannot exclude the possibility that at least some of the upper-airway reflexes inhibiting $\dot{V}E$ in the newborn may be a carry-over phenomenon from the fetal life, at a time when stimuli from the upper airways probably contribute to the sustained inhibition of breathing. Whatever the physiological meaning may be, it is important to know that, although hidden under normal circumstances, these reflexes exist, and in the developing organism are potentially capable of powerful ventilatory inhibition.

Interaction with Hypoxia. As is the case with other pulmonary vagal reflexes (see "Vagal Control during Hypoxia," above), it would be of interest to know, in the case of the laryngeal reflexes, what effect hypoxia might have on the inhibition of breathing. The premise is similar to that outlined for the pulmonary vagal reflexes (see Fig. 4.14); since in newborns during hypoxia the ventilatory "drive" decreases, owing to the concurrent hypometabolism, could this result in an increase in the relative efficacy of the upper-airway inputs inhibiting breathing? The available data, from piglets, lambs, and infants, give conflicting answers.

In anesthetized piglets (Fagenholz et al., 1979) and in 1-week-old conscious lambs (Grogaard et al., 1986), the apnea induced by H_2O placed directly on the laryngeal mucosa was not affected by peripheral chemodenervation. In most of the infants studied during normoxia and during breathing 15% O_2, the short apnea provoked by water instillation in the pharynx was slightly prolonged by hypoxia, although in a few cases the prolongation was pronounced (Wennergren et al., 1989). A prolongation of the apnea by hypoxia was also observed in anesthetized piglets (Lanier et al., 1983). In contrast, other studies have indicated that the carotid chemoreceptor activation during hypoxia limits the reflex apnea. In fact, in conscious lambs the apnea was reduced during acute hypoxia (Sladek et al., 1993b) or under pharmacological stimulation of the carotid body with a β-agonist (Grogaard et al., 1986). In addition, apnea was enhanced in lambs that had been chronically hypoxic since birth and thus had a decreased carotid-body sensitivity (Sladek et al., 1993b). Recently, in studies of conscious lambs, Milerad and Sundell (1999) confirmed that the duration of the apnea provoked by water instillation in the pharynx was slightly reduced by hypoxia. The authors also pointed out, however, that immediately following the apnea the inspiratory "drive," evaluated indirectly from the airway pressure profile seen during an inspiratory effort, decreased during hypoxia. This would mean that the duration of the apnea provides only a limited picture of the effects of interaction between laryngeal stimuli and hypoxia on the control of breathing. Indeed, several variables could confound the results. Because CO_2 level influences the magnitude of the laryngeal reflex (Litmanovitz et al., 1994), the duration of apnea is influenced not only by the hypoxic activation of the peripheral chemoreceptors but also by the degree of hypocapnia produced during the hypoxic hyperventilation, and therefore by the central chemosensitivity. The rate of progression of the hypoxia and hypercapnia during apnea is determined by the metabolic rate, which is affected by hypoxia to different degrees, depending upon species, age, and the normoxic metabolic level (see Chapter 5, under "Acute Neonatal Hypoxia"). Therefore, a better interpretation of the complex interaction between hypoxia and the inhibitory inputs of laryngeal origin would benefit from data on metabolic rate during normoxia and hypoxia.

Somatic and Chest Wall Reflexes

Somatic stimuli are thought to facilitate breathing, and could attenuate the vagal inhibitory inputs (Trippenbach and Flanders, 1999). The maternal licking of the newborn at birth, commonly observed in many species, is an instinctive behavior that could serve several purposes, perhaps including stimulation of breathing (Faridy, 1983). In infants, rubbing of the extremities reduced the apneic episodes (Kattwinkel et al., 1975), and there were fewer episodes of periodic breathing and apneas in preterm infants having skin-to-skin contact ("kangaroo care") with their mothers (Anderson, 1991). In decerebrate kittens, by contrast, although soft touching activates inspiration, intense cutaneous stimulation like pinching the ventral region of the thorax and skin can actually inhibit inspiration and provoke apnea (Khater-Boidin et al., 1994). Stimuli from the muscles and joints can have both facilitative and inhibitory effects on $\dot{V}E$. For example, in adult animals, phrenic activity was either increased or depressed by stimuli originating from the intercostal muscles (Sant'Ambrogio and Remmers, 1985; Shannon, 1986). In newborns, the compliance of the chest wall is high, relative to that of the lungs (see Fig. 3.10), and the distortion during inspiration is considerable. Some squeezing and deformation of the chest must be a common event in pups of large litters, in which huddling and very close interaction with the mother and littermates is essential for protection against heat loss. In newborns, if chest wall reflexes were as operational as they are in the adult, the continuously changing inputs from the skin and the chest could contribute greatly to the variability of their breathing pattern.

In studies of infants, an inverse relationship was found between the degree of chest distortion and the inspiratory time TI; in addition, vibration applied to the intercostal space reduced TI (Hagan et al., 1977), an effect also obtained by compression in some areas of the chest (Knill and Bryan, 1976). Occasionally in infants, especially prematures, the duration of the inspiratory effort after occlusion of the airways at end-expiration can be shorter than the open-airway control Ti (Knill et al., 1976; Thach et al., 1978; Gerhardt and Bancalari, 1981), hence a response opposite to that of the vagal compensation to a load discussed previously. All these findings were interpreted as supporting the existence of inspiratory-inhibitory reflexes originating from the distorted chest, but offered no clear indication of the exact source of the inputs.

Experimental data from animal studies could offer more specific and more mechanistic insights, but the few studies in this area to date have failed to yield uniform information. In newborn kittens, muscle and skin receptors are operational, although they are less active than in adults and rapidly adapting (Ekholm, 1967). Electrical stimulation of the intercostal nerves in kittens resulted in modifications of the inspiratory and expiratory time qualitatively similar to those reported in adult cats (Trippenbach, 1985). After vagotomy, end-expiratory occlusion of the airways in kittens could result in either a reduction (Trippenbach et al., 1981) or a prolongation of Ti (Trippenbach, 1985). Both findings could be attributed to inputs from the chest wall, which would be distorted during the effort, although in anesthetized and vagotomized kittens the breathing pattern was not significantly affected by chest compression (Trippenbach et al., 1982). A shortening of Ti during an inspiratory effort with airways occluded at end-expiratory was observed also in a quadriplegic infant (Thach et al., 1980); this seems a clear indication that an intact intercostal-phrenic pathway is not an essential requirement for the shortening of Ti during inspiratory loading.

Thus, information regarding the ventilatory reflexes originating from the chest wall in newborns is extremely fragmentary. As in adults, electrical inputs from the intercostal nerves can decrease Ti, but this does not imply that paradoxical responses to respiratory loading, such as the shortening of Ti during airway occlusion, are necessarily of chest wall origin. The core question, whether or not inputs from chest-wall structures in newborns hinder or facilitate the stability of the breathing pattern, remains unanswered.

Central Organization of Respiratory Neurons

Most studies of the properties of the central respiratory neurons have been performed on adult cats. Studies on the development of these neurons have adopted in vivo and in vitro preparations on a number of species, such as piglets, rabbits, rats, and kittens. The data on newborns are usually interpreted by making the results from adult cats the standard reference. In vitro preparation employed brain stems, with or without the spinal cord and in brain stem slices, obtained from newborn opossums or, mostly, from newborns of rats and other rodents. Despite technical problems related largely to the oxygenation of the in vitro preparations, these approaches have provided a wealth of new and detailed information, the main results of which have been reviewed by Hilaire and Duron (1999). A unifying interpretation of the neonatal network and of its developmental processes, however, remains complicated by the lack of an accepted model for respiratory rhythmogenesis.

Medullary units having respiratory modulations have been found in areas homologous to those where they are typically found in adults, between the *nucleus ambiguus* and the *nucleus tractus solitarius.* This has been the case even in the opossum, a marsupial that, because of its altricial characteristics, offers a rare opportunity to study the very early aspects of the development of respiratory control (Farber and Lawson, 1991). As has been proposed for the operation of a neural network of respiratory rhythmogenesis in adults, in the immature opossum and newborns of other species, neurons with the required characteristics, that is, inspiratory, postinspiratory, and expiratory, have been identified (Fig. 4.18).

In vitro preparations from fetal rats and mice have revealed rhythmic phrenic bursts only from embryonic day 16, out of the 21–22 days of gestation, i.e., 2–3 days after the phrenic axons have reached the primordial diaphragm (Fig. 4.19). If ~13 days is the minimum time needed to establish adequate phrenic-diaphragm connections, then animals born after a very short gestation, like some marsupials, would at birth have little control of their respiratory muscles. This could explain why a Julia Creek dunnart born after about 13 days of gestation, could produce, during the first days after birth, only generalized skeletal muscle twitches, which would therefore make pulmonary ventilation an impossible task; in this newborn its gas exchange is almost exclusively through the skin (Frappell and Mortola, 2000) (see Figs. 2.1 and 2.2). Once established, the fetal pattern of phrenic and respiratory neuronal firing is characteristically much more variable than that in the newborn (Hilaire and Duron, 1999).

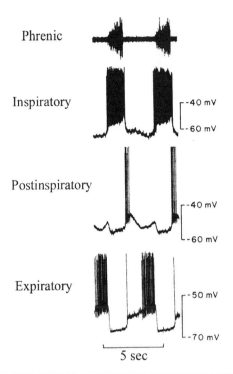

Figure 4.18. Combined representation of the activity of three neurons located in the ventral respiratory group of the medulla of the piglet, recorded intracellularly. Although the recordings originated from experiments on different animals, they are shown here combined, in relation to the activity of the phrenic nerve, shown at top. The three different patterns with respect to the timing of the phrenic burst (i.e., inspiratory, postinspiratory, and expiratory) are clearly distinguishable. From Farber and Lawson, 1991.

The activity pattern of the phrenic nerve and diaphragm of newborns during inspiration is often an abrupt and rapid burst with modest numbers of spikes, especially in newborns of altricial species, thus differing from the abundantly active ramp-shape firings of the adult (Hilaire and Duron, 1999). It has been suggested that with few spikes per breath the effects of feedback information could produce rather coarse changes in the amplitude and duration of the inspiratory burst (Farber and Lawson, 1991). If this is so, the great variability of the breathing pattern, a most common observation in conscious newborns of various animals, may not

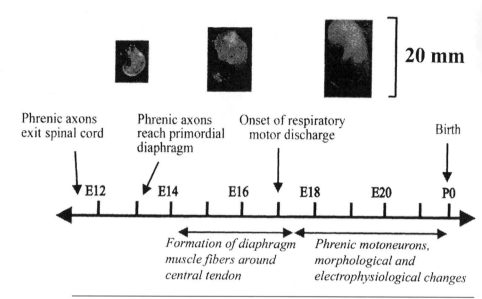

Figure 4.19. Time course of the events in the embryogenesis of phrenic nerve-diaphragm interaction in fetal rats. Photographs are of embryos at about 13, 16, and 18 days. Slightly modified from Greer et al., 1999.

necessarily originate only from the intrinsic properties of the respiratory rhythm generator, but could reflect a combination of its properties with those of the peripheral and central afferents acting on it (Hilaire and Duron, 1999). This peripheral-central interaction could contribute to the apparent paradox that, even if the afferent neural information is weak (see "Airway Receptors," above), vagal reflexes in newborns are at least as pronounced as those in adults, and laryngeal reflexes can be extremely powerful.

Breathing during Feeding

In mammals, with the exception of some diving species (dolphins, whales), the upper airways represent a pathway common to both the respiratory system and the digestive system; hence, appropriate coordination between the two functions is an absolute requirement during the neonatal period. Feeding periods occur more frequently in newborns than in adults, especially in the small species. An extreme case is that of the developing pouch-young marsupial, which remains continuously attached to the maternal nipple for weeks (Tyndale-Biscoe and Renfree, 1987). At the same time, some of the defense mechanisms protecting the upper airways, for example cough, are less pronounced in the newborn (see Fig. 4.13). Lack of coordination could lead to an excess of air in the digestive tract, with overdistention of the abdominal viscera and decreased efficiency of the digestive processes. Conversely, foreign particles drawn into the airways during sucking and swallowing, or during regurgitation and vomiting, would have obviously devastating effects on pulmonary gas exchange, and are often the cause of infection as well.

In the fetus, sucking exercises and the swallowing of amniotic fluid begin in the early phases of gestation, and the important role of the larynx in controlling fetal pulmonary volume has already been discussed (see Chapter 1, under "The Respiratory System before Birth"). In the newborn, sucking and breathing can occur simultaneously, although experimental observations have been reported only for the human infant and the lamb (Dreier et al., 1979; Harding and Titchen, 1981; Shivpuri et al., 1983; Mathew et al., 1985). The breathing pattern at such moments, however, differs from that during nonsucking periods (Mathew, 1988). The combination of breathing and sucking in newborns is favored by the higher anatomical location of the larynx within the neck organs, in combination with other anatomical characteristics of the developing upper airway (Noback, 1923; Crelin, 1973; Bosma, 1988). The higher position of the larynx reduces the size of the oropharyngeal region, thus favoring the opposition of the dorsal portion of the tongue to the soft palate, and therefore effectively separating the oral cavity from the naso- and oropharyngeal areas. As a result of this separation, the newborn can generate the negative pressures required for suction without involving other portions of the conductive airways.

The high location of the larynx would also permit a lateral passage of the liquid bolus toward the pharynx. Indeed, it was thought that, like sucking, swallowing too could occur simultaneously with breathing activity. In reality, at least in infants and lambs, the respiratory cycle is interrupted whenever swallowing occurs (Gryboski, 1969; Harding et al.,

EMG diaphragm

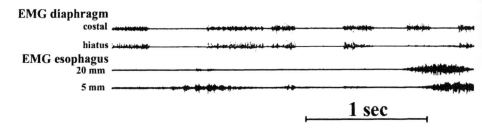

Figure 4.20. Multiple electromyographic (EMG) recordings in a lamb before and during swallowing. From the top, EMG of the costal fibers of the diaphragm, of the vertebral fibers of the diaphragm a few mm from the hiatus esophageus, and of the esophageal muscle fibers at 20 mm and 5 mm above the hiatus esophageus. At the start of the caudal contraction of the esophagus, the diaphragmatic fibers of the hiatus, but not those of the costal regions, decrease their activity. Modified from Harding and Titchen, 1981.

1978a; Wilson et al., 1981), not dissimilarly from what occurs in the adult, although the coordination between cessation of breathing activity and initiation of swallowing is not always perfect. The implications of the interruptions of breathing induced by swallowing on gas exchange should depend on the frequency of swallowing with respect to breathing rate. In the lamb, because of its higher rate of swallowing, gas exchange is affected more than it is in human infants (Harding et al., 1978a; Shivpuri et al., 1983). As soon as the bolus has entered the esophagus, inspiratory activity can safely resume. When the costal fibers of the diaphragm are active, the diaphragmatic vertebral region remains inactive in response to stimulation of vagal fibers from the lower esophagus, to favor the passage of the bolus through the esophageal hiatus (Fig. 4.20). Indeed, this is a good example of the differential control of the two diaphragmatic regions, made possible by the separate embryologic origins and motor innervations of the costal and vertebral portions of the diaphragm (Langman, 1977; Sant'Ambrogio et al., 1963).

Regurgitation of a small quantity of ingested material is common in the neonatal period, and continues later in life in ruminants. In the lamb the association between regurgitation and rumination begins at ~2–3 weeks after birth. Regurgitation represents another example of fine coordination between diaphragmatic regions, abdominal muscles, and upper-airway muscles (Mortola and Fisher, 1988).

Interspecies Comparisons

Newborns of all species tested, from the most altricial pouch-young opossum to the very precocial lamb, respond to an inspiratory load with increased inspiratory activity. Whether or not this compensatory response, which is a pulmonary reflex mediated by the vagus nerve, varies systematically with species size, or is quantitatively different from that of the adult, is still unclear. In newborns, the inhibition of breathing following lung inflation may be stronger in smaller species, as is the case among adults. The slowly adapting stretch receptors of the airways are also operational in the most altricial newborns, and even in fetuses, but their activity is typically modest compared to that of older animals. Rapidly adapting receptors, by contrast, seem to be underrepresented. Although studied in only a few species, the laryngeal area can be the source of powerful reflexes inhibitory to breathing. Liquid in the larynx of the newborn cat, dog, sheep, pig, or opossum can produce long and irreversible apneas, depending on its chemical composition. Ventilatory reflexes of somatic origin (from the skin, muscles, or joints) have been described in newborns of several species; and cutaneous touch stimuli, including those involved in mother-pup interaction, could facilitate breathing. Neurons having the key characteristics required for the generation of a respiratory rhythm have been identified in preparations from neonates of several species. In all of them the activity pattern of the phrenic nerve or diaphragm is invariably an abrupt and rapid burst with a modest number of spikes; this characteristic could make the inspiratory pattern more prone to coarse responses when confronted by peripheral stimuli.

Clinical Implications

The ability to defend tidal volume against mechanical loads imposed on the respiratory system, such as can occur with changes in posture or upper-airway obstruction, is a fundamental necessity for the infant's survival. Disturbances in the compensatory mechanisms for respiratory loading, which are based largely on reflexes of vagal origin, can lead to breathing irregularities and hypoxemia. During artificial ventilation, airway reflexes can explain a breathing pattern that is either in phase and synchronized with, or "fighting" against, the ventilator. Activation of the pulmonary vagal C-fibers could occur with interstitial edema, and reflexively cause rapid breathing, as occurs in the transient tachypnea syndrome of infants with delayed pulmonary fluid reabsorption. Inspiratory inhibition from lung expansion (the Hering-Breuer reflex) is enhanced by hyperthermia. This is particularly important during hypoxia, because hypoxia reduces the metabolic level and lowers the set-point of thermoregulation; the former decreases the ventilatory stimulus and the latter increases the risks of hyperthermia. The combination hypoxia-hyperthermia therefore sets the stage for the relatively greater effectiveness of stimuli inhibitory to breathing, such as those from the lungs or the upper airways, leading perhaps to breathing irregularities, apneas, or sudden death. In newborns, stimuli from the upper airways, specifically from the laryngeal region, can have powerful inhibitory effects on breathing. These reflexes can be triggered when foreign material accidentally reaches the laryngeal area, as during sucking, swallowing, or regurgitation, or with gastroesophageal reflux, especially in newborns, in which some of the typical defense mechanisms, including coughing and sneezing, are less powerful than in adults. In neonates, the skin is the source of stimuli that can facilitate breathing; perhaps this contributes to the more regular pattern observed in infants that are kept in direct skin-to-skin contact with the mother.

Summary

1. Compensation against mechanical loads is a fundamental necessity for the protection of tidal volume. Vagal reflexes are the core mechanisms for compensation; they include a prolongation and an increase in the activity of the inspiratory muscles, and are operational in the newborn. The extent of quantitative differences during development or between newborns and adults is still unclear.

2. When mechanical loads are sustained, the protection of tidal volume and ventilation is based on chemosensitivity as well. In newborns, a drop in metabolic rate may represent an additional mechanism.

3. Vagotomy results in a slower and deeper pattern in the newborn, as it does in the adult, whereas it has no appreciable implications for the breathing movements of the fetus.

4. Vagal innervation is not essential for the normal postnatal development of the lungs, but vagal innervation can affect bronchomotor tone, surfactant secretion, and the mechanical properties of the lungs, through changes in breathing pattern.

5. A sustained increase in lung volume inhibits inspiration and facilitates expiration (the Hering-Breuer inflation reflex) in all newborns studied, including those of the most altricial species.

6. At birth, the percentage of myelinated fibers in the vagus and superior laryngeal nerves is low. In general, the pattern and discharge characteristics of the airway receptors in newborns resemble those of the adults, most differences being quantitative rather than qualitative. The slowly adapting receptors have lower firing rates, and the rapidly adapting "irritant" receptors are scarce. The latter could contribute to the high threshold and decreased effectiveness of some airway defense and protective mechanisms, like sneeze and cough. Unmyelinated vagal C-fibers are numerous, but the threshold for their activation is higher than in adults.

7. Lung inflation during hypoxia can result in a greater vagally mediated inhibition of breathing than occurs in normoxia, especially if in combination with hyperthermia. In hypoxia, presumably, the decrease in metabolic rate lowers the "drive" to breathe, effectively increasing the relative strength of reflexes inhibitory to breathing.

8. Under special experimental conditions, it can be demonstrated that there are powerful ventilatory inhibitory reflexes from the upper airways, mostly from the larynx. These reflexes, which can be triggered by chemical and physical stimuli, are much more pronounced in newborns than in adults. The physiological role of these reflexes in resting conscious conditions, including their implications for the regulation of the breathing pattern and the changes in their effectiveness during hypoxia, are questions still to be resolved.

9. In newborns, somatic stimuli (from the skin, muscles, or joints) are likely to facilitate breathing, although reflexes from the chest can also shorten the inspiratory time. The exact role of these reflexes, and whether they hinder or facilitate the stability of the breathing pattern, remains unclear.

10. Both in vivo and in vitro preparations have indicated that the neurons having the characteristics thought to be required for central rhyth-

mogenesis in the adult are present in newborns. Experimental observations have been limited to only a few of the smallest species; but in these, at least, the output pattern of the respiratory generator is typically an abrupt and rapid burst with fewer spikes than in adults. Hence, peripheral information, even if of limited magnitude, could have coarse effects on the inspiratory pattern, and is potentially the source of the greater ventilatory variability seen in newborns, as opposed to adults. Presumably, the characteristics of the peripheral-central interaction could contribute to the apparent paradox that some neonatal reflexes are extremely powerful despite the rather limited neural afferent activity.

11. In newborns, feeding is a much more frequent and more prolonged event than it is in adults, and its coordination with breathing is crucial. Despite the paucity of observations (to date, only on lambs and infants), it seems clear that sucking can proceed simultaneously with breathing, a synchrony favored by the more cephalic location of the larynx in newborns than in adults. During swallowing, breathing is interrupted. As the bolus reaches the lower esophagus, the breathing-related diaphragm contractions continue with the selective relaxation of the periesophageal diaphragmatic fibers.

Chapter 5

Changes in Temperature and Respiratory Gases

Among the highest priorities for survival in the newborn are the protection of cellular oxygenation and temperature. The systems controlling these functions are highly interdependent. Suffice it to mention that the maintenance of body temperature is an energetically expensive function responsible for a high percentage of oxygen consumption, and that, conversely, the use of oxygen is temperature-dependent. In addition, the respiratory apparatus is a common effector for the operation of both thermoregulatory and respiratory systems. In fact, pulmonary ventilation is an important route of heat and water dissipation, and this function can often conflict with the task of regulating blood oxygenation and carbon dioxide.

Hence, the separate discussion of these two topics, the responses to changes in temperature and oxygenation, should not distract from recognizing the importance of their interdependence. The integration between temperature regulation and oxygenation in the newborn mammal assumes an importance possibly even greater than it does in the adult.

Changes in Ambient or Body Temperature

Adult mammals have the ability, known as *homeothermy*, to maintain body temperature (Tb) within narrow limits against changes in ambient temperature (Ta), via regulatory mechanisms of thermal regulation. These include the control of heat loss and the increase in heat production via endogenous mechanisms, or *endothermy*. Homeothermy does not exclude the possibility that Tb may vary, as in fact happens during sleep, torpor, or fever, which are all cases of change in the thermoregulatory set point. Furthermore, all mammals studied present circadian oscillations in Tb, commonly of about 1–1.5°C.

Metabolic and Behavioral Responses

Newborn mammals vary greatly in homeothermic and endothermic characteristics. Their Tb is not particularly different from the values commonly measured in adults (i.e., between 36 and 38°C), and there is a tendency for small species to exhibit lower values (Fig. 5.1). Tb can change substantially with environmental conditions, however, which in small species with large litters are often dependent on maternal behavior. In these cases, when pups are left unattended even for short periods of time Tb often plummets, because of their very limited thermogenic abilities (Hissa and Lagerspetz, 1964; Hissa, 1968).

Brown Adipose Tissue. Many newborns, including the human infant, do little or no shivering in response to cold, and their thermogenic response is therefore almost entirely based on the heat produced by the brown adipose tissue (BAT), the mechanism thus known as *nonshivering thermogenesis* (Brück, 1998). Marsupials develop in a very special habitat, the constant temperature of the maternal pouch, which is close to the maternal Tb (Hulbert, 1988). Not only is their specific metabolic rate low and constant during development (see Fig. 2.6), but their thermogenic capacity is almost nil. In fact, the absence of BAT is believed to be a general characteristic of marsupials, although this view is currently under review (Hulbert, 1988; Hope et al., 1997). Of the placental species so far examined, very few seem to pursue little or no BAT thermogenesis. Among them is the piglet, which can shiver in the first hours after birth, and, perhaps, the golden hamster.

BAT is the only animal tissue assigned the specific task of generating heat, and its physiology and biochemistry have been extensively studied during the last 35 years (Nedergaard and Cannon, 1998). Its bodily location and abundance vary greatly among species and with age, but common locations are around vessels, in many core and superficial regions of the thorax and abdomen, and, typically, in the easily identifiable symmetrical pads of the dorsal interscapular area (Smith and Horwitz, 1969), a deposit that furnishes one of the largest sites of BAT (Hull and Segall, 1965) (Fig. 5.2). Compared to the white adipose tissue, BAT is characterized by rich blood perfusion, which is responsible for its brownish

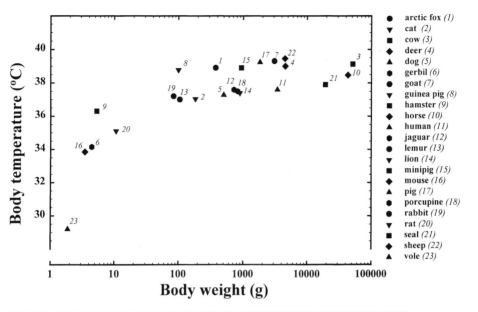

Figure 5.1. Body temperature of newborn mammals, as a function of the species' body mass. Values represent the average from one or more studies. Data for arctic fox are from[1]; cat[2-5]; cow[6-13]; deer[1,14]; dog[3,14-16]; gerbil[1,17]; african goat[14]; guinea pig[14,18-20]; hamster[1]; horse[21-22]; human[25-26]; jaguar[14]; lemur[14]; lion[24]; minipig[14]; mouse[1, 17]; pig[1, 28-29]; porcupine[14]; rabbit[1]; rat[16,23, 31-33]; seal[1]; sheep[30,34]; vole[27]. [1]Mortola and Lanthier, 1996; [2]Mortola and Rezzonico, 1988; [3]Pedraz and Mortola, 1991; [4]Mortola and Matsuoka, 1993; [5]Rohlicek et al., 1996; [6]Adams et al., 1993; [7]Reeves and Leathers, 1964; [8]Lekeux et al., 1984; [9]Kiorpes et al., 1978; [10]Bisgard et al., 1974; [11]Reeves and Leathers, 1967; [12]Donawick and Baue, 1968; [13]Peters et al., 1975; [14]Mortola et al., 1989; [15]Rohlicek et al., 1998; [16]Saiki and Mortola, 1996; [17]Waldschmidt and Muller, 1988; [18]Fewell et al., 1997; [19]Clark and Fewell, 1996; [20]Adamsons, 1969; [21]Ousey et al., 1991; [22]Stewart et al., 1984; [23]Bertin et al., 1993; [24]Parer et al., 1970; [25]Polgar and Weng, 1979; [26]Frappell et al., 1998; [27]Ru-Yung and Jinxiang, 1987; [28]Waters et al., 1996; [29]Rosen et al., 1993; [30]Sidi et al., 1983; [31]Mortola and Dotta, 1992; [32]Saiki and Mortola, 1994; [33]Saiki and Mortola, 1995; [34]Moss et al., 1987.

color, as well as sympathetic innervation, many small droplets of cytoplasmic triglycerides, and numerous mitochondria with dense cristae. The hallmark of BAT, however, is the presence of the mitochondrial uncoupling protein *thermogenin* (UCP) (Klingenberg, 1990; Nedergaard and Cannon, 1992). Although its molecular function is not fully under-

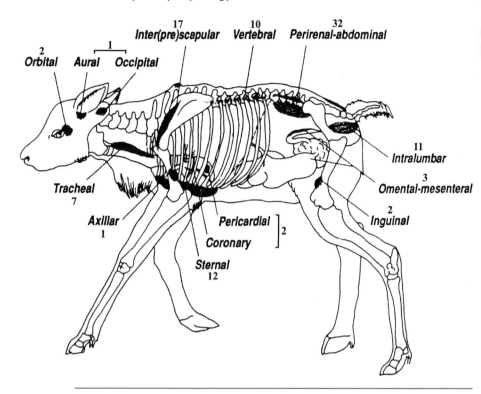

Figure 5.2. Anatomical locations of brown adipose tissue (BAT) in the newborn reindeer (body weight 4.3 kg). The total BAT wet weight was 48 grams, or 1.1% of body weight. Numbers indicate the percentage distributions at the several locations. Modified from Soppela et al., 1992.

stood, thermogenin behaves physiologically as a mitochondrial membrane carrier or channel; by letting H^+ leak through the inner mitochondrial membrane, it short-circuits the ATP-synthase, uncoupling the electron transport chain and $\dot{V}o_2$ from phosphorylation.

Thermogenesis and Heat Dissipation. The condition known as *Thermoneutrality* is the range in Ta over which Tb is maintained with minimal $\dot{V}o_2$ in normoxic conditions. Hence, because over the thermoneutral range $\dot{V}o_2$ is steady at its minimal value, the only avenue for maintaining Tb despite an increase in Ta is through heat-dissipation mechanisms. The magnitude of the thermoneutral range can therefore be interpreted as an index of the functional effectiveness of the mechanisms controlling heat loss. When these are no longer sufficient to protect Tb against the increase in Ta, hyperthermia intervenes. The lowest Ta of the thermoneutrality range is the *lower critical temperature;* below this value thermogenic mechanisms intervene to protect Tb, as indicated by the rise in $\dot{V}o_2$. Eventually, thermogenesis in the cold cannot compensate for the increasing gap between Tb and Ta, and Tb begins to fall, which in turn leads to a drop in $\dot{V}o_2$ because of the Q_{10} effect (Fig. 5.3, top). The Q_{10} (Arrhenius) factor expresses the change in reaction velocity for a 10°C change in T,

$$Q_{10} = (A'/A'')^{[10/(T'-T'')]} \qquad (Equation\ 5.1)$$

where A′ and A″ are the enzymatic activities, or reaction velocities, at the corresponding temperatures T′ and T″.

Hypoxia severely curtails the thermoregulatory mechanisms, as will be discussed later (see "Acute Neonatal Hypoxia," below). In adults, thermoregulation remains operative during slow-wave sleep, but during active (REM) sleep thermogenesis is depressed, resulting in a limited control of Tb in cooling conditions (Parmeggiani, 1990). Whether this may also occur in newborns has been studied in the lamb, but the results are conflicting (Berger et al., 1989; Fewell et al., 1990b).

In many newborns, especially of the smallest species, both heat-dissipation and heat-production mechanisms are very limited, with the result that thermoneutrality is almost undefinable (Fig. 5.3, bottom). These animals have such limited thermocontrol that their Tb behaves much as that in ectotherms. The small size, hence the large ratio of body surface to mass, favors heat loss to such a point that thermogenesis, although functional, is ineffective. For example, for 2-day-old rats at a Ta of 25°C to maintain Tb at 36°C would require a steady $\dot{V}o_2$ of ~250 ml · kg^{-1} · min^{-1}, which is 4 to 5 times higher than the summit metabolism that these animals can reach (Mortola and Dotta, 1992). Although body mass is undoubtedly an important issue, it is not the only factor determining the homeothermy ability in young animals. Newborns of some small but precocious species, such as the guinea pig (~80 grams), the coypu (~200 grams), and the pig (~1300 grams), have as good Tb control at birth as the lamb (~4000 grams), a species born with excellent thermogenic capacity (Fig. 5.4). By contrast, newborns of larger but altricial species such as humans (~3.5 kg) and other primates are far less able to maintain

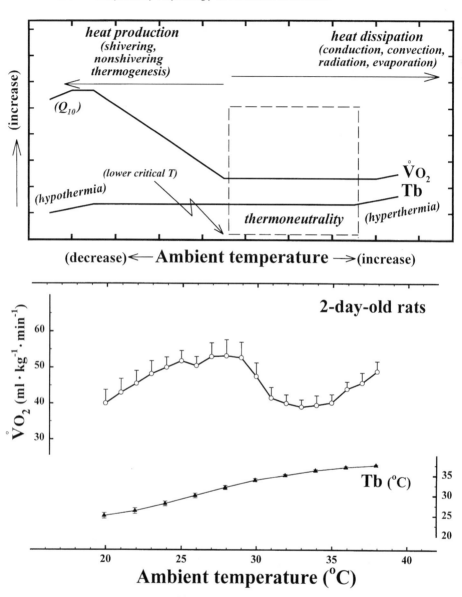

Figure 5.3. *Top:* Schematic representation of the changes in oxygen consumption ($\dot{V}O_2$) and body temperature (Tb) as a function of ambient temperature in mammals. The thermoneutrality range (*dashed box*) is determined by heat-dissipation mechanisms. For further definitions, see text. *Bottom:* Oxygen consumption ($\dot{V}O_2$) at different ambient temperatures in newborn rats. The drop in

homeothermy. With regard to the smallest species it has been proposed (Alexander, 1975) that the poikilothermic state of their newborns reduces their energetic demands, and is therefore an evolutionary response to the limited amount of milk available. Indeed, earlier calculations indicated that homeothermy is an impossible task for newborn mammals lighter than 2.5 g, because they cannot assimilate food fast enough to compensate for the heat loss they face (Pearson, 1948).

Behavioral Thermocontrol. Newborn mammals, like adults, when exposed to cold combine autonomic and behavioral thermogenesis. The latter, which includes interaction with the mother, positioning within the nest, postural changes, and huddling among siblings, could represent for them a mechanism of thermocontrol even more important than the automatic thermogenesis (Cabanac, 1972). Newborn mammals with sufficient locomotory control in a thermal gradient show thermotaxis toward a temperature that optimizes their Tb with minimal metabolic effort (Leonard, 1974; Fowler and Kellogg, 1975; Hull et al., 1986; Fewell et al., 1997). During overheating, the human infant adopts a stretched and extended posture to favor heat dissipation (Harpin et al., 1983); loud crying, by attracting the parents' attention, is also considered an important behavioral response to cold (Brück, 1998).

Huddling (i.e., close, shifting contact with siblings and parents) as a means to thermoregulate is a dynamic behavior finely controlled by extensive sensory information (Alberts, 1978a, 1978b). By reducing the total surface-to-volume ratio of the pups, huddling decreases the relative surface of heat loss. This behavior, in precocial newborns like the piglet that have efficient homeothermic capacity, implies a decreased need for thermogenesis, resulting therefore in a lowering of $\dot{V}o_2$. Differently, in an altricial newborn with poor thermocontrol, huddling limits the fall in Tb, and therefore limits the Q_{10} effect on $\dot{V}o_2$, with the result that each of the newborns can produce a more pronounced thermogenic effort (Fig. 5.5).

Newborns in Special Habitats. Both precocial and altricial species can be found in habitats presenting extreme climatic conditions, and their survival strategies differ markedly (Blix and Steen, 1979). In the arctic

body temperature (Tb) with the decrease in Ta occurs despite the thermogenic response to cold. Bars are 1 SEM. From data of Mortola and Dotta, 1992.

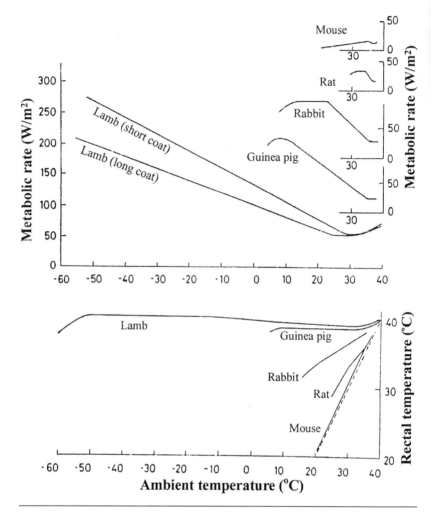

Figure 5.4. Metabolic rate (*top*) and rectal temperature (*bottom*) at different ambient temperatures in some newborn mammals (the vertical scale at left refers only to lambs). 1 Watt = 2.99 ml O_2 STPD/min. Dashed line is the line of identity. In the smallest newborns in the cold, despite the thermogenic effort, body temperature drops almost in proportion with ambient temperature. Slightly modified from Alexander, 1975.

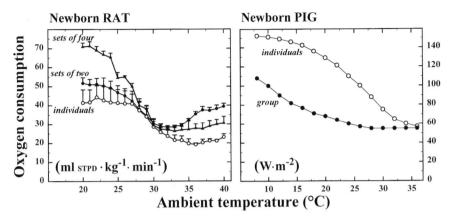

Figure 5.5. The effects of huddling on the metabolic response to changes in ambient temperature. In small newborn animals with limited thermogenic capabilities, such as the rat (*left panel*), the drop in body temperature is greater when they are in isolation ("individuals"); this lowers their thermogenic effort (oxygen consumption, $\dot{V}o_2$) via the Q_{10} effect. By contrast, in large and precocial species, such as the pig (*right panel*), the greater cold stimulus when they are in isolation further stimulates their thermogenic response, increasing $\dot{V}o_2$. 1 Watt = 2.99 ml O_2 STPD/min. Data on rat from Saiki and Mortola, 1996; on pig, modified from Mount, 1979.

polar regions the newborn northern fur seal (~5.5 kg) is exposed to an average Ta of 7°C, with strong winds and rain. Peripheral vasoconstriction, dense pelage, and liberal quantities of subcutaneous fat limit its heat loss, but these defenses would not be sufficient to maintain Tb without good thermogenic abilities and high $\dot{V}o_2$, which are made possible by a maternal milk of extremely high caloric value. The Canadian caribou, often born in areas still covered by snow, at birth exhibits an excellent thermogenic capacity, with potentials for a maximal $\dot{V}o_2$ in the cold five times the resting value (Hart et al., 1961). The musk-ox of the Canadian Northwest, even on the first day after birth, appears to tolerate temperatures as low as −25°C without obvious discomfort, presumably because of a pelt having excellent insulation and large amounts of BAT (Blix and Steen, 1979). The newborns of some whales that are permanent residents of the arctic seas do not seem to be afforded special insulation, and they must rely on shivering and BAT thermogenesis, in addition to the heat generated by swimming. These mammals, living permanently in chilly waters, give birth to pups of great size, spanning from ~80 kg for the newborn narwhal to probably ~1 ton for the newborn bowhead whale (Blix and Steen, 1979).

At the opposite end of the spectrum are altricial species like the polar bear. The bear cubs, born small (~700 grams) with limited thermogenic capacity and insulation, depend for survival uniquely on the care and sheltering provided by the mother. This dependency implies, as it does in many other altricial species, a major maternal investment in the rearing of the litter, since the mother is forced to remain in physical contact with the pups almost all the time. When the polar bear cubs finally leave the snow den their body weight is ~10 kg and their fur insulation is such that their lower critical T is an astonishing −30°C (Blix and Lentfer, 1979). The young of ermines, born in well-insulated nests, maintain Tb values lower than those in newborns of most other species. Like lemmings, the newborn ermines have very limited thermogenic capacity but an extraordinary ability to survive even when cooling is so severe as to stiffen their bodies and induce respiratory arrest (Segal, 1975).

Developmental Changes. In many eutherian mammals, the ability to maintain homeothermy improves quickly in the postnatal period. In the smallest species, like the rat (Fig. 5.6, left), this burgeoning ability is explained partly by the rapid body growth and increasingly effective insulation properties of the skin. In larger species, however, the improvement in thermogenesis occurs at a pace much faster than that of body growth (Fig. 5.6, right). BAT usually regresses with growth, somewhat in parallel with the appearance of shivering thermogenesis, although in some species, such as the rabbit, it may be found in large quantities even in the adult. In humans it is estimated that nonshivering thermogenesis (NST) prevails over shivering for the first 3–6 months of age (Hull and Smales, 1978). In rats the capacity for NST is small at birth, but shivering begins to be apparent as a response to cold at about 2 weeks (Janský, 1973; Sundin and Cannon, 1980). Because behavioral thermogenesis is absent during the first 3–4 days (Fowler and Kellogg, 1975), in the first few days after birth the rat's control of Tb is almost exclusively dependent on parental care.

The developmental changes conducive to the control of thermoregulation are accompanied by small, yet consistent, changes in Tb. In kittens Tb rises ~1°C during the first 2 postnatal months (Bonora and Gautier,

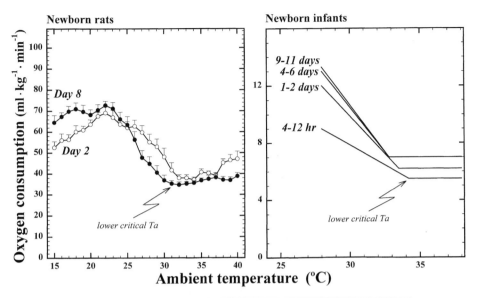

Figure 5.6. Changes in oxygen consumption for a progressive decrease in ambient temperature (Ta) in newborn rats (*left*) and infants (*right*) at different ages. In the younger pups and infants, thermogenesis occurs at higher Ta (i.e., the lower critical Ta has higher values than in adults), because of the narrow thermoneutral region. Also, summit metabolism is lower and not maintained as well as at older ages. Bars are 1 SEM. Data on rats from Mortola and Naso, 1998; on infants modified from Hey, 1969.

1987). In dogs, Tb increases ~3°C in the first 2 months, eventually reaching a value slightly higher than that of the adults (Mueggler et al., 1979) (Fig. 5.7). In smaller species, the postnatal increase in Tb can be even larger because of the lower values during the first postnatal days. In rats, for example, Tb can average only ~33°C during the first day but can reach ~37.5° by 1 month of age (Bertin et al., 1993). The postnatal increase in Tb, however, does not necessarily occur gradually and uniformly. Rather, it is possible to recognize a period of rapid changes in Tb and homeothermic ability that occurs between the postnatal poikilothermic phase and the later, adult-type homeothermy (Ru-Yung and Jinxiang, 1987). Among rodents the period of rapid change occurs earlier in the precocial species, such as the spiny mouse, than in the altricial species, such as the pallid gerbil (Waldschmidt and Müller, 1988).

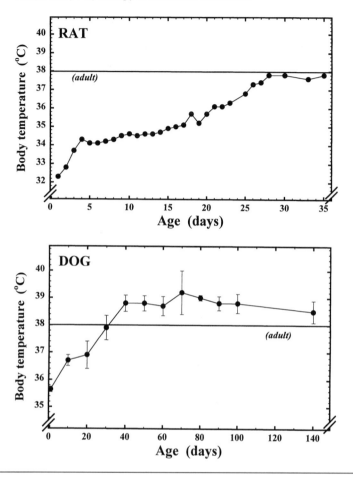

Figure 5.7. Postnatal changes in body temperature in rats (*top*) and dogs (*bottom*). Rats were raised at the constant ambient temperature of 28°C. For the dogs, each value is the mean of four animals; bars represent 1 SD. Data on rat from Bertin et al., 1993; on dog, from Mueggler et al., 1979.

In marsupials, the development of endothermy is a slow process (Hulbert, 1988). For example, even at 60–70 days of life the North American opossum is not capable of increasing its thermogenesis, whereas by 90–95 days it can protect Tb but only for values of Ta not lower than 10–15°C. As mentioned earlier, whether or not BAT or BAT-like tissues capable of NST are present in marsupials is still unclear.

Ventilatory Control at Different Temperatures

The $\dot{V}E$ responses to cold or warm conditions are examples of the conflict between gas exchange and thermoregulation. In fact, in the cold the increased thermogenic effort calls for an increase in $\dot{V}E$, which further increases heat loss through the airways. At high ambient temperatures, metabolic rate is low, and the hyperpnea, a common response appropriate for heat dissipation, can lead to respiratory alkalosis. Against cold or warm challenges, adult mammals find various forms of compromise, largely related to the possibility of dissociating alveolar from pulmonary ventilation (Mortola and Frappell, 2000). In newborns the responses can differ qualitatively from those of adults, because of their difficulties in maintaining homeothermy, as mentioned in the preceding section. This implies that in newborns, much more than in adults, the ventilatory responses to changes in ambient temperature can also include responses to the corresponding changes in body temperature.

Cold Conditions. The increase in $\dot{V}O_2$ during cold exposure could be met by an increase in $\dot{V}E$, an increase in pulmonary O_2 extraction, or any combination of the two, as becomes apparent from an application of the Fick equation. In fact,

$$O_2 \text{ consumption} = O_2 \text{ delivery} \cdot O_2 \text{ extraction fraction, or}$$
$$\dot{V}O_2 = (\dot{V}E \cdot FIO_2) \cdot [(FIO_2 - FEO_2)/FIO_2]$$
$$(Equation \ 5.2)[1]$$

where FIO_2 and FEO_2 are the inspired and expired fractional O_2 concentrations. In many adult mammals, changes in $\dot{V}O_2$ during thermogenesis are matched closely by changes in $\dot{V}E$, thereby maintaining the same $\dot{V}E/\dot{V}O_2$ and O_2 extraction as in normothermia, and only a few species deviate from this pattern (Mortola and Gautier, 1995; Mortola and Frappell, 2000).

[1] This equation assumes that the values of the inspired and expired ventilation are equal (i.e., that the respiratory quotient RQ = 1). In the more general case of RQ < 1, the expired ventilation is less than the inspired ventilation, by a magnitude that can be computed from the inspired-expired difference in nitrogen concentration (Otis, 1964). The $\dot{V}O_2$ error when neglecting to correct for RQ < 1 is proportional to the concentration of the inspired oxygen (FIO_2). In the worst case, RQ = 0.7 in normoxia (FIO_2 = 21%), $\dot{V}O_2$ is underestimated by ~6.3%, whereas with FIO_2 = 100% the error is 30% (Frappell et al., 1992a).

In newborn lambs, a decrease in Ta was met by proportional increases in $\dot{V}O_2$ and $\dot{V}E$, with no changes in Tb, blood gases, or pH (Sidi et al., 1983; Andrews et al., 1991). In 11-day-old dogs at 20°C, $\dot{V}O_2$ was 70% higher than at 30°C, with no changes in blood gases, although in this species Tb in the cold dropped by ~ 1°C (Rohlicek et al., 1998). In newborn rats, a decrease in Ta from 33–36 to 23–25°C increased $\dot{V}E$ less than it increased $\dot{V}O_2$, while Tb decreased by 3–6°C (Saiki and Mortola, 1996; Merazzi and Mortola, 1999a). Hence, despite the paucity of information, it seems probable that in the cold, in newborns as well as in adults, the changes in $\dot{V}E$ follow approximately those in $\dot{V}O_2$, as long as Tb is not greatly decreased. When Tb drops, $\dot{V}E$ may increase disproportionately less than $\dot{V}O_2$, probably because of the depressant effect of the low Tb on $\dot{V}E$ (Gautier, 1996). A limited hyperpnea in the cold would reduce the pulmonary heat loss but could jeopardize acid-base homeostasis.

Warm Conditions. Because in warm conditions the Tb − Ta difference is reduced, radiation, convection, and conduction lose their effectiveness, and evaporation becomes the most important heat-dissipation mechanism. Changes in breathing pattern, by mismatching $\dot{V}E$ from $\dot{V}A$ and maximizing dead-space ventilation, play a fundamental role in the control of heat loss in all adult homeotherms, and it seems likely that this should also apply to newborns, although specific relevant information is surprisingly scanty. Four-day-old rats maintained at 35–36°C exhibited a higher incidence of breathing irregularities than they did at 25°C (Cameron et al., 2000), and warm environmental temperatures are thought to favor apneic episodes in human infants (Daily et al., 1969; Perlstein et al., 1970; Berterottière et al., 1990). Lambs in warm conditions increase breathing frequency (Andrews et al., 1991), a tachypnea reminiscent of panting or of the thermal polypnea of adult mammals and birds. At high Ta the $\dot{V}E$ response to CO_2 was decreased in the newborn cat (Watanabe et al., 1996a) and rat (Merazzi and Mortola, 1999a), but the mechanisms for this effect remain speculative.

During neonatal fever, it is accepted that breathing rate increases (O'Dempsey et al., 1993), but thorough studies of respiratory control are not available. Injection of pyrogens in kittens increased both Tb and $\dot{V}O_2$, but did not elicit obvious changes in chemosensitivity (Watanabe et al., 1998).

Changes in Ta and Tb have profound implications for the neonate's resistance to anoxia and for the $\dot{V}E$ response to hypoxia, since both are highly dependent on metabolic level; these responses, and their interaction with changes in Tb, will be discussed in detail below (see "Acute Neonatal Hypoxia"). The strength of pulmonary vagal reflexes, and specifically of the $\dot{V}E$ inhibition to sustained lung inflation, was found to be more substantially reduced in newborn rats at Ta = 24°C than at 36°C; this difference could be attributed to the higher metabolic rate in the cold, and therefore the greater chemical stimuli offsetting the vagal inhibition. The effects of the combination of hypoxia and hyperthermia on the pulmonary reflexes inhibiting $\dot{V}E$, and the possible implications for neonatal breathing irregularities and apnea, were discussed in Chapter 4, under "Vagal Relexes."

Circadian Oscillations. Many physiological parameters in animals oscillate regularly over time, and the rhythms of Tb and $\dot{V}O_2$ have been among the most studied (Refinetti and Menaker, 1992). These rhythms have an endogenous origin controlled by a biological clock located at the level of the suprachiasmatic nucleus (Van den Pol and Dudek, 1993). The natural period of the endogenous clock can be measured in free-running conditions (i.e., in the absence of any external cue), and usually differs slightly from 24 hours; the daily light-dark cycle is believed to be the most powerful time synchronizer (Moore, 1997).

Like adults, newborns show daily oscillations in Tb and $\dot{V}O_2$, although very few species other than the rat have been investigated in this regard (Hellbrügge, 1960; Recabarren et al., 1987; Lee and Zucker, 1988; Davis and Gorski, 1988; Spiers, 1988; Nuesslein and Schmidt, 1990). The persistence of these oscillations in pups that were separated from their mother and artificially reared proves that the rhythm has an endogenous origin and is not necessarily cued by maternal behavior (Mumm et al., 1989; Redlin et al., 1992). Measurements in 6-day-old rats indicated that $\dot{V}E$ was higher in the evening than in the morning, which accords with differences in $\dot{V}O_2$ and Tb (Saiki and Mortola, 1995). Whether these circadian differences apply to other species and are accompanied by differences in $\dot{V}E$ control has not been investigated.

O_2 and CO_2 Chemoreception

The ability to respond to changes in tissue oxygenation and to altered levels of CO_2 are fundamental necessities for homeostasis. Pulmonary ventilation is the common pathway for O_2 and CO_2 transport, and hence plays a fundamental role in oxygenation and acid-base control. Most control mechanisms are reflex in nature, and their sensors are represented by chemoreceptors in the peripheral circulation and in the brainstem.

Peripheral Chemoreceptors

The physiology of the peripheral chemoreceptors, anatomically known for several centuries, has been progressively elucidated since the 1920s, when their involvement with the response to hypoxia was first demonstrated. Although the modalities of the transduction of arterial hypoxia into neural activity are still the object of debate, their basic physiological characteristics are well established (Fidone and Gonzalez, 1986; Lahiri, 1997). In many species the peripheral O_2 sensors are the carotid and aortic bodies, located respectively at the level of the carotid bifurcation and at the aortic arch. The former sense hypoxia whenever a drop in arterial O_2 content is accompanied by a decrease in the corresponding partial pressure (PaO_2), but are not activated when the arterial O_2 content decreases with no changes in PaO_2, as in hypoxia due to carbon-monoxide poisoning or anemia. The aortic bodies have a low hypoxic sensitivity and small reflex effects on $\dot{V}E$. At the level of the carotid body, hypercapnia interacts with hypoxia in enhancing the activity of the chemoreceptors, whereas in the aortic bodies the interaction is minimal. In adult animals during resting air-breathing, the peripheral chemoreceptors undergo some tonic activity that reflexively stimulates $\dot{V}E$. Indeed, bilateral carotid body denervation usually reduces $\dot{V}E$ and increases arterial PCO_2. Hence, acute hyperoxia, by suppressing the normoxic sensory activity, results in an abrupt drop in $\dot{V}E$ (Dripps and Comroe, 1947; May, 1957; Dejours, 1957). Chronic hyperoxia, especially if occurring during the neonatal period, can have long-lasting depressant effects on chemoreceptor activity and its sensitivity to hypoxia (Hanson et al., 1989a; Ling et al., 1997; Erickson et al., 1998).

  Response to Acute Hypoxia. The immediate increase in $\dot{V}E$ during hypoxia is the effect of the activation of the peripheral chemoreceptors, and in newborns, as will be discussed later, this increase is small in comparison to that at older ages (Hanson et al., 1989b; Williams et al., 1991). The $\dot{V}E$ decrease immediately upon the onset of hyperoxia was observed to be small in the newborn lamb (Purves, 1966b, 1966d; Bureau and Bégin, 1982), the newborn rat (Hertzberg et al., 1990), and the human infant

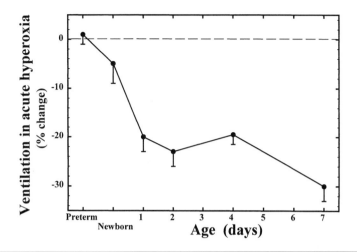

Figure 5.8. Reflex assessment of the activity of peripheral chemoreceptors in rat pups. Ventilation ($\dot{V}_E$) during acute hyperoxia is expressed as percentage change from the normoxic value (0%, dashed line). *Preterm* refers to newborn rats artificially delivered close to term. Hyperoxia is thought to silence the peripheral chemoreceptors; hence, the decrease in $\dot{V}_E$ during the first few seconds of a sudden hyperoxic exposure is a functional assessment of their level of activity during normoxia. The activity of the peripheral chemoreceptors evaluated by this method progressively increases in the days after birth. Bars are 1 SEM. Slightly modified from Hertzberg et al., 1990.

(Brady et al., 1964; Hertzberg and Lagercrantz, 1987; Parks et al., 1991), becoming more pronounced in the following days (Fig. 5.8). Both sets of findings could therefore suggest that the peripheral chemoreceptors are less functional in the newborn.

Single- or multiple-fiber recordings from the peripheral cut end of the sinus nerve confirmed this impression. In the sheep fetus, the only species investigated prenatally, carotid sinus activity was present and increased with hypoxia (Blanco et al., 1984a), a finding that differs from that of an earlier report (Biscoe et al., 1969), but the response occurred at much lower PaO_2 values than in adults. Further, the aortic chemoreceptors are functional in the fetal lamb (Dawes et al., 1969). In the neonatal period, activity of the carotid sinus nerve has been recorded in kittens (Schwieler, 1968; Hanson et al., 1989a; Blanco et al., 1984b; Marchal et al., 1992; Bairam et al., 1993; Carroll et al., 1993), lambs (Biscoe and

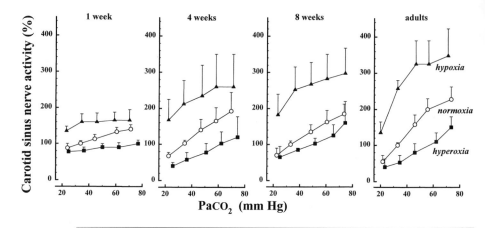

Figure 5.9. Activity of the carotid sinus nerve in kittens at various postnatal ages, as a function of the arterial partial pressure of carbon dioxide ($PaCO_2$). Data are expressed as a percentage of the baseline activity in normoxia-normocapnia. Open symbols, normoxia; filled squares, hyperoxia (>300 mm Hg); filled triangles, hypoxia (40–50 mm Hg). Bars are 1 SD. Modified from Carroll et al., 1993.

Purves, 1967; Blanco et al., 1984a), piglets (Mulligan, 1991), and rats (Kholwadwala and Donnelly, 1992; Bamford et al., 1999). All studies indicated an increase in activity with hypoxia, although at any value of PaO_2 the activity was in general less than that in older animals (Fig. 5.9). In kittens, the PaO_2 activity curve shifts in the neonatal period in a process of resetting that gradually increases sensitivity to hypoxia (Blanco et al., 1984a, 1988; Kumar and Hanson, 1989). Because the level of oxygenation interferes with this process (Hanson et al., 1989b), and because the time course of the postnatal changes in blood gases can differ among species (see Fig. 1.9), it is probable that the time period required for the carotid body to reset from the fetal to the adult pattern also varies among species. For similar reasons, one may expect differences in the process of postnatal resetting of the chemoreceptors for species living in hypoxic environments, as at high altitude or in burrows, but no comparative data are available.

Whether or not during hypoxia the increased chemoreceptor activity is sustained or rapidly adapts to the prehypoxic level has been the object of much attention. In fact, a rapid adaptation toward the normoxic value could be considered a mechanism for the poor hypoxic hyperpnea of many newborns, a characteristic of the neonatal $\dot{V}E$ response to hypoxia to be discussed in more detail later. Almost all measurements of carotid sinus nerve afferent activity in young or newborn animals have indicated that the hypoxic excitation was sustained during hypoxia, even when the ventilatory output was declining (Blanco et al., 1984b), which accords with what is observed in adults (Vizek et al., 1987; Tatsumi et al., 1992). Some adaptation during hypoxia has been noticed in kittens, but it has not been a consistent finding, and even when adaptation took place the activity still exceeded the normoxic activity (Marchal et al., 1992; Bairam et al., 1993; Carroll et al., 1993). In one study in rats, the adaptation of the carotid-body activity to hypoxia was observed in both newborns and adults, but to a lesser degree in newborns (Kholwadwala and Donnelly, 1992).

It should be noted that the recording technique employed in these studies almost invariably requires severing the carotid sinus nerve, therefore eliminating the efferent control of the chemosensory cells. The developmental aspects of this efferent control, which is thought to be inhibitory to afferent activity via inhibitory neuromodulators such as dopamine and noradrenaline, are unclear. In the early postnatal hours the level of the inhibitory neuromodulators is high (Hertzberg et al., 1990), but in cats the effect of dopamine on chemosensory activity did not differ between newborns and adults (Bairam et al., 1993). The potential problem inherent in recording from the nerve was circumvented in piglets by recording from the petrosal ganglion, which is the sensory ganglion of the carotid sinus afferents, thus maintaining the carotid sinus nerve intact (Mulligan, 1991); when this preparation was employed, the carotid sensory activity appeared to adapt to hypoxia. Yet even this piece of information does not accord with the hypothesis that the adaptation of the carotid body may explain the poor hyperpneic response to hypoxia in the newborn. In fact, the newborn piglet happens to be one of those few species in which neonates exhibit minimal or no ventilatory adaptation during acute hypoxia (Davis et al., 1988a).

Response to Hypercapnia and Asphyxia. In newborns, as in the adult, the carotid afferents respond to changes in arterial PCO_2 (Mulligan, 1991; Marchal et al., 1992; Carroll et al., 1993). The effect of combined hypoxia and hypercapnia on carotid sinus activity has been studied in lambs (Blanco et al., 1984a), kittens (Carroll et al., 1993), piglets (Mulligan, 1991), and rats (Bamford et al., 1999). The results are generally consistent in indicating that the effect of interaction between hypoxia and hypercapnia on chemoreceptor activity is small in the newborn and, as is true of the response to hypoxia, becomes progressively more apparent with age (see Fig. 5.9).

Carotid Body Denervation. In adult animals, because the peripheral chemoreceptors are active during normoxic breathing, denervation of the carotid and aortic bodies reduces $\dot{V}E$ and increases $PaCO_2$. In one study of the sheep fetus, breathing activity was found to be reduced by chemodenervation (Murai et al., 1985). Other studies, however, indicated that chemodenervation in the fetus did not affect breathing activity, nor had it any effect on the establishment of breathing at birth (Jansen et al., 1981; Rigatto et al., 1988b).

In newborn lambs, carotid denervation resulted in a decrease in $\dot{V}E$ and a rise in $PaCO_2$, with periods of irregular breathing (Purves, 1966c, 1966d). Major disturbances in breathing after bilateral carotid-body denervation were also observed in newborn cats (Schwieler, 1968) and rats (Hofer, 1984, 1986). In lambs, the arousal response to hypotension, hypoxia, hypercapnia, or upper-airway obstruction was blunted after carotid-body denervation (Horne et al., 1989; Fewell et al., 1989a, 1989b, 1990a). Further, the febrile response to bacterial pyrogen was eliminated by carotid denervation (Fewell, 1992). Several authors have called attention to the fact that some lambs and piglets died in the weeks after chemodenervation (Bureau et al., 1985; Carroll and Bureau, 1988; Donnelly and Haddad, 1990); because some of these deaths were "unexpected," the authors suggested a parallelism between chemodenervated animals and the sudden infant death syndrome (SIDS), raising the possibility that SIDS may be a manifestation of peripheral chemoreceptor insufficiency.

It should be stressed, however, that reports on the neonatal tolerance to chemodenervation are highly variable. For example, in 1–3-day-old goats, chemodenervation produced only transient breathing irregularities, and the alterations were considered quite similar to those that can be observed in adult goats after chemodenervation (Lowry et al., 1999a). Donnelly and Haddad (1990) reported that more than half of their piglets carotid-denervated at 3–9 days of age died within a week from surgery, whereas Côté et al. (1996) had no fatalities in various groups of piglets chemodenervated between 4 and 22 days of age. A very recent study on 5–15-day-old piglets indicated that $\dot{V}E$ was only minimally affected by chemodenervation; in fact, the authors commented that the outcome of the chemodenervation was related to the modalities of the neck surgery and dissection performed (Lowry et al., 1999b). Thus, the possibility that chemodenervation may result in a more dramatic respiratory insufficiency in newborns than in adults, and perhaps create a life-threatening condition specific to the newborn, seems to be based on unconvincing anecdotal evidence. In addition, sectioning of the carotid sinus nerve also eliminates the baroafferents, a component of the blood pressure control system from the carotid sinus, thus complicating the interpretation of the effects of the chemodenervation.

Central Chemosensitivity

Many experiments on adult mammals have indicated that after thorough peripheral chemodenervation the $\dot{V}E$ response to hypoxia may be greatly reduced or even abolished, but the $\dot{V}E$ responses to CO_2 and acidosis remain. These responses are due to the activation of chemosensory areas in the brainstem, or *central* chemoreceptors. Estimates of the relative contributions of the central and peripheral chemoreceptors to the $\dot{V}E$ response to CO_2 are quite variable, ranging between 30% and 80%, even when experiments are performed without the use of anesthesia, which has major confounding effects on the results (Nattie, 1983; Cherniack and Altose, 1997). The exact location of the central chemosensory areas is still a matter of debate; hence, the mechanisms of central chemoreception are also largely hypothetical (Nattie, 1995). Nevertheless, it is accepted that the cerebrospinal fluid (CSF) or brain extracellular fluid may represent the microenvironment of the central chemosensitive regions better than the blood does.

In the newborn sheep, goat, and dog, the CSF pH is slightly more acidic and the CSF PCO_2 slightly higher than those in the arterial blood (Hodson et al., 1968; Bureau et al., 1979; Nattie and Edwards, 1980). The same is true for the human infant (Krauss et al., 1972). In the rat the difference in pH between CSF and arterial blood is almost nil in the first postnatal week and increases in the following weeks (Johanson et al., 1988).

In one study in newborn lambs and goats (Bureau et al., 1979), a progressive metabolic acidosis obtained by HCl infusion was accompanied by a continuous drop in CSF pH. Because the level of hyperpnea remained stable despite the progressive systemic and central acidosis, the authors suggested that both the peripheral and central chemoreceptors were functionally immature. Most other studies, however, have indicated that the newborn CSF pH remained constant during metabolic acidosis, suggesting that the neonatal blood-brain barrier was impermeable to the diffusion of hydrogen and bicarbonate and was performing, at least qualitatively, as in adults. In newborn rats, CSF pH was better protected against systemic metabolic acidosis than against metabolic alkalosis (Johanson et al., 1988). Newborn puppies had the ability to control CSF pH by regulating bicarbonate against asphyxia or metabolic acidosis (Nattie and Edwards, 1980), which agrees with earlier measurements in lambs (Hodson et al., 1968; Herrington et al., 1971).

In the conscious lamb fetus, ventriculocisternal perfusion with mock CSF stimulated fetal breathing movements when pH was low, and depressed it when pH was high (Hohimer et al., 1983). This demonstration, which accords with the observation of a fetal breathing response to hypercapnia, even after peripheral chemodenervation (Woodrum et al., 1977), indicates the presence of functional central chemosensory areas in the fetus.

In the newborn, experiments aiming to evaluate the role of the central chemoreceptors in breathing have been performed on an isolated rat brainstem preparation superfused with mock CSF (Issa and Remmers, 1992), by perfusion of the medullary surface with mock CSF in anesthetized rabbits and guinea pigs (Wennergren and Wennergren, 1980, 1983), and by superfusion of the ventral medullary surface of anesthetized kittens with mock CSF of varying pH (Whittaker et al., 1990). The results have been consistent in demonstrating a CO_2 and pH sensitivity. Hence, in both the fetus and the newborn the central chemoreceptors are responsive to stimuli in a way qualitatively similar to that of the adult. In lambs the $\dot{V}E$ response to CO_2 was only minimally altered by peripheral chemodenervation (Purves, 1966a). In piglets that had been anesthetized and artificially ventilated, it was calculated, from an analysis of the time course of a squared change in alveolar PCO_2, that the central chemoreceptors contributed $\sim70\%$ of the $\dot{V}E$ response to CO_2 (Wolsink et al., 1991). Hence, these studies in lambs and piglets indicate that the central chemoreceptors are by far the major contributors to the $\dot{V}E$ response to CO_2, although it would be of interest to extend these observations to other species more altricial at birth.

Changes in Oxygenation

Hypoxia and hyperoxia depart in opposite directions from normoxia, the normal state of tissue oxygenation. One may therefore expect that the regulatory mechanisms aiming to reestablish normoxia should determine opposite reflex responses on $\dot{V}E$, that is, an increase during hypoxia and a drop during hyperoxia. Indeed, this would be so if we considered strictly the main feedback loops, that is, those initiated by the activation of the peripheral chemoreceptors discussed in the preceding section. In reality, many other variables concur in determining the $\dot{V}E$ response to hypoxia and hyperoxia, and their importance can vary over time. The result is that the ventilatory responses to hypoxia and hyperoxia are not necessarily reciprocal. In fact, in newborns hyperpnea is quite common during hyperoxia; during hypoxia, moreover, hyperpnea can be minimal or absent. The fine details of how this all comes about are still unclear, but changes in metabolic rate are certainly an important factor.

O_2 Transport

The hemoglobin (Hb) of the fetus shows high O_2 affinity, or low P_{50},[2] a characteristic important for adequate O_2 transport in conditions of low PaO_2. After birth the newborn PaO_2 increases rapidly (see Fig. 1.9), rendering this characteristic no longer necessary. In fact, because of the high metabolic requirements of the neonate, rapid unloading of O_2 at the tissue level would be hindered by an Hb with high O_2 affinity. Following birth the affinity of Hb for O_2 declines[3] (Fig. 5.10), mostly because of the

[2] P_{50} is the partial pressure of O_2 at which 50% of Hb is O_2-saturated. The greater the affinity of Hb for O_2 (i.e., the more displaced to the left is the PO_2-Hb dissociation curve), the lower the P_{50}. Because the Hb curve is affected by temperature and pH, P_{50} is usually measured at body temperature and pH = 7.4.

[3] The opposite pattern, however, occurs in some marsupials, such as the Tammar wallaby, the brushtail possum, and, possibly, the Virginia opossum, but not in the fat-tailed dunnart (Baudinette et al., 1988; Holland et al., 1994).

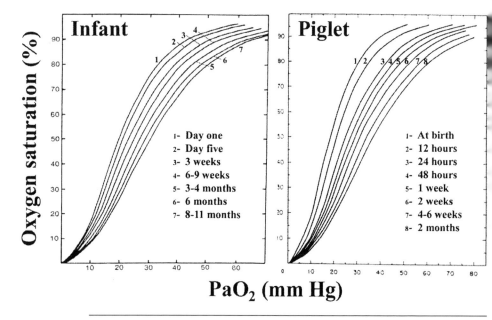

Figure 5.10. Arterial oxygen pressure (PaO_2)-hemoglobin O_2 saturation curves of human infants (*left*) and piglets (*right*) at different postnatal ages. Curves were obtained at pH = 7.4. Data on piglet from Delivoria-Papadopoulos et al., 1974; on infant, from Delivoria-Papadopoulos and McGowan, 1998; both slightly modified.

replacement of the fetal Hb with the adult Hb, which is accompanied at least in some species by an increase in erythrocyte 2,3-DPG (Fig. 5.11). This molecule (2,3-diphosphoglycerate), a phosphate compound produced in the process of erythrocyte glycolysis, is important in regulating Hb function and its affinity for O_2, which it does by lowering intracellular pH. In human infants P_{50} during the first year of life increases from ~19 to ~30 mm Hg, and fetal Hb declines from 77% of the total Hb to less than 2% (Delivoria-Papadopoulos and McGowan, 1998).

Figure 5.11. (*opposite*) Average values of hematocrit, hemoglobin, P_{50}, and *2,3-DPG* in dogs (Labrador retriever) at various postnatal ages. *Cross-hatched areas* indicate the range of adult values. P_{50} was measured at pH = 7.4 and at the body temperature of the animal. Bars are 1 SD. Modified from Mueggler et al., 1979, 1980.

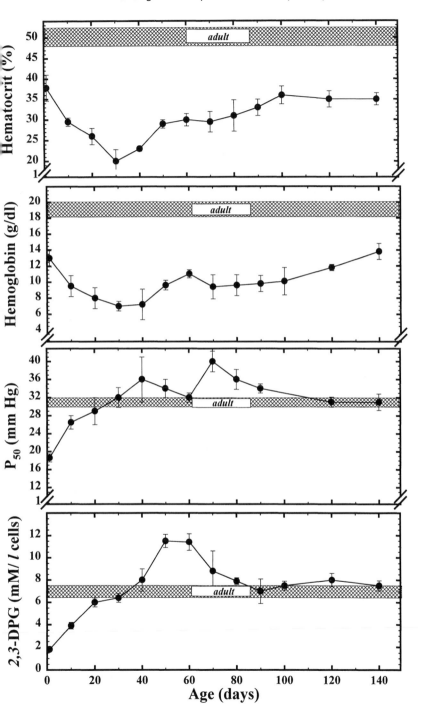

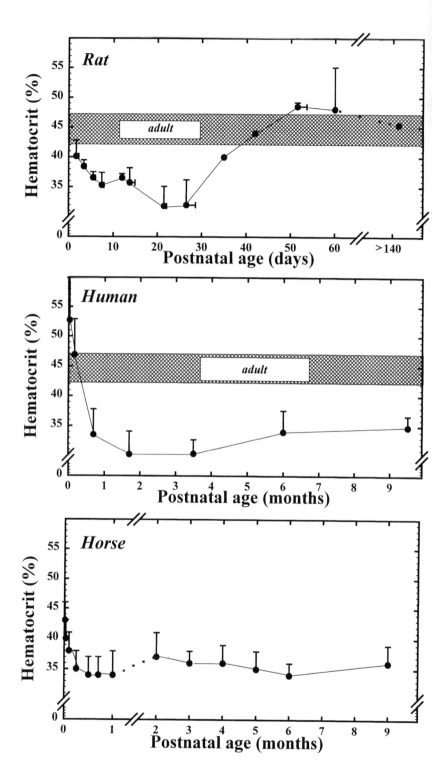

As Hb is changing its functional characteristics, during the first post-natal days or weeks, hematocrit (Hct) and hemoglobin concentration progressively decrease ("postnatal anemia") (Figs. 5.11 and 5.12). These changes probably reflect not only changes in erythrocyte turnover rate (erythropoiesis and erythrolysis), but also hemodilution and vascular expansion.

In adult mammals the Hb concentration is ~15 ml/dl, with little variation among species. With full O_2 saturation, this corresponds to ~20 ml O_2/100 ml blood (1 gram of Hb fully saturated holds ~1.34 ml O_2). Because the total mass of circulating blood is a nearly constant fraction of the animal's weight (60–80 ml/kg) (Prothero, 1980), it follows that the amount of O_2 in the arterial blood is a fixed proportion of the animal weight. Interspecies differences in metabolic rate are therefore matched not by differences in arterial O_2 content but by differences in cardiac output and O_2 delivery (oxygen delivery = cardiac output · arterial O_2 content). In addition, the unloading of O_2 at the tissue level differs among adults of various species, because P_{50} is inversely proportional to body mass (Schmidt-Neilsen and Larimer, 1958; Lahiri, 1975).

To what extent the adult interspecies relationships pertinent to O_2 transport may apply to newborns is difficult to establish, not only because of the paucity of data but also because, as mentioned previously, some of these parameters change rather rapidly during the postnatal period. The data available for Hct, Hb, and P_{50} (Table 5.1) do not suggest any significant trend for systematic changes with the body mass of the newborns. To a small extent, some of the interspecies differences can be explained by the source of the blood sample (arterial versus venous) and the modality of collection. In human infants, clamping of the umbilical cord is known to increase the values of Hct and Hb.

Prenatal Hypoxia

Many studies have examined the effects of hypoxia on fetal growth during gestation. Maternal hypoxia was produced either by exposing animals to high altitude or placing them in chambers where barometric pressure

Figure 5.12. (*opposite*) Postnatal changes in hematocrit values in the rat (*top*), human (*center*), and horse (*bottom*). Bars represent 1 SD. Data on human from Delivoria-Papadopoulos and McGowan, 1998; on horse, from Harvey et al., 1984.

Table 5.1. Parameters pertinent to blood oxygen transport in newborn mammals of various species

Species	W	Age	Hct	Hb	P_{50}	Reference(s)
Hamster	8	7	25			1
Rat	12	5	36	10		1, 14
Rabbit	62	1	42	13	21	11, 12, 21
Guinea pig	70	1	48	14		14
Cat	244	12		7		1
Dog	460	6	38	12	20	1, 2, 17
Deer	620	1	31	9		6, 19
Baboon	760	1	49			20
Lion	890	8	29	12	39	5
Pig	1400	4	30	10	26	7, 18
Human	3500	4	54	18	19	3, 15
Goat	3490	1	35	11	20	13, 22
Sheep	4150	3	41	12	34	8, 13, 16, 22
Cow	33100	1	34	11		10
Horse	49050	3	35	13	26	4, 9, 23, 24

Note: Values refer to animals within 2 weeks after birth, at the body weight indicated, and combine data from several publications. W, body weight, g; Hct, hematocrit, %; Hb, hemoglobin concentration, ml/dl; P_{50}, partial pressure of O_2 at 50% Hb saturation, mm Hg. When not available from the original references, body weight appropriate for the animal's age was derived from other sources. Numbers in final column: [1]Published and unpublished data, 1985-present; [2]Mueggler et al., 1979; [3]Delivoria-Papadopoulos and McGowan, 1998; [4]Littlejohn, 1975; [5]Parer et al., 1970; [6]Rawson et al., 1992; [7]Rosen et al., 1993; [8]Moss et al., 1987; [9]Medeiros et al., 1971; [10]Adams et al., 1993; [11]Bortolotti et al., 1989; [12]Harris et al., 1983; [13]Huisman et al., 1969; [14]Porcellini et al., 1966; [15]Matoth et al., 1971; [16]Battaglia et al., 1970; [17]Dhindsa et al., 1972; [18]Delivoria-Papadopoulos et al., 1974; [19]Chapple et al., 1991; [20]Brans et al., 1991; [21]Riegel and Ruhrmann, 1964; [22]Riegel et al., 1961; [23]Harvey et al., 1984; [24]Lavoie et al., 1990.

(Pb) was decreased (hypobaric hypoxia), or by reducing the fractional concentration of inspired O_2 at constant Pb (normobaric hypoxia). The results, in the rat (e.g., Van Geijn et al., 1980; Grauw et al., 1986; Larson and Thurlbeck, 1988; Gleed and Mortola, 1991; Cheung et al., 2000), sheep (Jacobs et al., 1988), guinea pig (Gilbert et al., 1979), and rabbit (Chang et al., 1984), have uniformly indicated a decrease in fetal growth. Birth weight is reduced also in infants born at altitudes above 2000 m, or at Pb less than ~590 mm Hg; the birth weight loss averages 65 grams for every 500 m of additional altitude (Fig. 5.13).

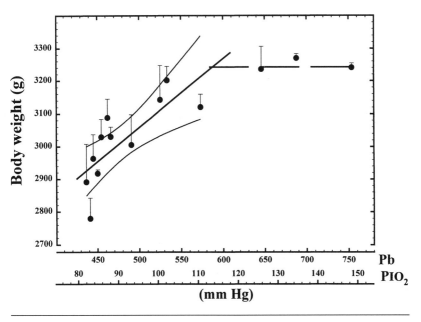

Figure 5.13. Birth weight at altitude. Data are from infants of various communities in Peru, in the region between Lima (50 m altitude) and Cerro de Pasco (4330 m), and in surrounding villages. At altitudes above ~2000 m, corresponding to a barometric pressure (Pb) of ~590 mm Hg and a dry inspired O_2 pressure (PIO_2) ~115 mm Hg, birth weight decreased on average 105 grams every 50 mm Hg drop in Pb, or about 65 grams for every additional 500 m altitude between 2000 and 4500 m. Bars are 1 SEM. Oblique line is the linear regression through the data points that fall significantly below the average birth weight at the highest Pb, which is indicated by the broken horizontal line. Curved lines are 95% CI of the linear regression. From Mortola et al., 2000.

Acute experiments in sheep indicated that maternal PaO_2 needs to drop from 91 to 53 mm Hg, and fetal O_2 delivery by about 30%, before the $\dot{V}o_2$ of the fetus is reduced (Rudolph, 1984; Edelstone, 1988). Hence, it may seem surprising that blunting of fetal growth in both humans and animals is manifest even when the maternal hypoxia is not severe. Several experimental findings, however, suggest that chronic and acute hypoxia have substantially different effects on the fetus. First, in chronic hypoxia a reduction in fetal $\dot{V}o_2$ can occur for levels of oxygenation that normally would have no effects in acute conditions (Anderson et al., 1986). Uterine blood flow of pregnant women at 3100 m altitude was only two-thirds of that of pregnant women at 1,600 m, and infant birth weight showed a positive correlation with uterine blood flow (Zamudio et al., 1995). Hence, in chronic hypoxia the redistribution of maternal blood flow may worsen the O_2 delivery to the fetus, probably limiting its O_2 supply. It has also been noticed that the level of maternal hemoglobin concentration and the degree of hyperventilation correlated positively with the infant's birth weight (Moore et al., 1982, 1986), which are additional indications that in chronic hypoxia the fetal $\dot{V}o_2$ may be determined by the supply of O_2.

The lungs of newborn rats after gestation in hypoxia were smaller, with lower total DNA and protein content (Faridy et al., 1988; Larson and Thurlbeck, 1988; Gleed and Mortola, 1991; Cheung et al., 2000). In addition, both prenatal and perinatal hypoxia diminished the postnatal process of alveolar septation, an effect reflected in fewer alveolar attachments to bronchioli (Massaro et al., 1989, 1990). The implications that these changes have for neonatal breathing and pulmonary functions have been only marginally examined. Lambs born from ewes made anemic during gestation had higher $PaCO_2$ levels, presumably because of abnormal lung functions (Moss and Harding, 2000). In rats, after gestation in hypoxia, the compliance of the respiratory system (Crs) was decreased at postnatal day 3 and increased at day 50 (Cheung et al., 2000). Crs had higher values in infants at high altitude than in infants born at sea level (Mortola et al., 1990a). Perhaps these discrepancies are related to species differences in the timing of the process of alveolar formation, which can occur entirely after birth, as in rats, or be essentially completed before birth, as in precocial species like the guinea pig and the pig (Collins et al., 1986; Winkler and Cheville, 1985). In the rat the transformation of thick-walled saccules into thin-walled alveoli begins around the fourth postnatal day, and accelerates by the second week, whereas in humans an important component of the process of alveolar formation occurs before birth (Burri, 1974; Langston et al., 1984; Brody and Thurlbeck, 1986). Hence, in the human but not in the rat, prenatal hypoxia can compromise the process of alveolar formation during fetal life, sufficiently that the consequences for pulmonary recoil pressure and compliance can be already apparent at birth.

The neonate's thermogenic capacity was found to be compromised by a gestation in hypoxic conditions, both in infants (Frappell et al., 1998) and in newborn rats, with a drop in BAT tissue and uncoupling protein (Mortola and Naso, 1998). Whether these differences in thermoregulation are accompanied by differences in ventilatory control is still unknown. In neonatal rats following hypoxic gestation, $\dot{V}E$ measured in normoxia was slightly increased and $\dot{V}O_2$ decreased; this hyperventilation, which was not corrected by hyperoxia, probably reflected the persistence of fetal circulation with pulmonary shunts (Gleed and Mortola, 1991). In infants at high altitude the values of $\dot{V}E$ were not too different from those measured in sea-level infants (Lahiri et al., 1978; Cotton and Grunstein, 1980; Mortola et al., 1992b). Furthermore, acute hyperoxia had small and insignificant effects on $\dot{V}E$, and acute exposure to hypoxia evoked similar $\dot{V}E$ responses in low-altitude and high-altitude infants (Lahiri et al., 1978). Presumably, the onset of air breathing at birth represents a sudden increase in oxygenation, or relative hyperoxia (see Chapter 1, under "Birth, a Hyperoxic Event"), not only at sea level but also at high altitudes. In such a case, the effect of high-altitude hypoxia on breathing would be apparent not immediately at birth but only at a later time, once the process of resetting of the peripheral chemoreceptors is completed (see "Peripheral Chemoreceptors," above).

Acute Neonatal Hypoxia

Both newborn and adult mammals respond to hypoxia with a *hyperventilation,* defined as an increase in $\dot{V}E$ above the metabolic requirements. Hence, by application of the alveolar gas equation for CO_2 (Equation 2.5), hyperventilation implies a reduction in the alveolar partial pressure of CO_2 ($PaCO_2$), whereas the term *hyperpnea* indicates an increase in the absolute level of $\dot{V}E$ above its resting normoxic value. It follows from these definitions that the degree of hyperventilation is not necessarily related to the degree of hyperpnea, a distinction that is important for an understanding of the $\dot{V}E$ response to hypoxia in newborn mammals.

A sudden occurrence of hypoxia provokes an increase in $\dot{V}E$ followed by a gradual decline over the next several minutes; by 15–30 min $\dot{V}E$ is usually stable, although its level relative to normoxia can be quite variable among subjects and species. Many factors, some stimulatory and others inhibitory, are known to influence the level of hypoxic $\dot{V}E$, and their relative importance may differ with the magnitude of hypoxia and the animal's age. The activation of the peripheral chemoreceptors and the hyperpnea-induced respiratory alkalosis are rapidly occurring mechanisms, respectively stimulating and inhibiting $\dot{V}E$ during hypoxia. The metabolic level, moreover, is an important determinant of the level of $\dot{V}E$, and its drop during hypoxia, quite common in newborns, reduces $\dot{V}E$. A decrease in body temperature often accompanies the drop in metabolic rate, and this too can lower $\dot{V}E$. Changes in blood pressure and hormonal levels, including catecholamines, affect the magnitude of the $\dot{V}E$ response, either directly or via changes in metabolism. In severe hypoxia, the acidosis represents an additional $\dot{V}E$ stimulus, although profound acidosis eventually depresses metabolic rate and $\dot{V}E$.

Hence, given the complex interplay of these and other factors, it is not surprising to find an enormous variation in the degree of the $\dot{V}E$ response to hypoxia both among and within species. Equally, it is not surprising that the early $\dot{V}E$ response often shows a biphasic pattern (i.e., a sudden rise followed by a gradual decline). Indeed, any response is expected to have a biphasic pattern when it results from the interaction of two or more opposing events each occurring with a different time course. In the case of the $\dot{V}E$ response to hypoxia, even if one considers only two factors, the stimulatory effect from the carotid body and the inhibitory effect of hypometabolism, a biphasic pattern is to be expected, because the former has a faster time course than the latter (Mortola, 1996; Powell et al., 1998).

Hypoxic Hyperpnea. In newborns of many mammals, hypoxic hyperpnea is low in comparison to that of the adults. In several species, in fact, including the human infant, the hypoxic $\dot{V}E$ level can be below the normoxic value. In kittens, hypoxic hypopnea occurs for at least one month after birth (Fig. 5.14), and in human infants at 2 months of age, $\dot{V}E$ during 15% O_2 did not differ from the normoxic value (Cohen et al., 1997). Some exceptions are found in the most precocial species, such as the pig (Davis et al., 1988a; Rosen et al., 1993), and an unusually high $\dot{V}E$ response to hypoxia was also observed in the newborn opossum (Farber et al., 1972). As will be discussed later, the low hypoxic hyperpnea of many newborns is largely explained by the decrease in their metabolic level. In other words, the magnitude of the hyperpnea does not represent the degree of hyperventilation. Several other mechanisms could theoretically be responsible for the low hypoxic hyperpnea, but several considerations suggest that their contribution is small, if present at all.

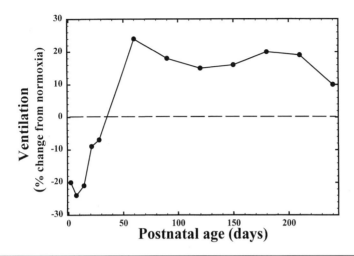

Figure 5.14. Ventilation ($\dot{V}_E$) during hypoxia (inspired O_2 concentration 11%) in cats at various ages, expressed as a percentage change from the normoxic value. During the first postnatal month, the $\dot{V}_E$ level in hypoxia is less than in normoxia. Redrawn from Bonora et al., 1984.

For example, one of the most obvious possibilities would be an adaptation of the peripheral chemoreceptors, but experimental measurements in adults and newborns do not support this view. In fact, as discussed previously, peripheral chemoreceptors are less active at birth than at older ages, but, once activated, their firing rate does not show any major adaptation over time. A related possibility, that of a central filtering of the inputs from the peripheral chemoreceptors (Eldridge and Millhorn, 1986), seems unlikely. In fact, during hypoxia the low levels of $\dot{V}_E$ are entirely accounted for by the low V_T, while breathing rate remains elevated (Fig. 5.15), a pattern of breathing that may be modified by the state of arousal in some species (Haddad et al., 1982) but not in others (Bonora et al., 1984). This suggests that peripheral inputs are not "gated" centrally, and that even during the decline of hyperpnea, information about the arterial hypoxia continues to be transmitted from the peripheral chemoreceptors and is recognized centrally. As in the adult, so also in newborns, the alveolar and arterial hypocapnia accompanying the hypoxic hyperpnea blunts the increase in $\dot{V}_E$, but this effect cannot explain the absence of hyperpnea. Vagal inputs inhibitory to breathing, although

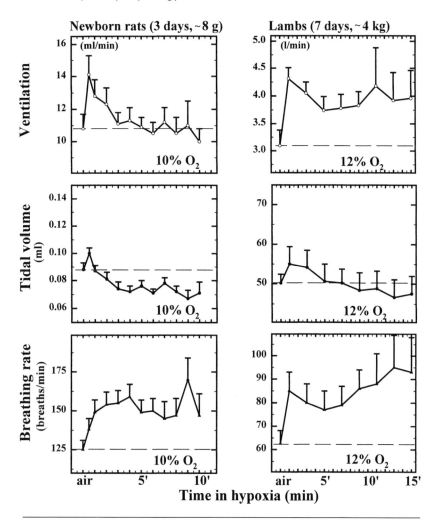

Figure 5.15. Biphasic pattern of the acute ventilatory response to hypoxia. Average values of pulmonary ventilation (V̇E), tidal volume (VT), and breathing frequency (+ 1 SEM) during the first 10–15 min of hypoxia in newborns rats and lambs. *Dashed lines* are values during normoxia. In hypoxia, the biphasic pattern of V̇E is often due to the biphasic pattern of VT, whereas breathing rate retains the high value of the immediate response. Data on rats from Saetta and Mortola, 1987; on lambs, from Bureau et al., 1984.

believed to be more effective in newborns than in adults, are not involved, because the pattern of the $\dot{V}E$ response to acute hypoxia remains unaltered after vagotomy (Trippenbach et al., 1985a; LaFramboise et al., 1985).

The hypothesis that during acute hypoxia the increase in the mechanical impedance of the respiratory system may pose a limit to both tidal volume and $\dot{V}E$ found some support from measurements in young monkeys (LaFramboise et al., 1983) but not in newborn lambs (Côté et al., 1988). In hypoxic kittens, tidal volume (VT) was found to be low despite persistently high activity of the diaphragm electromyogram (EMG) (Rigatto et al., 1988a), which would be compatible with the hypothesis of an increase in mechanical impedance. It was later shown, however, that the persistence of the diaphragm activity was due only to the diaphragm tonic end-expiratory component of the EMG, whereas the phasic volume-related EMG component changed in parallel with VT (Bonora et al., 1992). The hypothesis of fatigue of the respiratory muscles, too, is not a very realistic one, since newborns can maintain high levels of $\dot{V}E$ for many days, as has been observed, for example, in newborn rats in chronic hypercapnia (Rezzonico and Mortola, 1989).

Hypoxia can have a depressant effect on breathing, an effect usually apparent in carotid body denervated animals (Tenney and Ou, 1977a, 1977b). The mechanisms of this depression, which is more marked in anesthetized preparations, are still elusive. This aspect of the effect of hypoxia on the brain could contribute to the progressive decline in $\dot{V}E$ after the immediate rise, and it is often considered a likely mechanism for the poor hyperpneic response to hypoxia in newborns of many species (Edelman et al., 1991, 1997). In rabbit pups, transection at the level of the pons resulted in a greater $\dot{V}E$ response to hypoxia, an effect not seen when the transection was at a more rostral location (Schwieler, 1968; Martin-Body and Johnston, 1988). Hence, brain structures at the approximate level of the pons appear to have an inhibitory effect on breathing during hypoxia, as is shown to be the case in the fetus (see Chapter 1, under "Fetal Breathing Movements"). Whether or not this $\dot{V}E$ inhibition from pontine structures applies also to the newborn's response to hypercapnia, as it does in the fetus, would be of interest to know, since in intact newborns only the $\dot{V}E$ response to hypoxia, and not that to hypercapnia, is characteristically small. The possibility that a central neural mechanism is responsible for the poor hypoxic hyperpnea of the newborn needs to be reconciled with the fact that the $\dot{V}E$ response to hypoxia varies depending on arterial PCO_2. In fact, the hyperpnea can be modest or absent during normocapnic hypoxia, but pronounced when hypoxia occurs with hypercapnia (see "Acute Conditions," below).

Implicit in the concept of a central depression of breathing in the hypoxic newborn is the idea that the ventilatory output during hypoxia does not fulfill the metabolic requirements, which should lead to an elevation of alveolar and arterial PCO_2. But the experimental observations do not indicate the occurrence of CO_2 retention. On the contrary, since the earlier observations in infants (e.g., Brady and Ceruti, 1966; Rigatto and Brady, 1972) and kittens (Schwieler, 1968), numerous measurements in newborns of several species have indicated that PCO_2 during hypoxia is similar to or below the normoxic value, even when the hypoxic $\dot{V}E$ is below the normoxic level. The latter observations emphasize that the drop in metabolic level is an important aspect of the hypoxic response.

Hypoxic Hypometabolism. Many newborns decrease metabolic rate during hypoxia; notwithstanding great variability, a trend toward greater decrease can be recognized, across many species, in newborns having high normoxic $\dot{V}O_2$ (Fig. 5.16).

Animals can allow their internal state to fluctuate with changes in external conditions, thus conforming to the environment, or can protect the internal state against external challenges via mechanisms of regulation. The question has been raised whether the phenomenon of hypoxic hypometabolism is an expression of conformity forced on the animal by the limitation in O_2 supply, but clearly it is not. A drop in $\dot{V}O_2$ can occur even with very mild levels of hypoxia, as for example during breathing 15–18% inspired O_2 in some of the first studies of the subject by Hill (1959) and Taylor (1960b) on newborn kittens and rats. Similarly, Cross et al. (1958) indicated that even during very modest hypoxia, as during breathing 15% O_2, human infants significantly dropped their metabolic rate. In cases of mild hypoxemia the level of lactic acid in the blood increases very little, suggesting that the reduction in aerobic metabolism is not compensated by anaerobic energy production. Indeed, on return to normoxia, $\dot{V}O_2$ rises to the normoxic level, not above it, indicating that no anaerobic debt had been contracted during the hypoxic drop in $\dot{V}O_2$ (Hill, 1959; Adolph and Hoy, 1960; Fahey and Lister, 1989; Frappell et al., 1991) (Fig. 5.17). Dur-

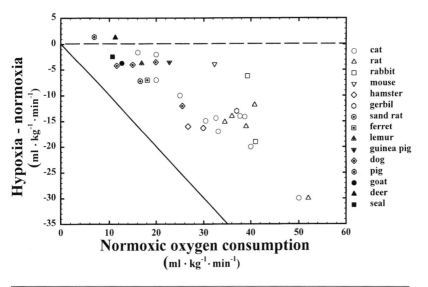

Figure 5.16. Changes in oxygen consumption ($\dot{V}o_2$) during hypoxia (~10% in-spired O_2) in newborns of several species. The difference between hypoxic and normoxic $\dot{V}o_2$ is represented as a function of the normoxic $\dot{V}o_2$/kg. The drop in $\dot{V}o_2$ tends to be greater the larger the normoxic $\dot{V}o_2$ value. This applies across species, as shown in the figure, as well as within a species, when normoxic $\dot{V}o_2$ is increased by a cold stimulus. *Dashed line,* no hypoxic-normoxic difference; *oblique line,* line of identity. Data for cat are from[1–7,12]; rat[2,3,8,9]; rabbit[10,13]; mouse[3]; hamster[3]; gerbil[3]; sand rat[3]; ferret[3]; lemur[3]; guinea pig[3]; dog[3,4,11]; pig[3]; goat[3]; deer[3]; seal[3]. [1]Frappell et al., 1991; [2]Mortola and Dotta, 1992; [3]Mortola et al., 1989; [4]Pedraz and Mortola, 1991; [5]Mortola and Matsuoka, 1993; [6]Rohlicek et al., 1996; [7]Hill, 1959; [8]Mortola, 1991; [9]Matsuoka and Mortola, 1995; [10]Trippen-bach, 1994; [11]Rohlicek et al., 1998; [12]Mortola and Rezzonico, 1988; [13]Blatteis, 1964.

ing hypoxic hypometabolism it is still possible to raise $\dot{V}o_2$, for example pharmacologically, or with a cold stimulus (Rohlicek et al., 1998). In other words, for the same value of hypoxemia (whether arterial or ve-nous) the magnitude of hypometabolism can vary depending on the con-comitance of other metabolic stimuli (Fig. 5.18). This means that the level of $\dot{V}o_2$ measured during hypoxic hypometabolism is not necessarily the maximum that the animal can reach for that degree of hypoxia. Var-ious arguments and experimental evidence thus support the view that hy-poxic hypometabolism is not a forced response but a "choice" (i.e., a phe-

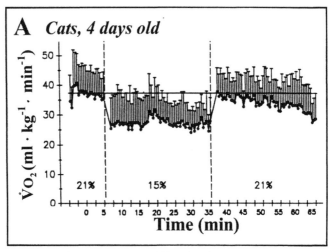

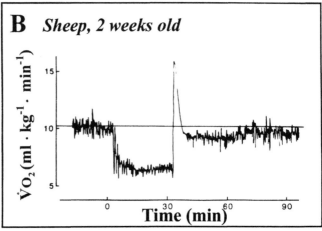

Figure 5.17. Decrease in oxygen consumption ($\dot{V}_{O_2}$) in hypoxia induced by a decrease in inspired O_2 from 21% to 15% in kittens (*panel A*, dots are mean values, bars are 1 SD) or by a decrease in cardiac output in one lamb (*panel B*). In either case, on termination of the hypoxia, $\dot{V}_{O_2}$ returns to normoxic value (*horizontal lines*) with no (kitten) or little (lamb) overshoot. The fact that on return to air breathing, the $\dot{V}_{O_2}$ overshoot (i.e., the time integral of $\dot{V}_{O_2}$ in excess of the pre-hypoxic value) is minimal or absent indicates that during the hypoxia there was little or no O_2 debt accumulation (i.e., that anaerobic energy sources did not compensate for the lower aerobic metabolism). *Above:* modified from Frappell et al., 1991; *below:* modified from Fahey and Lister, 1989.

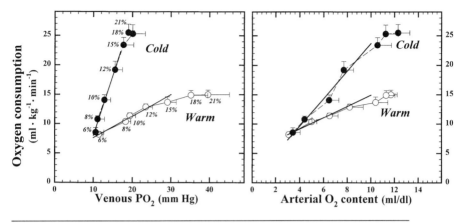

Figure 5.18. Hypoxic hypometabolism and blood oxygenation in newborn dogs. The animals were breathing progressively lower concentrations of O_2 (indicated as percentages from 21 to 6), in warm (○) or cold conditions (●). As venous and arterial oxygenation decreased, oxygen consumption ($\dot{V}O_2$) also dropped. The *straight lines* are through the data points of hypometabolism (i.e., with $\dot{V}O_2$ below the normoxic value). Bars are 1 SEM. Hypometabolism in warm conditions occurred at levels of oxygenation that could sustain higher $\dot{V}O_2$ levels during cold exposure. From Rohlicek et al., 1998.

nomenon of *regulated conformity*). Of course, this raises the question, what are the functions that the hypoxic newborn "chooses" to depress, and via what mechanisms?

Depression of Thermogenesis. The magnitude of hypoxic hypometabolism is roughly proportional to the resting $\dot{V}O_2$ of the animal (Mortola, 1996; Gautier, 1996), both in newborns (see Fig. 5.16) and in adults (Frappell et al., 1992b). Among adults, hypoxic hypometabolism is more marked in small species, which, relative to body mass, have higher $\dot{V}O_2$ than larger species. Moreover, animals have a stronger hypometabolic response to hypoxia when their normoxic $\dot{V}O_2$ is increased during cold-induced thermogenesis. In fact, hypoxia during cold can decrease $\dot{V}O_2$ even in those species, such as humans, that normally would not lower metabolic rate in warm conditions. These observations and much other direct experimental evidence on several species have indicated that a major factor in the drop in $\dot{V}O_2$ during hypoxia is the inhibition of thermogenesis (Mortola and Gautier, 1995), which is in full support of the early suggestions by June Hill (1959). Hence, the greater hypometabolic response to hypoxia in newborn and young animals is consistent with the notion that their thermogenic needs, and $\dot{V}o_2/W$, are usually greater than those in adults (see Fig. 2.4). In addition to the depression of thermogenesis, in young animals, cell repair, tissue growth, and differentiation are curtailed by hypoxia and contribute to the lowering of $\dot{V}o_2$; the consequences of the depression of these functions become apparent with prolonged or chronic hypoxia (see "Chronic Neonatal Hypoxia," below).

In newborns, heat production is mostly the result of nonshivering thermogenesis of the brown adipose tissue (BAT), and is depressed in hypoxia (Fig. 5.19) by mechanisms the nature of which is still unclear. Chronic hypoxia decreases both the mass of BAT and the concentration of the mitochondrial uncoupling protein thermogenin (Mortola and Naso, 1997, 1998), but these effects are apparent only after several days; hence, thermogenin reduction is too slow a process to be of any impor-

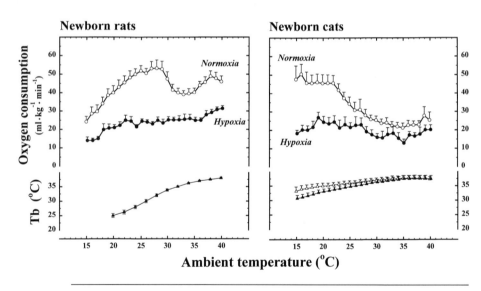

Figure 5.19. Oxygen consumption at different ambient temperatures (Ta) in 2-day-old rats (*left panel*) and kittens (*right panel*) during normoxia and hypoxia (10% inspired O_2 concentration). Hypoxia abolished the thermogenic response to cold. In newborn rats, the drop in body temperature (Tb) with the decrease in Ta was the same whether in normoxia or hypoxia. Bars are 1 SEM. From the data of Mortola and Dotta, 1992.

tance in acute hypoxia. Blood flow to BAT is reduced by hypoxia in warm conditions, but not in the cold, which is when $\dot{V}_{O_2}$ drops the most, indicating that changes in BAT perfusion are not the primary mechanism for the hypoxic depression of thermogenesis (Mortola et al., 1999b). The fact that hypercapnia (breathing 2–5% CO_2) does not depress $\dot{V}_{O_2}$ (Saiki and Mortola, 1996) indicates that hypometabolism is not an undifferentiated response to chemoreceptor stimulation. Indeed, the hypometabolic response to hypoxia also occurs after chemoreceptor denervation, or with experimental anemia or carbon monoxide poisoning. The latter conditions reduce the arterial oxygen content with no changes in PaO_2, hence without activation of the carotid bodies (Mortola and Gautier, 1995; Gautier, 1996). Hypoxia could also act on BAT adipocytes, either directly or via neural control through the hypothalamus, but no experimental information is available.

Huddling, the other common means of defense of the newborn against cold, is also depressed in hypoxia (Fig. 5.20), thus aggravating the decrease in body temperature (Tb) (Mortola and Feher, 1998). This fact is in line with the notion, which has emerged from experiments in adults of some species (e.g., Dupré et al., 1988; Gordon and Fogelson, 1991), that hypoxia not only decreases thermogenesis but also lowers the preferred ambient temperature and the set point for thermoregulation. Considering that in hypoxia all forms of thermogenesis are suppressed— whether shivering, nonshivering, or behavioral heat control, as well as the febrile response to pyrogen (Ricciuti and Fewell, 1992; Fewell, 1992)— the existence of a central coordination of the hypoxic effects on these processes seems very probable, and the hypothalamus is the most likely candidate for the site of their integration.

Newborns, in comparison to adults, have less ability to control heat dissipation (see "Metabolic and Behavioral Responses," above). Hence, it is often observed that during hypoxia, as thermogenesis is suppressed, Tb drops (see Fig. 5.19). The advantages of a low Tb in the hypoxic newborn, which favors a further drop in $\dot{V}_{O_2}$ via the Q_{10} effect and improves the affinity of hemoglobin for O_2, have often been discussed, and have been the object of several experiments (e.g., Miller et al., 1964; Miller and Miller, 1969; Dupré et al., 1988; Pedraz and Mortola, 1991; Rohlicek et al., 1996). During hypoxia, artificially increasing Tb to the normoxic value evokes responses like hyperpnea and a drop in systemic vascular

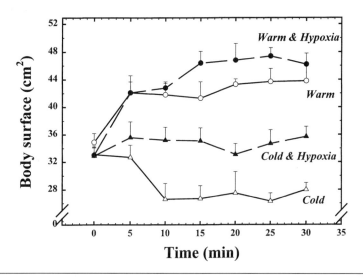

Figure 5.20. Hypoxia and behavioral thermogenesis. Body surface area is computed from video images of five rat pups in warm (33°C) or cold (15°C) conditions, during normoxia or hypoxia (10% inspired O_2). At time 0 the rats were grouped close together. In warm conditions they spread out, increasing the collective body surface. In the cold they huddled, decreasing the exposed surface area. Hypoxia reduced huddling. From Mortola and Feher, 1998.

resistance (Fig. 5.21), which would indicate the newborn's attempt to dissipate heat. In particular, the reduction in systemic vascular resistance, by redistributing blood to the peripheral circulation, could jeopardize the actions of the defense mechanisms against hypoxia (Rohlicek et al., 1996). In short, it is evident that in the hypoxic newborn the drop in Tb is part of a regulated strategy. For if Tb is artificially forced toward the normoxic value, the newborn interprets it as *relative* hyperthermia, and reacts to it in a way that could be counterproductive for the delivery of O_2 to the essential organs.

A decrease in temperature increases the pH of electrochemical neutrality. Because constancy of the net charge of proteins, and specifically of the imidazole group of histidine, is believed to represent an important goal of homeostasis ("alphastat regulation"; Reeves, 1977), one would expect that, for the same level of hypoxia, the greater the drop in Tb the greater the increase in $\dot{V}E/\dot{V}O_2$, to meet the condition of neutral pH for

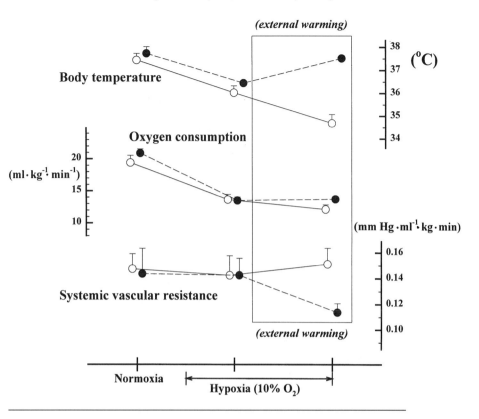

Figure 5.21. Effects of body warming during hypoxia. When kittens 12 days old were exposed to hypoxia (10% inspired O_2) for 30 min, both their oxygen consumption ($\dot{V}o_2$) and their body temperature (Tb) dropped. Then, some kittens (●, *broken lines*) were artificially warmed to increase their Tb to the normoxic value. The external warming had no effect on $\dot{V}o_2$, whereas it decreased systemic vascular resistance, implying a redistribution of blood toward the peripheral vascular districts. This presumably reflected the attempt of the kitten to dissipate heat in response to a Tb too high for its hypoxic condition. From Mortola, 1999, based on data of Rohlicek et al., 1996.

that Tb. On the other hand, a lowering in Tb is thought to have an inhibitory effect on $\dot{V}E$ (Gautier, 1996). The hypoxic newborn would offer an opportunity to explore the extent of alphastat regulation in a homeotherm, which, because of hypometabolism, lowers its Tb and therefore behaves as a poikilothermic animal. To my knowledge, experimental data addressing this issue are lacking.

Hypoxic Hyperventilation. As is apparent from the Fick equation as applied to the respiratory system (see Equation 5.2), and as is implicit in the alveolar gas equation for CO_2 (see Equation 2.5), hypoxic hyperventilation can be obtained with hyperpnea, hypometabolism, or any combination of the two. Indeed, in first approximation hyperpnea and hypometabolism are inversely proportional to each other, their relative magnitudes depending on the level of hypoxia and the animal's age. To an inspired O_2 concentration of 10% maintained for at least 15–30 min, which corresponds to values of arterial PO_2 (PaO_2, mm Hg) of 35–40, many adult animals respond preferentially with hyperpnea, but in small species hyperpnea is combined with hypometabolism (Frappell et al., 1992b). To the same degree of hypoxemia, newborns of many species respond preferentially with hypometabolism, and only the most mature newborns, such as those of the pig, the deer, and other precocial species, respond with hyperpnea (Fig. 5.22). Data from normal human infants are scarce, for obvious ethical reasons, but from the few measurements made since the seminal work of Cross et al. (1958), it is clear that infants, too, respond predominantly with hypometabolism, and with only a modest hyperpnea or none at all. With growth, hypometabolism gradually gives way to hyperpnea as the primary contributor to hypoxic hyperventilation (Fig. 5.23).

In a hypoxic newborn, hyperventilation can result from different combinations of changes in $\dot{V}E$ and $\dot{V}O_2$, depending on the starting metabolic level. For example, in newborn dogs in warm conditions (Fig. 5.24), hypoxia decreased $\dot{V}O_2$ only when PaO_2 was around 40 mm Hg, and had minimal effects on alveolar ventilation ($\dot{V}A$). In the cold, when $\dot{V}O_2$ was high because of thermogenesis, hypoxic hypometabolism was much more pronounced, the level of $\dot{V}O_2$ beginning to drop at a PaO_2 of about 55 mm Hg; in this case, $\dot{V}A$ decreased markedly during hypoxia. As is apparent from the decrease in $PaCO_2$, the degree of hyperventilation in warm conditions was identical to that in cold conditions (Rohlicek et al., 1998). This example emphasizes two points. First, as is also clear from numerous cases in adults (Mortola, 1996), the ventilatory response to hypoxia needs to be considered in light of the metabolic level. A correct interpretation is fundamental in the clinical setting, for example in making a decision regarding pharmacological interventions aimed at stimulating $\dot{V}E$ in a hypoxic infant. Second, the remarkable similarity in degree of

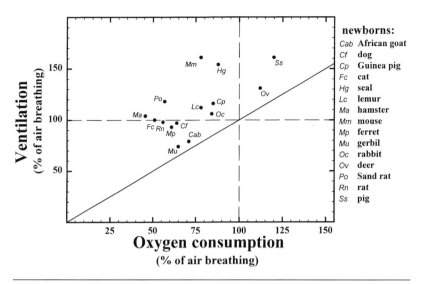

Figure 5.22. Hypoxic hyperventilation in newborns. Oxygen consumption ($\dot{V}o_2$) and ventilation ($\dot{V}e$) in several newborn species during 10% O_2 breathing for 10–30 min. Data, obtained at the ambient temperatures most common to the species, are expressed as a percentage of normoxia. The *oblique line* indicates the normoxic $\dot{V}e/\dot{V}o_2$ value. In hypoxia, all species hyperventilate (i.e., increase $\dot{V}e/\dot{V}o_2$), although some hyperventilate predominantly by decreasing $\dot{V}o_2$ (hypometabolism) and others by increasing $\dot{V}e$ (hyperpnea). From data of Mortola et al., 1989.

hyperventilation between warm and cold conditions, despite the very different $\dot{V}e$ and metabolic responses, raises the question, what mechanisms provide the link between metabolic rate and pulmonary gas convection? This is an unsolved problem, one that has long intrigued physiologists interested in muscle exercise, thermogenesis, and metabolic responses to various pharmacological interventions; the fact that the question also surfaces within the discussion of neonatal hypoxia emphasizes once more its importance for the understanding of respiratory control. The possibility of an involvement of the metabolically produced CO_2 has often been raised in adult studies (Mortola and Gautier, 1995), and has also been considered to be important for the spontaneous breathing activity observed during fetal life (Blanco, 1994). In newborns, data emerge indirectly. For example, in normocapnic kittens, hypoxia did not

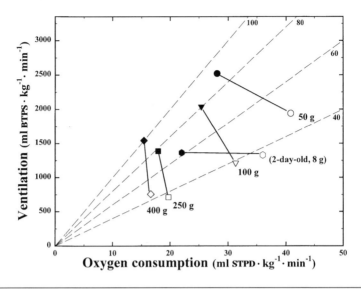

Figure 5.23. Hypoxic hyperventilation in growing rats. Values of oxygen consumption ($\dot{V}o_2$) and ventilation ($\dot{V}e$) in rats of several ages (weights indicated) during normoxia (*open symbols*) and hypoxia (10% inspired O_2 for 15–30 min, *filled symbols*). N = 10 per group. *Dashed lines* are constant $\dot{V}e/\dot{V}o_2$ at 40, 60, 80, 100. All animals were studied at ambient temperature (Ta) of 23–25°C except the newborns (2 days old, 8 grams), studied at Ta of 33°C. Hyperventilation is obtained mostly by hypometabolism in the newborn, and by hyperpnea in the adult. From the data of Mortola et al., 1994, and Matsuoka and Mortola, 1995.

increase $\dot{V}e$; but in hypercapnic kittens, the same degree of hypoxia promoted an adult-type hyperpnea. One interpretation of these results is that the added inspired CO_2 compensated for the hypoxic drop in CO_2 metabolically produced (Mortola and Matsuoka, 1993). Whether in the newborn the hypoxic reduction in metabolic level, in addition to affecting the absolute level of $\dot{V}e$, may also reduce the sensitivity to ventilatory stimulants has not been specifically addressed by experiments.

Chronic Neonatal Hypoxia

The major disadvantages of hypometabolism are apparent when hypoxia becomes a chronic condition. In fact, the processes of tissue growth and repair, as well as thermogenesis, are reduced. Hence, body growth is severely blunted (Fig. 5.25).

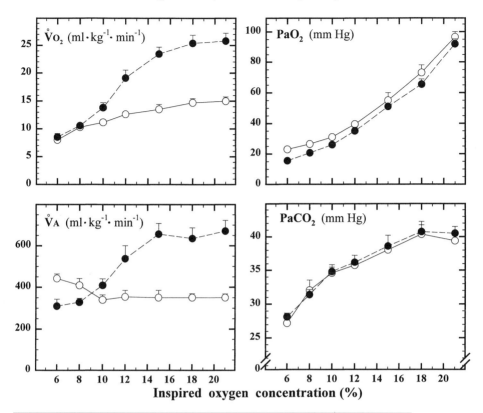

Figure 5.24. Hypoxic response of newborn dogs in cold (20°C, ●) or warm (30°C, ○) conditions. Bars are 1 SEM. In the cold, hypoxia (from 21% to 6% O_2, in 15-min steps) determined a marked hypometabolism and hypopnea. In warm conditions, the hypometabolism was more modest and accompanied by normopnea or hyperpnea. But despite the major differences in metabolic and ventilatory responses, the hyperventilation in warm conditions was very similar to that in cold conditions, as indicated by the similar drop in arterial $PaCO_2$. From the data of Rohlicek et al., 1998.

Implications for Body and Lung Growth. Most experimental studies of neonatal chronic hypoxia are performed by exposing to normobaric or hypobaric hypoxia an entire litter, with the mother; hence, maternal loss of appetite, decreased lactation, and reduced care for the pups are confounding factors that can aggravate the effects of hypoxia on the fetus and the newborn (Moore and Price, 1948). Attempts to minimize these ma-

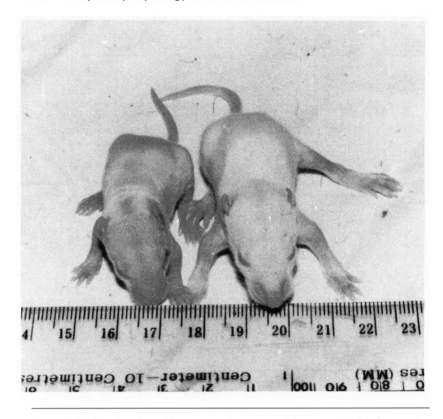

Figure 5.25. Twelve-day-old rat pups: the one at right was raised in normoxia, the one at left in hypoxia (FIO_2 = 11%). The chronically hypoxic rat is visibly smaller and growth-retarded.

ternal factors have been made, for example by artificial rearing of the pups, or by daily rotation of the dams for any given hypoxic litter, but they may introduce other confounding issues. The results are almost universally consistent in finding that hypoxia retards body growth, although few experiments have been performed on species other than the rat. Blunting of body growth also occurs in children with cyanotic heart diseases (Prader et al., 1963; Gingell et al., 1981) or living at high altitude (Monge and León-Velarde, 1991). In the semifossorial hamster, but not the rat, metabolism dropped during the acute phase of hypoxia, but was eventually maintained at the normoxic level during chronic hypoxia, with the result that body growth was almost unaltered (Frappell and Mortola, 1994).

The consequences of prolonged hypoxia on the structural and functional development of the individual organs depend on their relative oxygen needs and the extent of the redistribution of blood flow. For example, under these conditions the delivery of O_2 to brain, lungs, heart, and diaphragm is not as impaired as it is to other organs (Côté and Porras, 1998). Several reports have indicated that lung mass not only is not compromised, but could actually increase under hypoxia (Cohn, 1939; Pepelko, 1970; Bartlett and Remmers, 1971; Tucker and Horvath, 1973; Cunningham et al., 1974; Bartels et al., 1979; Sekhon and Thurlbeck, 1995). In part, the earlier results, which were based on measurements of wet weights, may have been exaggerated by fluid accumulation, since interstitial pulmonary edema can develop in some hypoxic conditions (Bartlett and Remmers, 1971). More recent measurements of proteins and DNA content, however, have confirmed that the lungs are among the protected organs, although whether or not hypoxia actually stimulates lung growth is still not clear, and may depend on the degree and duration of hypoxia (Piazza et al., 1988; Mortola et al., 1990b; Sekhon and Thurlbeck, 1995, 1996). Recently, the possibility has been raised that the effects on lung growth may differ depending on whether the hypoxia occurred under normobaric or hypobaric conditions (Sekhon et al., 1995; Sekhon and Thurlbeck, 1995, 1996).

The effects of prolonged hypoxia on the structural and functional development of the internal organs, and whether or not these effects are compatible with survival, depend on the duration and severity of hypoxia, and also on the timing of the hypoxic episode. In the developing rat the chances of surviving hypoxia in the perinatal period are much greater if hypoxia does not occur during the first four postnatal days (Fig. 5.26). Presumably, prolonged hypoxia has more devastating consequences in the first postnatal days because of the many changes occurring at that time, of which those of the cardiovascular system, including the pulmonary circulation, are the best known (see Chapter 1, under "Implications for Pulmonary Circulation"). Numerous experiments, conducted mostly with rats but also with calves, lambs, and piglets, have demonstrated that perinatal hypoxia impairs the functional development of the pulmonary circulation. In calves, perinatal hypoxia halts the normal fall of pulmonary artery pressure and resistance, contributing to the establishment of pulmonary hypertension (Reeves and Leathers, 1967; Sten-

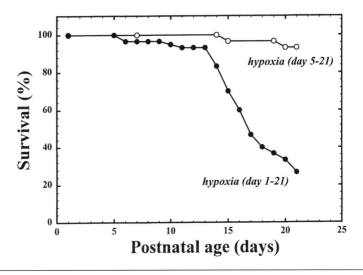

Figure 5.26. Chronic hypoxia in newborn rats. Neonatal rats were exposed to hypobaric hypoxia (inspired PaO_2 ~90 mm Hg) from either postnatal day 5 (○) or postnatal day 1 (●; day 0 = day of birth). The former group had a much greater survival rate, at least up to weaning age (day 21).

mark et al., 1987; Inscore et al., 1990; Durmowicz et al., 1993). Structural alterations probably responsible for the persistence of pulmonary hypertension were seen in piglets exposed to hypoxia at birth, but were less in evidence when the exposure had begun a few weeks after birth (Allen and Haworth, 1986). In piglets, the pulmonary hypertension resulting from neonatal chronic hypoxia was attributed to a decreased vascular nitric oxide production (Fike and Kaplowitz, 1996). Infants and children at high altitude present mild pulmonary hypertension and increased vascular resistance, possibly manifestations of the delayed adaptation of the vascular bed to postnatal life (Sime et al., 1963).

    In addition to its role in cardiovascular adaptation, at birth the respiratory function must cope with the rapid increase in metabolic rate and the gradual resetting of the carotid chemoreceptors from fetal to postnatal PaO_2 values (see "Peripheral Chemoreceptors," above); chronic hypoxia can interfere with this process (Hanson et al., 1989b). Changes in the compliance and resistance of the lungs have been observed in 2-week-old calves raised in hypoxia (Inscore et al., 1990). The implications of perinatal hypoxia for alveolar formation have been well described, and the existence of a critical period of high sensitivity to hypoxia in the perinatal period has also been recognized (Massaro et al., 1989; 1990). Thus, hypoxia can interfere with many variables during the first postnatal days, and there are therefore many reasons for considering the first postnatal days a very vulnerable period, a period probably more sensitive than any other to the challenges of chronic hypoxia.

Implications for Adult Respiratory Function. The long-term effects of chronic neonatal hypoxia have been studied extensively, with particular attention to neurological development. An intriguing aspect during the posthypoxic period is the progressive recovery in body weight. The mechanisms underlying the phenomenon of "catch-up growth," observed in experimental animals after termination of the neonatal hypoxia (e.g., Okubo and Mortola, 1988, 1989) and in children after surgical correction of cyanotic heart defects (Prader et al., 1963; Gingell et al., 1981), are unknown.

Experiments concerned with the long-term effects of neonatal hypoxia on the respiratory system have been performed mostly on adult rats that had been exposed to hypoxia in the neonatal period. In general, these studies have indicated the persistence of pulmonary vascular hypertension, or changes in the vascular response to a new hypoxic episode (Rabinovich et al., 1981; Tucker et al., 1984; Hakim and Mortola, 1990; Herget and Hampl, 1990; Tucker and Penney, 1993), changes in lung mechanics (Okubo and Mortola, 1989), and modifications in various aspects of respiratory control (Okubo and Mortola, 1990). Patients studied at least one year after correction of the tetralogy of Fallot, a congenital cyanotic cardiac defect, continued to present a blunted $\dot{V}E$ response to hypoxia (Sørensen and Severinghaus, 1968). A later report (Edelman et al., 1970) did not support this conclusion, but was based on only three patients whose congenital cardiac defects may or may not have caused continuous hypoxemia before the surgical correction. The blunted $\dot{V}E$ response to hypoxia in older animals with chronic hypoxia in the neonatal period (Eden and Hanson, 1987; Okubo and Mortola, 1990; Sladek et al., 1993a; Matsuoka et al., 1999) is a phenomenon reminiscent of what is observed in humans after many years of life at high altitudes (Forster et al., 1971; Lahiri, 1984; Weil, 1986). Chronic hypoxia later in life, however, did not result in blunting of the adult response to hypoxia (Okubo and

Mortola, 1990). Animal experiments, therefore, considered globally, underline the importance of early postnatal hypoxia in shaping the later properties of the respiratory function. They also support the view that the characteristics of humans living at high altitudes are not necessarily genetic traits, but could be the long-term effects of hypoxia experienced early in life.

Hyperoxia

Hyperoxia is a common event in the clinical setting. Physiologically, it occurs in mammals only at birth, when, in the transition from the intrauterine life to postnatal air-breathing, PaO_2 can rapidly increase two- to threefold (see Chapter 1, under "Birth, a Hyperoxic Event"); its implications for the resetting of the peripheral chemoreceptors were mentioned previously.

Measurements in newborn mice (Mortola and Tenney, 1986), rats (Mourek, 1959; Taylor, 1960b; Frappell et al., 1992a; Dotta and Mortola, 1992a), lambs (Alexander and Williams, 1970), and infants (Mortola et al., 1992a, 1992b) have all indicated an increase in $\dot{V}o_2$ during acute hyperoxia. In adults the results are mixed, and in adult men only $\dot{V}o_2$max (but not resting $\dot{V}o_2$) is increased by hyperoxia (Mortola and Gautier, 1995). In the early postnatal period the increase in metabolic rate is more apparent in cold conditions (Fig. 5.27), when hyperoxia can elicit greater muscle activity and, in older animals, shivering, thus affording better protection of body temperature (Tb). The mechanisms for the rise in $\dot{V}o_2$ are unclear, although the involvement of thermogenic mechanisms suggests that the site of action may be hypothalamic. It is of interest that in newborn rats the maximal $\dot{V}o_2$/W attained in the cold is still substantially below the levels that the adults can reach in normoxia, and not sufficient to maintain their Tb. The fact that even with abundant O_2 supply the newborn rat cannot maintain homeothermy suggests the presence of a limitation on the rate of maximal O_2 utilization independent of O_2 availability, contrary to what was previously believed (Thompson and Moore, 1968). The absent or limited capacity for shivering, a characteristic of many newborns, is probably the main reason for the limitation on O_2 use and heat production.

As mentioned earlier, the immediate effect of hyperoxia is a decrease in $\dot{V}E$ because of the elimination of the normoxic activity of the periph-

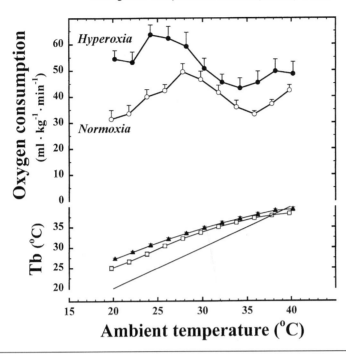

Figure 5.27. Hyperoxia and nonshivering thermogenesis. Oxygen consumption at different ambient temperatures (Ta) in 2-day-old rats during normoxia and hyperoxia (100% inspired O_2 concentration). Hyperoxia shifted the curve upward and specifically increased the thermogenic response to cold, resulting in a modest increase in body temperature, Tb (*oblique continuous line* represents the condition Tb = Ta). Bars are 1 SEM. Redrawn from Dotta and Mortola, 1992a.

eral chemoreceptors. When hyperoxia was maintained for several minutes in newborn mice, V̇E remained below the normoxic value (Mortola and Tenney, 1986). But in infants numerous studies have indicated an increase in V̇E during sustained hyperoxia (Cross and Warner, 1951; Brady et al., 1964; Davi et al., 1980; Mortola et al., 1992a, 1992b). The differences between the responses of newborn mice and those of human infants probably reflect differences in the balance of the many variables involved in the V̇E response to hyperoxia (Mortola and Gautier, 1995); generalizations therefore cannot be made until further data are obtained on additional species.

Although newborns are much better equipped to resist hyperoxia than adults (see Chapter 1, under "Birth, a Hyperoxic Event"), prolonged hyperoxia can have serious cytotoxic effects on many tissues. In the clinical setting, prolonged hyperoxia in infants can result in major pulmonary lesions (bronchopulmonary dysplasia) (Frank et al., 1998). In rats, prolonged hyperoxia during the neonatal period resulted in a blunted response to acute hypoxia during adulthood, which was not observed if the hyperoxic exposure occurred later in life (Ling et al., 1996). This phenomenon of blunting could be attributed specifically to a derangement of the normal postnatal development of the carotid bodies (Hanson et al., 1989a; Ling et al., 1997; Erickson et al., 1998), and is very similar to what is observed with neonatal chronic hypoxia, as discussed in the preceding section.

Hypercapnia

Hypercapnia, seen in the diseased state of respiratory insufficiency, can also occur during accidental rebreathing or during breathing in an environment presenting increased levels of CO_2. In many cases, an increase in arterial PCO_2 is accompanied by a drop in PaO_2 (i.e., a combination of hypercapnia and hypoxia defined as *asphyxia*).

Acute Conditions

As is the case in hypoxia, in hypercapnia the primary response is hyperventilation, but contrary to what occurs under hypoxia, the metabolic component of the hyperventilation is small. In fact, for levels of inspired CO_2 between 1 and 5%, the changes in $\dot{V}O_2$ are often negligible (Várnai et al., 1970; Saiki and Mortola, 1996). Hence, hypercapnic hyperventilation is essentially proportional to the degree of the hyperpneic response (Fig. 5.28, left panel), and can be present even in those newborn animals that shows minimal or absent $\dot{V}E$ response to hypoxia (Watanabe et al., 1996b).

For the same level of inspired CO_2, newborns of larger species tend toward slightly greater hyperventilatory responses than do those of smaller species (Mortola and Lanthier, 1996), perhaps because larger species are usually more precocial than the smaller ones. In adult mammals, a comparative study that compiled data from various sources suggested that the $\dot{V}E$ response to hypercapnia was greater in species of small size (Williams et al., 1995), whereas in adult marsupials studied under

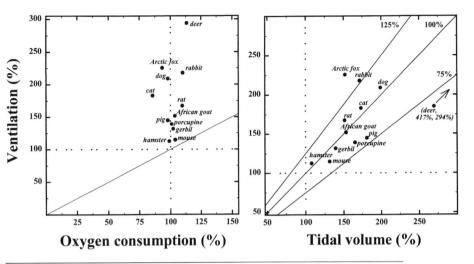

Figure 5.28. Hypercapnic hyperventilation in newborns. *Left:* Oxygen consumption ($\dot{V}O_2$) and ventilation ($\dot{V}E$) in newborns of several species during 5% CO_2 breathing for 30–60 min. Data, obtained at the ambient temperatures most common to the species, are expressed as a percentage of the air-breathing values. The oblique line indicates the $\dot{V}E/\dot{V}O_2$ value in air. In hypercapnia all species hyperventilate almost exclusively by increasing $\dot{V}E$, with minimal changes in $\dot{V}O_2$, a pattern quite different from the response to hypoxia (e.g., Fig. 5.22). *Right:* Average values of tidal volume (VT) and ventilation during 5% CO_2 breathing, expressed as a percentage of the air values. *Oblique lines indicate* breathing rate (f), as a percentage of the f value in air. The increase in $\dot{V}E$ during hypercapnia usually results from increases in VT, with little change in f. From Mortola and Lanthier, 1996.

the same experimental conditions the hypercapnic $\dot{V}E$ response was found to be size-independent (Frappell and Baudinette, 1995).

Unlike the condition of sustained hypoxia, in which the breathing pattern is rapid and often shallow (see Fig. 5.15), the hypercapnic hyperpnea of many newborns, including the human infant, is characterized by an increase in tidal volume (e.g., Brady and Dunn, 1970; Haddad et al., 1980; Guthrie et al., 1985; Martin et al., 1985; Saetta and Mortola, 1987; Mortola and Lanthier, 1996) (Fig. 5.28, right panel). In newborn rats and kittens, the rapid, shallow pattern seen during hypoxia was reversed into a deep, normofrequency pattern when the inspired CO_2 was increased (Saetta and Mortola, 1987; Mortola and Matsuoka, 1993). The rapid, shallow pattern during hypoxia and the slow, deep pattern in hypercapnia can be substantially modified when the hypoxic or hypercapnic conditions are made more severe, and may not occur at all in premature newborns or under anesthesia. To a lesser extent these patterns also occur in adult animals, so long as they are not anesthetized, and have been interpreted as differences between the central effects of hypoxia and those of hypercapnia on the termination of inspiration (Gautier, 1976).

During asphyxia, hypoxia and hypercapnia combine their effects, resulting in the maximal degree of hyperventilation. This positive $\dot{V}E$ interaction has been observed in the newborns of several species, such as the opossum (Farber et al., 1972), rat (Saetta and Mortola, 1987), lamb (Purves, 1966a), cat (Mortola and Matsuoka, 1993), and human (Cross et al., 1954; Brady and Dunn, 1970). When in combination with hypercapnia, hypoxia does not cease to reduce metabolic rate, but the hypoxic tachypnea is replaced by a slower and deeper pattern (Saetta and Mortola, 1987; Mortola and Matsuoka, 1993).

Unusual $\dot{V}E$ interactions between oxygenation and hypercapnia, hypoxia decreasing the $\dot{V}E$ response to CO_2 and hyperoxia increasing it, have been reported in premature monkeys (Guthrie et al., 1985) and premature human infants (Rigatto et al., 1975). In both cases, hypoxia and hyperoxia may have caused substantial changes in metabolic rate, with parallel effects on $\dot{V}E$. The possibility that the hypoxic- hypercapnic ventilatory interaction may differ qualitatively between premature and full-term newborns should also be considered, although there are no data from which to speculate on the basis for such an eventuality.

The common notion that newborn mammals, as opposed to adults, are extremely resistant to acute episodes of severe hypoxia or asphyxia finds support in numerous experimental studies on many species, although the details of the mechanisms are still unclear (e.g., Glass et al., 1944; Mott, 1961; Miller et al., 1964; Adolph, 1969, 1973; Guntherot, 1974; Duffy et al., 1975; Thach et al., 1991). During anoxia or asphyxia, a brief hyperpneic response is usually followed by apnea of variable duration, succeeded by gasping. Postnatal age and ambient temperature are important, since with their increase the resistance to anoxia or asphyxia rapidly decreases (Fig. 5.29). The lack of O_2, more than the severity of the acidemia, influences the duration of the apnea, although the acidemia can further drop metabolic rate, and reduction of metabolism is consid-

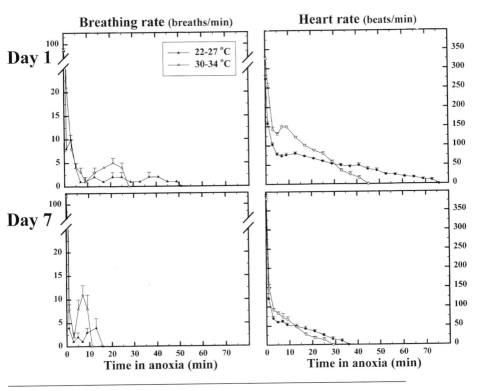

Figure 5.29. Changes in breathing rate (*left panels*) and heart rate (*right panels*) during anoxia in rats at day 1 and day 7 after birth. The period of respiratory and cardiac activities in anoxia is substantially longer in the younger rats, and is increased by the cold temperatures (22–27°C). Bars are 1 SEM. Modified from Saiki and Mortola, 1994.

ered to be the important variable in the developing brain's resistance to anoxia (Guntheroth and Kawabori, 1975; Duffy et al., 1975; Vannucci and Duffy, 1976; Duffy and Vannucci, 1977). Even with daily episodes of anoxia during the first postnatal week, young rats did not later present detectable abnormalities in body growth, metabolism, or breathing pattern (Saiki and Mortola, 1994; Saiki and Matsumoto, 1999).

Chronic Conditions

Studies on chronic hypercapnia or asphyxia in newborn animals are very few. Newborn rats maintained in 7% CO_2 for the first postnatal week increased V_T and $\dot{V}_E$ for the whole period, with no significant changes in breathing rate (Rezzonico and Mortola, 1989). Two days after return to air, their $\dot{V}_E$ remained elevated, and they presented a blunted response to a new acute episode of hypercapnia. Lung mass was unchanged, a finding that has some bearing on interpretations of increased lung mass during chronic hypoxia (see "Chronic Neonatal Hypoxia," above). In fact, the negative results during chronic hypercapnia would exclude the possibility that the increase in lung mass during chronic hypoxia is due to prolonged hyperpnea and the associated mechanical stresses on the lung parenchyma. A similar conclusion had been reached by Bartlett after exposing young rats to 5% CO_2 for 20 days (quoted in Bartlett and Remmers, 1971) and by Lechner et al. (1987) from studies on young guinea pigs in 5% CO_2 for 4 weeks. The fetuses of pregnant rabbits chronically exposed to 8% CO_2 had larger alveoli, which the authors (Nagai et al., 1987) speculated may have resulted from increased fetal breathing movements; but even in this case the lung mass of the fetuses was not increased. The compliance of both air-filled and liquid-filled lungs was higher in adult rats that had experienced chronic hypercapnia in the neonatal period, but not in those that had had chronic hypercapnia at later stages of development (Rezzonico et al., 1990). This finding is reminiscent of what is observed in adult rats chronically hypoxic in the neonatal period, and could indicate that prolonged hyperpnea during the early postnatal period, albeit not affecting lung mass, could influence the normal development of the lung-tissue constituents.

Conditions Confronting Burrowing Mammals

A typical situation of increased environmental CO_2 with a drop in O_2 is seen in the habitat offered by burrows, a situation that many semifossorial species (e.g., ground squirrels, prairie dogs, and woodchucks) experience intermittently and a few fossorial species (e.g., the moles) experience throughout life. The hypoxia and hypercapnia in the burrow are usually not much elevated, although large values have occasionally been measured; for example, in the burrow of the golden hamster, O_2 and CO_2 concentrations averaged respectively 15% and 5.7%, with CO_2 peaks up to 10% (Kuhnen, 1986). Burrowing animals typically exhibit hypoxemia and respiratory acidosis and a blunted $\dot{V}_E$ response to hypercapnia, whereas their $\dot{V}_E$ response to hypoxia may not differ greatly from that of surface dwellers (Boggs et al., 1984; Tenney and Boggs, 1986). To what extent these characteristics are genetic traits or reflect neonatal development in the asphyxic environment of the burrow could be experimentally addressed either by studying adult fossorial species raised in normoxia-normocapnia, or by studying adult surface dwellers raised in asphyxia. Very few data are available. In golden-mantled ground squirrels raised for over one year under simulated burrow conditions, respiratory control was found to be similar to that of those raised in room air conditions (Milsom, 1992). Adult hamsters growing in normoxia presented some characteristics typical of burrowers, but no marked differences were observed in their newborns. Hence, it was concluded that the characteristics of this species reflect genetic traits expressed during postnatal development, irrespective of environmental conditions (Holloway and Heath, 1984; Walker et al., 1985; Mortola, 1991).

Interspecies Comparisons

Body temperature tends to be slightly less in newborns than in adults, particularly in newborns of small species, which have limited thermogenic capacity and a delayed development of endothermy. Marsupials differ from eutherian mammals in developing endothermy very slowly and, possibly, in not having brown adipose tissue, presumably because thermogenesis is not an important requirement during development in the pouch. Behavioral thermocontrol, including huddling, is otherwise very important in newborns. For precocial eutherians, such as the piglet, huddling reduces heat loss, and therefore the amount of cold-induced thermogenesis. But in small altricial species like the rat, huddling in the cold limits the fall in body temperature, which results in higher thermogenesis because of the Q_{10} effect on $\dot{V}o_2$.

Peripheral chemoreceptors have been found to be functional in newborns of all species investigated, and the central chemoreception also seems operational in newborns. Hence, newborns respond to hypoxia, hypercapnia, and asphyxia with hyperventilation, as do adults; but hypoxic hyperventilation is mostly achieved through a reduction in metabolic rate, rather than in hyperpnea, and the distinction is more pronounced in altricial species than in precocial species. Hypoxic hypometabolism is due largely to a depression of all thermoregulatory mechanisms, but with chronic hypoxia there is also a blunting of tissue differentiation and organ growth. Prenatal hypoxia can have different effects on the postnatal respiratory functions, depending on the characteristics of prenatal lung development; in fact, in some altricial species alveolar formation is almost entirely a postnatal event, whereas in precocial newborns it is largely a prenatal process. Growth retardation in chronic hypoxia has been observed in many species; but it may not occur in those adapted to the hypoxic life of burrows or a high-altitude environment. Some of the peculiarities in the structure and functions of the respiratory system of species adapted to hypoxic environments could be acquired postnatally, and do not necessarily need to be attributed to the genetic makeup of the species.

Hemoglobin affinity for oxygen is strong in the fetus, but declines postnatally when its fetal form is replaced by the neonatal form, although this may not be the case in several marsupial species. Blood parameters related to O_2 transport (Hct, Hb, and P_{50}), although variable among newborns, do not seem to change systematically with species size.

Clinical Implications

Infants respond to cold mostly by behavioral means and nonshivering thermogenesis. Hypoxia abolishes these responses and resets the control of body temperature around a value lower than that in normoxia, as part of a hypometabolic strategy aimed at protecting oxygen delivery to the essential core organs. Hence, body temperature drops, and this drop, together with the hypometabolism, contributes to the extraordinary resistance of the newborn to hypoxia, asphyxia, and even anoxia. Artificial warming of the hypoxic infant can in fact present a counterproductive hyperthermic stimulus. In hypoxia, the hypometabolic response largely explains why the ventilatory level may not be higher than that in normoxia; in reality, this seemingly low ventilatory response to hypoxia can be appropriate for the infant's metabolic needs, and does not necessarily justify administration of respiratory stimulants. It is nevertheless conceivable that when metabolic rate is low, inputs normally inhibitory to breathing assume a relatively greater power, with the effect of further inhibiting pulmonary ventilation. These physiological events could create the setting for the establishment of a vicious circle leading to breathing irregularities and apneas, and perhaps constitute one reason for onset of the sudden infant death syndrome.

Chronic hypoxia reduces body growth, with differential effects on various organs. Prenatal hypoxia (e.g., gestation at high altitude, maternal smoking, anemia) results in small-for-gestational-age infants; at high altitude, infants are born at term, but their weight is low, approximately in proportion to the drop in barometric pressure. The lungs are highly compliant, presumably because of a derangement in the process of alveolar septation; pulmonary hypertension is also a characteristic of newborns after a hypoxic gestation.

As with prenatal hypoxia, postnatal hypoxia has implications for the mechanical properties of the lungs, pulmonary circulation, and aspects of respiratory control, including the ventilatory response to hypoxia. Animal experiments suggest that the early postnatal phases are a particularly critical period for sensitivity to sustained hypoxia. If the causes of the hypoxia are removed, as in infants following correction of cyanotic congenital heart diseases, body growth resumes at a pace even more rapid than normal ("catch-up growth"), but long-term effects on various aspects of the respiratory function can persist into adulthood.

Infants, like newborns of most other species, are well equipped against hyperoxia, presumably because birth itself is the only physiological incidence of an increase in environmental oxygen. Sustained hyperoxia, however, has deleterious effects on many organs, and bronchopulmonary dysplasia is one of the best known manifestations of oxygen toxicity.

The infant's postnatal anemia (decrease in hematocrit and hemoglobin during the early postnatal period) is a physiological process and is common in many other species as well.

Summary

1. Newborn mammals have limited abilities to maintain homeothermy, and in small species body temperature (Tb) is often below the adult value. Thermoregulation is based largely on behavioral means and oxygen consumption of brown adipose tissue (BAT) in nonshivering thermogenesis; the latter is characteristic of the neonatal period, although in some species it remains prominent even in adulthood.

2. Even with moderate cold exposure, thermogenesis may not be sufficient to maintain Tb. A drop in Tb favors a decline in $\dot{V}o_2$ via the Q_{10} effect, thus aggravating the hypothermia.

3. With postnatal growth, the lower critical temperature of the neonate decreases and thermoneutrality widens, both changes reflecting an improvement in the control of heat loss. In addition, Tb increases, and is better protected against changes in ambient temperature. In marsupials, which are so thoroughly altricial, the development of an effective endothermic capacity is a very slow process.

4. As is the case in adults, in newborns the increase in $\dot{V}o_2$ in the cold is largely accompanied by an increase in ventilation ($\dot{V}E$). Very little is known about neonatal ventilatory control in hyperthermia, a condition characterized by great irregularity of the breathing pattern.

5. Circadian oscillations in metabolism and Tb are seen in newborns, but whether or not they are governed by the same mechanisms employed by the adult is still unclear.

6. The peripheral chemoreceptors are active in both the fetus and the newborn, but their response to hypoxia is less than that in adults, and in the first days after birth they undergo a process of resetting that gradually increases sensitivity to hypoxia.

7. Carotid body denervation in newborns can lead to hypoventilation and irregular breathing; there is no strong evidence that the effects of these alterations are any worse than those in the adult.

8. A few measurements have indicated that the neonatal blood-brain barrier performs, at least qualitatively, as in adults. The central chemosensory areas are active, and promote ventilatory responses comparable to those of the adults.

9. Both hematocrit and hemoglobin (Hb) concentration decline in the weeks after birth, which are marked by a variable period of transient anemia. The Hb affinity for O_2, very high in the fetus, gradually drops postnatally, mostly because of replacement of the fetal- with the adult-type Hb and the increase in erythrocyte 2,3-DPG.

10. Prenatal hypoxia affects the body and lung growth of the fetus, with functional implications even at birth.

11. Hyperventilation is the neonate's most immediate response to acute hypoxia, but in newborns hyperventilation is achieved mostly by a decrease in metabolic rate (hypometabolism) rather than by hyperpnea.

Hypometabolism is a regulated process largely determined by inhibition of thermogenesis in all its forms, whether shivering, nonshivering, or behavioral heat control, and involves a resetting of the thermocontrol at a lower value of Tb.

12. During hypoxia, if Tb is artificially forced toward the normoxic value, the newborn can respond with mechanisms of heat dissipation. Among these, the redistribution of blood to the body periphery can reduce the O_2 delivery to the most vital organs.

13. A weak or absent hyperpnea in the hypoxic newborn is the expected response to the decrease in metabolic rate; therefore, it should not necessarily be regarded as an expression of inadequate ventilatory control. During hypoxia, however, the low metabolic rate can enhance the relative efficacy of inputs inhibitory to breathing.

14. The advantages of the hypometabolic strategy are many, and are at the basis of the extraordinary ability of newborn mammals to survive periods of severe hypoxia. Its disadvantages become apparent with chronic hypoxia, because the reduced growth of tissues and organs may be incompatible with survival, or can lead to long-lasting structural and functional alterations. The most critical period is in the first few days after birth.

15. Hypercapnia, like hypoxia, promotes hyperventilation. In this case, however, hyperventilation is due mostly to hyperpnea, with no important changes in metabolic rate. The breathing pattern in hypercapnia is deep and slow, whereas in hypoxia it is most often rapid and shallow. An interaction between the two stimuli results in the greatest level of hyperventilation, by combining hypoxic hypometabolism with hypercapnic hyperpnea.

16. Prolonged hypoxia and hypercapnia, singly or combined, even if occurring only in the neonatal period, can have long-lasting effects on various aspects of the adult respiratory function, because of the structural and functional alterations that result. It seems probable that some of the characteristics of animals living permanently in hypoxic habitats, such as at high altitude or in burrows, may be acquired during postnatal development and are not necessarily an expression of their genetic traits.

Passive Respiratory Mechanics: Some Applications to Measurements in Newborns

The respiratory system is in a passive mode when the respiratory muscles are relaxed (see Chapter 3, under "Terminology"). This is the case in a paralyzed subject, but it can also occur without paralysis immediately after a period of hyperventilation (i.e., when arterial PCO_2 is below the threshold of respiratory muscle activation). During spontaneous breathing, the passive mode can be induced by forcing the lung volume to remain elevated above the passive resting volume (Vr); in infants this can easily be achieved by occlusion of the outlet of a face mask at end-inspiration or during expiration. The maneuver establishes relaxation of the inspiratory muscles via the stimulation of the slowly adapting stretch receptors within the lungs and the vagally mediated Hering-Breuer inflation reflex (see Chapter 4, under "Vagal Reflexes"). Hence, in a subject that is spontaneously breathing, this brief period of artificially provoked muscle relaxation is sufficient for the computation of the passive mechanical properties of the respiratory system.

Measurements of passive mechanics can also be obtained during artificial ventilation. In studies of this sort, one should bear in mind that in infants the endotracheal tube often introduces an important nonlinear resistance, the correct value of which can be difficult to evaluate. Moreover, the upper airway often behaves as a nonlinear resistor. Hence, only in experimental animals breathing through a tracheostomy can the mechanical properties of the respiratory system be approximated to those of a linear system. When this is the case, the conceptual approach and the analysis of measurements are substantially simplified.

Artificial Ventilation

The change in volume (ΔV) of the lung is usually measured by integration of the respiratory airflow signal, which is obtained via a pneumotachograph placed in series between the tracheal cannula (or endotracheal tube) and the ventilator. The total pressure driving the respiratory system (Prs) during artificial ventilation can be conveniently measured at the airway opening (Pao) (Fig. A.1, right), and the values of Crs and Rrs are either separately computed or derived from the equation of motion.[1] In fact, because three of the five variables of the equation ($\dot{V}$, V, and Pao) are experimentally measured, it is possible to derive the two unknown constants mathematically.

Compliance

At end-inflation, $Crs = \Delta V/\Delta Pao$, where ΔV and ΔPao represent the difference between the two zero-flow points, at the beginning and the end of inflation. From the record of esophageal pressure (Pes), C_L is similarly computed as $\Delta V/(\Delta Pao - \Delta Pes)$, from which chest-wall compliance $Cw = 1/(1/Crs - 1/C_L)$. The measurement of C_L obtained in this way is defined as *dynamic* compliance, $C_L dyn$ (see Chapter 3, under "Inspiration").

One approach to verifying the accuracy of the Pes measurement is to compare the changes in Pao and Pes during an effort against occluded airways (Asher et al., 1982; Baydur et al., 1982). Because during the effort $\Delta V = 0$, the reasoning is that ΔP_L is also nil, and the change in Pes should equal that of Pao. Hence, when during an inspiratory effort against closed airways ΔPes equals ΔPao, it is commonly accepted that Pes is representative of mean pleural pressure (Ppl). In newborns, however, the distortion of the chest may be great, and the distortion in normal open-airway breathing is undoubtedly not the same as that in the inspiratory occluded effort used for the test. Therefore the possibility exists that Pes is not representative of Ppl during open airway breathing, despite the fact that it may seem so during the occluded effort.

---

[1] $Prs = (1/Crs) \cdot V + Rrs \cdot \dot{V}$ (see Chapter 3, under "Terminology").

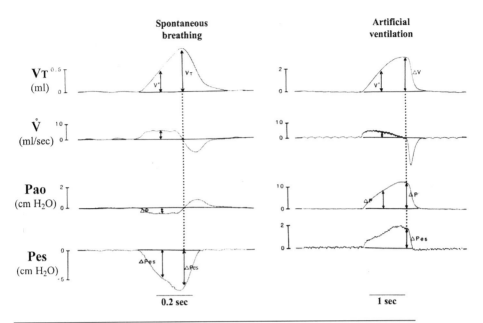

Figure A.1. Measurements of compliance and resistance in an anesthetized animal. Records are of the changes in lung volume (VT), airflow (V̇), pressure at the airway opening (Pao), and esophageal pressure (Pes) during spontaneous breathing (*left*) and artificial ventilation (*right*). *Dotted lines* join the points of zero V̇ at end-inspiration. Compliance is measured from the change in P and V between points of zero V̇. Resistance is often computed in the middle of inspiration, when the V̇ is maximal. Slightly modified from Okubo and Mortola, 1989.

Resistance

The total resistance of the respiratory system (Rrs) at any given time *during* the inflation can be calculated according to the following formula: Rrs = [Pao − Pel(rs)]/V̇, where Pel(rs) is the elastic pressure component (V/Crs) of Prs (Mead and Whittenberger, 1953), and V̇ is flow. The measurement is more accurate at the highest flow, which is usually in the middle third of inflation (Fig. A.2, A).

A *time-average* inflation resistance can be computed on the chart record by *planimetry* (Fig. A.2, B), as originally proposed by J. Hildebrandt (quoted in LaFramboise et al., 1983). The two Pao points, at the beginning and the end of inflation, are joined, and the area enveloped between this line and the actual Pao curve, divided by the inflation time (TI), corresponds to the average resistive pressure. The ratio between this value and the mean inflation flow (V/TI) is therefore the time-average inflation Rrs.

Alternatively, from the *equation of motion* of the respiratory system (Equation 3.1), it follows that

$$\text{Prs/V} = 1/\text{Crs} + \dot{\text{V}}/\text{V} \cdot \text{Rrs}. \qquad (Equation\ A.1)$$

Hence, in the plot of the Pao/V values (y-axis) against the corresponding isotime $\dot{\text{V}}$/V (x-axis[2]), the slope represents the total resistance of the respiratory system (Rrs), and the intercept is Ers (= 1/Crs) (Fig. A.2, *C*).

All the analytical approaches presented above can also be applied to the computation of the total pulmonary resistance (RL), from which Rw = Rrs − RL.

Spontaneous Breathing

In this procedure, the pneumotachograph is connected to a face mask, an endotracheal tube, or a tracheostomy tube. Newborn infants tolerate the face mask well, although one should consider the possibility of alteration of the breathing pattern owing to the stimulation of the trigeminal area (Fleming et al., 1982; Dolfin et al., 1983) and to the added dead space. The latter can be minimized with a bias flow of air (electronically offset) through a side port of the mask. Brief occlusions of the pneumotachograph outlet represent the basis for the measurements of Crs and Rrs described below. Average values for humans are presented in Table A.1.

Compliance

After occlusion of the airways, at end-inspiration or midexpiration, Pao rises to the value corresponding to the recoil pressure of the respiratory system at that volume. The time course of the increase in Pao depends on the relaxation time of the inspiratory muscles. The occlusion must be

---

[2] With this analytical approach, the data points at onset- and end-inflation are usually neglected, because of the low reading accuracy at these points.

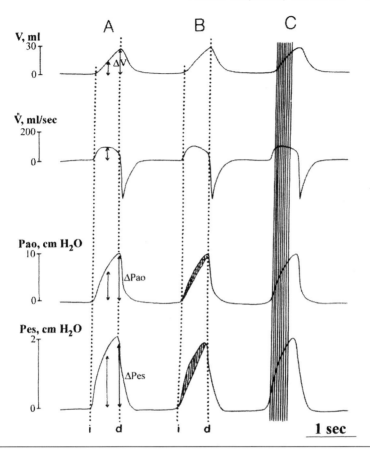

Figure A.2. Schematic representation of the changes in lung volume (V), airflow (V̇), pressure at the airway opening (Pao), and esophageal pressure (Pes) during artificial ventilation: *i* and *d* indicate the beginning of, respectively, inflation and deflation; *A, B,* and *C* represent the three analytical approaches to the computation of Rrs and RL described in the text. Adapted from Rezzonico et al., 1990, with modifications.

brief because the same vagal reflex inhibiting the inspiratory muscles can eventually promote the activation of the expiratory muscles (the Hering-Breuer expiratory promoting reflex; see Chapter 4, under "Vagal Reflexes"). A steady Pao plateau is indirect evidence of complete muscle relaxation, as indicated by electromyographic measurements in animals.

From the data points obtained from numerous occlusions at different

Table A.1. Average values of some parameters pertinent to respiratory mechanics in human infants

Parameter	Infant	Adult Male
Resting pulmonary ventilation ($\dot{V}_E$, ml $\cdot$ kg^{-1} $\cdot$ min^{-1})	310	90
Resting tidal volume (V_T, ml $\cdot$ kg^{-1})	7	7
Resting breathing rate (f, breaths $\cdot$ min^{-1})	45	13
Respiratory system compliance (Crs, ml $\cdot$ cm H$_2$O^{-1} $\cdot$ kg^{-1})	1.0	1.2
Lung compliance (CL, ml $\cdot$ cm H$_2$O^{-1} $\cdot$ kg^{-1})	1.2	2.4
Chest wall/lung compliance (Cw/CL)	5	1
Respiratory system resistance (Rrs, cm H$_2$O $\cdot$ ml^{-1} $\cdot$ sec $\cdot$ kg)	0.21	0.4
Total pulmonary resistance (RL, cm H$_2$O $\cdot$ ml^{-1} $\cdot$ sec $\cdot$ kg)	0.15	0.34
Functional residual capacity (ml $\cdot$ kg^{-1})	30	30
FRC-resting volume (FRC-Vr, ml $\cdot$ kg^{-1})	3	0
Respiratory system time constant (τrs, sec)	0.2	0.5

Note: These values can vary greatly among studies, especially in newborns, since they are highly dependent on the measurement conditions employed. This table thus serves chiefly to indicate the directions of the changes when comparing infants and adult men. Differences or similarities suggested by this table are not necessarily shared by other species.

lung volumes, Crs is calculated as the slope of the regression line of the ΔV-ΔPao relationship (Fig. A.3); the intercept on the V-axis represents the FRC-Vr difference (see Chapter 3, under "Expiration").

An alternate method for the measurement of Crs in spontaneously breathing infants consists of performing two brief occlusion maneuvers. The first occlusion is needed to establish a reference V and the corresponding static Pao; this occlusion is then released and a second occlusion is performed at a different V, within the same expiration, or in one of the immediately following breaths (Fig. A.4). Crs corresponds to the difference in volume between the two occlusions (ΔV) divided by the corresponding ΔPao. This method circumvents the potential problems introduced by the breath-by-breath oscillations in FRC.

As is the case during artificial ventilation, the simultaneous measurement of Pes during spontaneous breathing would permit the partitioning of Crs into its lung and chest-wall components. To this end, CL can be computed during the breathing cycle (CLdyn) or as the slope of the ΔV-ΔPL curve. CLdyn is usually less than CL (see Chapter 3, under "Inspi-

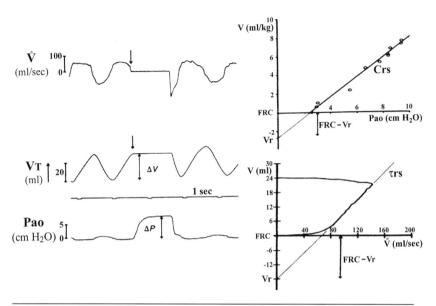

Figure A.3. Records of airflow (V̇), tidal volume (VT), and pressure at the airway opening (Pao) in an infant during spontaneous breathing. Time mark = 1 sec. At the arrow (end-inspiration), the airways are briefly occluded, to trigger relaxation of the respiratory muscles via the Hering-Breuer reflex (the muscle relaxation is suggested by the stability in Pao). From the V and Pao values of occlusions at various lung volumes it is possible to construct the Pao-V relationship, the slope of which represents the compliance of the respiratory system, Crs (*panel at top right*). The linear portion of the expiratory V̇-V curve following occlusion release (*bottom right panel*) represents the time constant of the respiratory system (τrs), and its extrapolation to zero V̇ is the dynamic elevation of the end-expiratory level (FRC-Vr differences). Slightly modified from Mortola, 1987.

ration"). From Crs and CL, chest-wall compliance Cw = 1/(1/Crs − 1/CL).

Time Constant and Resistance

Because respiratory-system compliance and resistance are approximately constant within the tidal volume range, both V̇ and V decrease during expiration, following closely an exponential curve. Therefore, a semilog representation of V̇, or V, during expiration, after reopening of the airway occluded at end-inspiration, yields a linear relationship, the reciprocal of

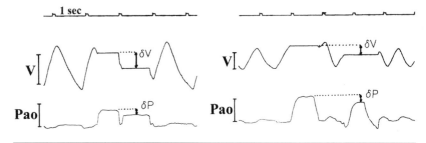

Figure A.4. Change in lung volume (V, inspiration upward; vertical bar = 20 ml) and pressure at the airway opening (Pao; vertical bar = 10 cm H_2O) in a spontaneously breathing infant. Time mark = 1 second. *Left:* An example of calculation of compliance of the respiratory system (Crs) from two occlusions within the same expiration. After the reference occlusion, the airways are briefly opened and occluded again at a lower V. *Right:* Another example from two airway occlusions performed one breath apart. After the reference occlusion, the airways are reopened and the second occlusion is performed during an inspiration, at a V lower than the reference volume. δV and δP indicate, respectively, the change in V and Pao between the two occlusions. Modified from Mortola et al., 1993.

the slope being τrs, the passive time constant of the respiratory system (Brody, 1954). Because the ratio of two exponential curves ($\dot{V}$ and V in time) is a constant, τrs can also be calculated from the slope of the V (y-axis)-$\dot{V}$(x-axis) relationship (McIlroy et al., 1963), after reopening the airway occluded at end-inspiration (Mortola et al., 1982b; Zin et al., 1982; LeSouef et al., 1984) (Fig. A.3, bottom right panel). From τrs, Rrs is calculated as τrs/Crs. Several factors, including respiratory muscle activation or alinearities in the mechanical properties of the upper airway, can alter the overall mechanical behavior of the respiratory system to the extent that the postocclusion $\dot{V}$ and V cannot be approximated by exponential curves. In these cases, the values of τrs and Rrs change with $\dot{V}$, and the $\dot{V}$ at which they are measured must be defined.

By extrapolation of the linear segment of the $\dot{V}$-V curve to zero flow, it is possible to estimate the dynamic elevation of the end-expiratory level (or FRC − Vr difference) (Mortola et al., 1982b, 1984a). In fact, on theoretical grounds, by recording V and $\dot{V}$, with only a single end-inspiratory occlusion and the following release, it should be possible to compute τrs, FRC − Vr, Crs, and Rrs (LeSouëf et al., 1984).

Comparisons and Normalization

Experimental observations are rarely limited to a lone subject. More commonly, they are performed on a sample of subjects considered representative of the total population according to some statistically defined probability. If the subjects sampled were identical clones, the value of the sample would be voided. Yet if their differences were too great, it would be difficult to combine the results and reach a meaningful conclusion. An appropriate balance between the size of the sample and its homogeneity is at the core of much experimental research in biology. Although statistical approaches to defining the size of the sample are in use, decisions about the homogeneity of the group and how to control for its inherent differences are usually left to the experience and common sense of the investigator.

Normalization

In some cases, direct inter-animal comparison of the raw data is physiologically meaningful. For example, in respiratory physiology, time-derived parameters (e.g., breathing rate, inspiratory and expiratory time, mechanical time constants, partial pressure of gases, etc.) can be readily related. In many cases, however, direct comparison of raw data determined by linear dimensions and their powers (length^2 = surfaces, length^3 = volumes) proves to be of minor interest. In respiratory physiology, inter-animal comparisons of lung surfaces, volumes, and related variables (e.g., alveolar or capillary surface, diffusing capacity, tidal volume, dead space, lung capacities, flows, compliance, resistance) are often made with reference to easily obtained parameters, such as body weight, body length or height, dry or wet lung weight (LW), or lung volume (V).

It is important to realize that, although each normalization procedure may yield some more or less meaningful information, none can provide the whole picture. For example, lung compliance (CL) can be normalized by body weight (W). This approach is practical and useful if the interest lies in examining the elastic properties of the lung with respect to the size of the organism that the lung has to serve, as represented by W. But the information may be misleading if the interest lies in comparing lung properties per se; in fact, normalization by W obviously includes the weight of anatomical structures (head, limbs) extraneous to the lungs, and not equally represented in animals of different species or sex or age. In this case, normalization by LW or V, rather than W, may be more appropriate. The former parameter (CL/LW) relates to the density of the organ; the latter (CL/V) has the dimensional units of 1/pressure, hence it is size-independent parameter of the lung's elastic characteristics.

The choice of the normalizing parameter may be crucial for the conclusion sought. For example, many lung variables when normalized by LW are smaller in newborns than in adults, whereas the opposite can occur after normalization by W, since LW is a rather large fraction of body mass in newborns and decreases with age. Often, coefficients or dimensionless parameters (e.g., dead space/tidal volume, resting volume/total lung capacity, chest wall-lung compliance ratio, Reynolds' number, exponential constant of the deflation P-V curve, ventilation/oxygen consumption,[1] etc.), or pressures (P_{50}, alveolar or arterial Po_2 and Pco_2) are desirable choices for comparisons among animals, because they have a clear and readily interpretable physiological value, with no need for normalization.

Body weight (W) as a normalizing parameter can be particularly misleading in some special cases. For example, the content of the digestive tract in adult ruminants can be up to 20% of W, complicating comparisons with nonruminant species and with neonates of ruminants *or* nonruminants. The bony shell of the armadillo and the thick blubber of the seal are other special examples of body parts that may complicate the use of W as a normalizing parameter.

[1] This is not really a dimensionless parameter, since ventilation is usually measured at BTPS conditions, and oxygen consumption at STPD conditions, but the ratio is a convenient indication of the ventilatory needs per unitary oxygen requirements, or *ventilatory equivalent.*

Allometry

A special case of inter-animal comparison by body weight (W) is the *allometric* [from Greek, "measure of what is different"] *analysis,* by which a morphological or physiological variable Y is related to the W of the animal according to the exponential function $Y = a \cdot W^b$. This is conveniently expressed by the log-transformed version $\log Y = \log a + b \cdot \log W$, where b is the slope of the function and a the antilog of its intercept. When b = 1, the variable Y increases in direct proportion to W, the proportionality factor being a. Positive slopes different from unity indicate that Y is not directly proportional to W, and therefore Y/W is not an interspecies constant; rather, Y increases disproportionately more (b > 1) or less (b < 1) than W (Fig. B.1). In these cases, inter-animal comparisons of the Y/W values need careful interpretation, because of the potential problems intrinsic to the normalization by W mentioned above.

Allometry as a scaling approach has been widely used in all fields of biology (Kleiber, 1961; Stahl, 1962; Schmidt-Nielsen, 1984; Peters, 1983), including respiratory physiology (Tenney and Remmers, 1963; Günther and De La Barra, 1967; Stahl, 1967; Leith, 1976). In this book, allometry is used to compare newborns of different species, and to compare newborns and adults of the same species. Comparison of the allometric relationships of newborns and adults of various species can be an analytical tool useful in separating the relative importance of changes in size versus changes in age in the modifications accompanying postnatal growth (Fig. B.2).

By use of simple algebraic operations the allometric approach allows predictions of relationships between variables that can be useful in the understanding of the functional design of the respiratory system. For example, the scaling of the passive respiratory time constant (τrs) in newborn mammals can be calculated on the basis of the individual allometric relationships of compliance (Crs) and resistance (Rrs). Since τrs = Crs $\cdot$ Rrs, and Crs $\propto W^{1.02}$ (see Fig. 3.9) and Rrs $\propto W^{-0.80}$ (see Fig. 3.16), the product of these two equations yields an exponent[2] $[1.02 + (-0.80)$

---

[2] The *ratio* of two exponential functions A and B is a new exponential function C, with an exponent equal to the *difference* between the two original exponents of A and B. The *product* of two exponential functions A and B is also a new exponential function C, with an exponent equal to the *sum* of the two original exponents of A and B.

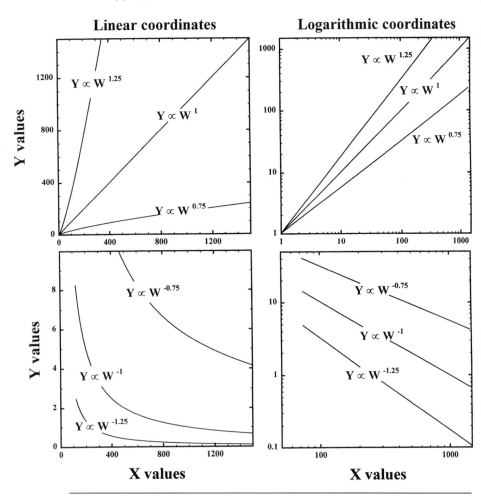

Figure B.1. Exponential functions represented on linear coordinates (*left*) and logarithmic (log$_{10}$) coordinates (*right*). The log-transformation gives a linear relationship, the slope of which is the original exponent. Even small differences between slopes of the log-transformed functions can indicate important differences between the functions of the raw data.

= 0.22] that is, slightly, yet significantly, higher than zero. Hence, from the product of the allometric equations of Crs and Rrs one can predict that τrs in newborns is not an interspecies constant but should increase slightly with the size of the species. This prediction can then be verified experimentally.

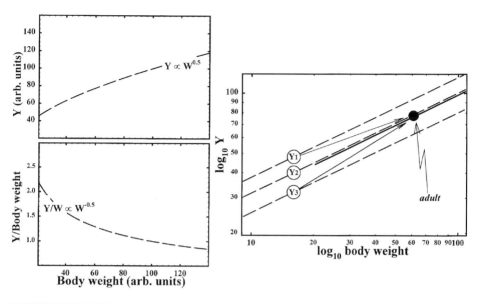

Figure B.2. Hypothetical representation of a variable Y scaling to body weight (W) with the same exponent, 0.5, in both adult and newborn species (Y ∝ a · $W^{0.5}$). Because the exponent is less than 1, a plot of the raw data is a curvilinear function with a concavity toward the body weight (W) axis (*top left*). When Y is represented normalized by W (Y/W), the values decrease with the increase in W (*bottom left*). *Right:* Logarithmic representations of the Y and W values. A larger Y/W value in the newborn than in the adult can be interpreted in different ways, depending on the relative displacement of the neonatal (*dashed line*), with respect to the adult, allometric curve (*solid line*). If the two curves were overlapping (i.e., same intercept, a), the higher Y/W in the newborn than in the adult of the same species can be attributed strictly to its smaller size. In this case, the value of the newborn Y2 would be similar to that of adults of other species of smaller size. If the neonatal allometric curve is displaced above the adult line (i.e., greater value of a, Y1), the larger Y/W of the newborn would reflect both size and developmental (age) differences. Finally, if the newborn function is displaced below the adult line (i.e., smaller value of a, Y3), the larger Y/W value of the newborn cannot be explained by its smaller size, but must be attributed to differences in development. In case 1 it can be predicted that within a species changes in Y during development (*arrows*) will be less than for the interspecies allometric curves (i.e., the function describing changes in Y during development will have an exponent < 0.5). In case 3, the opposite will occur.

Table B.1. Allometric relationships of newborn mammals

Y	X	Total No. of Species	a (value of Y at X=1)	b (slope)	Adult slope	Figure/section
Gestation (days)[1]	Ad W (kg)	228[1]	63	0.25		Fig. 1.3
Total litter mass (kg)[1]	Ad W (kg)	154[1]	0.109	0.75		Chapter 1, "Gestation and Birth"
Birth weight (kg)[1]	Ad W (kg)	164[1]	0.042	0.86		Fig. 1.4
$\dot{V}_{O_2}$ (ml/min)	Nb W (g)	33	0.025	0.92	(0.75)	Fig. 2.4
$\dot{V}_{O_2}$ (ml/min)[2]	Nb W (g)	(9 orders)	0.011	0.99	(0.75)	Fig. 2.5
$\dot{V}_{O_2}$ (ml/min)[1,2]	Nb W (g)	(8 orders)[1]	0.022	0.91	(0.75)	Fig. 2.5
$\dot{V}_{E_2}$ (ml/min)	Nb W (g)	23	1.22	0.91	(0.75)	Fig. 2.15
V_T (ml)	Nb W (g)	23	0.01	1.01	(1)	Fig. 2.15
f (breaths/min)	Nb W (g)	23	115	-0.09	(-0.25)	Fig. 2.15
Tracheal volume (ml)	Nb W (g)	11	0.00052	1.02	(1)	Fig. 2.16
Tracheal volume (ml)[1]	Nb W (g)	10[1]	0.00055	1.01	(1)	Fig. 2.16
$\dot{V}_{E_2}$ (ml/min)	Nb $\dot{V}_{O_2}$	21	39.9	1.04	(1)	Fig. 2.18
Diaphragm weight (g)[3]	Nb W (g)	10	0.00092	0.99	(1)	Fig. 3.2
Intercostals weight (g)[3]	Nb W (g)	9	0.00012	1.03		Fig. 3.2
Lung weight (g)[3]	Nb W (g)	9	0.00369	0.99	(1)	Fig. 3.7
C_L (ml/cm H_2O)	Nb W (g)	15	0.0024	1.01	(1)	Fig. 3.9

C_L (ml/cm H_2O)[1]	Nb W (g)	14	0.0020	1.03	(1)	Fig. 3.9
Crs (ml/cm H_2O)	Nb W (g)	15	0.0016	1.02	(1)	Fig. 3.9
Crs (ml/cm H_2O)[1]	Nb W (g)	13	0.0017	1.01	(1)	Fig. 3.9
Alveolar diameter (m)	Nb W (kg)	11	62	−0.02	(0.2)	Chapter 3, under "Mechanical Properties of the Lungs"
Vr (ml)	Nb W (kg)	7	27	0.98	(1)	Chapter 3, under "Static Coupling between Lungs and Chest Wall"
Rrs (cm $H_2O \cdot ml^{-1} \cdot sec^{-1}$)	Nb W (g)	11	25	−0.80	(−0.75)	Fig. 3.16
Rrs (cm $H_2O \cdot ml^{-1} \cdot sec^{-1}$)[1]	Nb W (g)	10	28	−0.82	(−0.75)	Fig. 3.16
R_L (cm $H_2O \cdot ml^{-1} \cdot sec^{-1}$)	Nb W (g)	10	26	−0.89	(−0.75)	Fig. 3.16
R_L (cm $H_2O \cdot ml^{-1} \cdot sec^{-1}$)[1]	Nb W (g)	9	36	−0.93	(−0.75)	Fig. 3.16
Derived relationships:						
τrs (sec) = Crs · Rrs	Nb W			0.22	(0.25)	Chapter 3, under "Resistance to Airflow"
τL (sec) = C_L · R_L	Nb W			0.12	(0.25)	Chapter 3, under "Resistance to Airflow"
$\dot{V}_E/\dot{V}_{O_2}$	Nb W			−0.01	(0)	Chapter 2, under "Ventilation-Metabolism Coupling"

[1] Eutherian species only
[2] By mammalian order
[3] Dry weight

Ad, adult. Nb, newborn. W, body weight. $\dot{V}_{O_2}$, oxygen consumption. $\dot{V}_E$, pulmonary ventilation. C_L, lung compliance. Crs, respiratory system compliance. Vr, passive resting volume of the respiratory system. Rrs, resistance of the respiratory system. R_L, total pulmonary resistance. τrs, time constant of the respiratory system. τL, time constant of the lung.

In some cases, the remaining exponent of allometric operations (residual mass index) does not differ significantly from zero (as in the case of $\dot{V}E/\dot{V}O_2$, or of the pulmonary surface area/$\dot{V}O_2$), indicating that the relationship between the variables is mass-independent.

Allometric Relationships of Newborn Mammals

The equation here is the log-transformed version of $Y = a \cdot X^b$,

$$\log Y = \log a + b \cdot \log X, \qquad (Equation\ B.1)$$

where Y is the dependent variable and X is body weight in kg. Hence, a is the antilog of the intercept (the value of the dependent variable at $X = 1$) and b is the exponent of the equation in its original exponential form. The equations are calculated from the mean values of each species, rather than from the total number of data points, to minimize biases toward those species for which larger sample sizes have been used (Table B.1). In the table, *adult slope* is the most commonly reported value for adult mammals; *section* indicates the text location where the relationship has been mentioned in this book.

Orders and Suborders
of the Class Mammalia

Subclass			
Order			
	Suborder		Common Names of Some Families and Species
	Monotremata (3)		platypus, echidnas
Metatheria (marsupials)			
	Diprotodontia (117)		wombats, potoroos, opossums, wallabies, koalas
	Didelphimorphia (60)		New World opossums
	Microbiotheria (1)		monito del monte
	Dasyuromorphia (60)		mulgara, Tasmanian devil
	Paramelemorphia (21)		bandicoots, bilbies
	Notoryctemorphia (2)		marsupial moles
	Paucituberculata (5)		rat opossums
Eutheria (placentals)			
	Artiodactyla (220)		camels, cows, sheep, goats, pigs, hippos, giraffes, deer, antelopes
	Carnivora (250)		dogs, bears, cats, weasels, ferrets, minks, martens, otters, civets, mongooses, pandas, foxes, skunks, ermines, coatis, racoons, badgers, wolverines
		Pinnipedia (33)	seals, sea lions, walruses
	Cetacea (79)		whales, porpoises, dolphins

(*continued*)

Subclass (*continued*)

Order	
Suborder	Common Names of Some Families and Species

Eutheria (placentals)

Chiroptera (950)	bats (of many families)
Dermoptera (2)	colugos
Edentata (29)	armadillos, sloths, anteaters
Hyracoidea (7)	hyraxes
Insectivora (370)	shrews, moles, tenrecs, hedgehogs
Scandentia (19)	tree shrews
Macroscelidea (19)	elephant shrews
Lagomorpha (66)	pikas, rabbits, hares
Perissodactyla (16)	rhinos, tapirs, horses, zebras
Pholidota (7)	pangolins
Primates (190)	apes (bonobos, chimpanzees, gorillas, orangutans), gibbons, humans, lemurs, marmosets, tamarins, colobus, baboons, loris, indri, galagos
Proboscidea (2)	African and Indian elephants
Rodentia (1700)	rats, mice, porcupines, capibaras, guinea pigs, beavers, squirrels, lemmings, voles, marmots, chinchillas, gerbils, muskrats, chipmunks, prairie dogs
Sirenia (5)	dugongs, manatees
Tubulidentata (1)	aardvarks

Note: Numbers in parentheses indicate approximate numbers of surviving species.

Glossary

See also Abbreviations, following the Preface.

abductor Any of various muscles (e.g., one of several that open the laryngeal cavity) that, upon shortening, move structures apart (*opposite:* adductor).

acidemia A reduction of blood pH below the normal value (*opposite:* alkalemia).

acidosis A condition of decreased pH in tissues and cells tending to cause *acidemia* (*opposite:* alkalosis). *Metabolic acidosis* is manifested primarily by a decrease in blood bicarbonate, *respiratory acidosis* primarily by an increase in blood partial pressure of carbon dioxide.

active In a mode characterized by muscle contraction (*opposite:* passive).

adenosine 5′-triphosphate (*ATP*) The principal carrier of chemical energy in the cells of animals.

adductor Any of various muscles (e.g., one of several that close the laryngeal cavity) that, upon shortening, move structures together (*opposite:* abductor).

adrenaline A hormone (also called *epinephrine*) secreted by chromaffin cells of the adrenal gland and by some neurons.

adrenergic fiber One of the nerve fibers secreting epinephrine (or adrenaline) as a synaptic transmitter.

afferent Sending information from a peripheral organ toward the central nervous system (*opposite:* efferent).

airflow A change in volume of air per unit of time.

air hunger *See* dyspnea.

airway opening In general, the mouth (at the level of the lips), the nose (at the level of the nostrils), or both.

airway spasm Contraction of the smooth muscle of the airways, induced by many conditions, including allergic reactions and cold air, and resulting in an increase in airway resistance.

alkalemia *See* acidemia.

alkalosis *See* acidosis. *Metabolic* alkalosis is manifested primarily by an increase in blood bicarbonate, *respiratory* alkalosis primarily by a decrease in blood partial pressure of carbon dioxide.

allantois The fluid-filled compartment adjacent to the umbilical cord, its tissue contributing to the formation of the placenta (*see* chorioallantoic).

allometric analysis Inter-animal comparison by body weight (*see* allometry).

allometry Comparison of functional or anatomical parameters as functions of body weight.

altricial Said of an animal species born at an early stage of development, therefore less mature at birth (*opposite:* precocial).

alveolar sac An anatomical structure from which alveoli are derived by the process of alveolar septation.

alveolar septation The division of alveolar sacs in the process of alveolar formation.

alveolar stability The ability of alveoli to retain air, despite their differences in size and therefore in the pressure generated at the air-liquid interface (*see* Young-Laplace relationship).

alveolar ventilation The quantity of gas entering or leaving the gas-exchange regions of the lungs per unit time.

anabolism *See* metabolism.

anaerobic Not utilizing oxygen; said of certain metabolic processes (*see* glycolysis).

anatomical dead space The total volume of those air-conductive passages that do not have the anatomical structures necessary for gas exchange.

anion An ion having a negative charge (*opposite:* cation).

anoxia The condition characterized by the absence of oxygen.

antagonistic contraction A contraction of muscles that yields opposing mechanical effects (e.g., the simultaneous contraction of *abductors* and *adductors*).

antioxidant A molecule or compound having the property of reducing the oxidative power of other molecules or compounds.

apnea A cessation of breathing movements in the end-expiratory position, characterized by muscle relaxation.

apneusis A cessation of breathing movements in the end-inspiratory position, brought on by a sustained contraction of the inspiratory muscles.

apposition area The lower region of the rib cage in which the diaphragm is apposed to the rib cage without interposed lung.

Arrhenius factor *See* Q_{10} factor.

artificial ventilation Ventilation of the lungs achieved by external means, usually with a mechanical ventilator.

asphyxia A combination of reduced oxygen and increased carbon dioxide, tending to cause unconsciousness.

atelectasis Alveolar collapse, brought on by a decrease in the elastic pressure of a region of the lungs (for example, in the case of *pneumothorax*).

atresia Obstruction of a normally opened pathway because of developmental defects.

β-agonist A molecule stimulating a beta-receptor of the sympathetic system.

baroafferent Sending neural information from baroreceptors to the central nervous system.

baroreceptor One of the receptors sensing blood pressure.

basal tone The minimal tone of a muscle.

Biot's respiration Breathing by alternating sequences of gasps and periods of apnea.

blood perfusion See perfusion.

body box See plethysmograph.

body surface pressure The pressure applied outside the body, usually equal to atmospheric pressure.

bradycardia A decrease in heart rate (*opposite:* tachycardia).

bradypnea A decrease in the rate of breathing (*opposite:* tachypnea).

breath holding Voluntary cessation of breathing.

bronchoconstriction A reduction in bronchial diameter because of smooth-muscle contraction (*opposite:* bronchodilatation).

bronchodilatation See bronchoconstriction.

bronchomotor tone The level of activity of the airway smooth muscle.

brown adipose tissue (*BAT*) A body tissue having the specific task of producing heat.

calorie A unit of heat; one calorie is the amount of heat needed to raise the temperature of 1 gram of water by 1°C.

carbon dioxide stores The total amount of carbon dioxide, in all its forms, stored in the body (dissolved and gaseous carbon dioxide, carbonic acid, carbonate, bicarbonate, carbamino compounds, carboxy-hemoglobin).

cardiac output See perfusion.

catabolism See metabolism.

cation An ion having a positive charge (*opposite:* anion).

cerebrospinal fluid (*CSF*) A fluid surrounding the brain and spinal cord.

C-fiber Any of various small-diameter, unmyelinated neural fibers having low conduction velocity.

chemodenervation Sectioning of the nerves of the peripheral chemoreceptors (the aortic depressor nerve and the carotid sinus nerve).

chemoreceptor A receptor sensing chemical stimuli.

chest wall Functionally, all the anatomical structures, outside the lungs, that move as the result of the breathing act.

Cheyne-Stokes respiration Breathing characterized by cycles of gradually increasing and decreasing tidal volumes, usually separated by brief periods of apnea.

cholinergic Said of nerve fibers that secrete acetylcholine as a synaptic transmitter.

chorioallantoic The placental layer formed by the fusion of the *chorion* and the *allantois*.

chorion The outermost layer of the early embryonic structures, contributing to the formation of the placenta (*see* chorioallantoic).

circadian oscillation A pattern having a period of approximately 24 hours.

collagen A fibrous protein rich in glycine and proline; a major component of the extracellular matrix and connective tissues.

colloid osmotic pressure *See* oncotic pressure.

compliance An index of the elastic characteristics of an organ, specifically the change in volume per unitary change in pressure (i.e., 1/elastance).

conductance Flow generated by a unitary change in pressure (i.e., 1/resistance).

convection Movement of molecules resulting from the movement of the fluid of which they are part.

corticoid Any of various hormones secreted by the adrenal cortex.

cyanosis Blueness (of a blue-purple shade) of the skin caused by excessive levels of poorly oxygenated hemoglobin.

cyanotic heart disease A cardiac disease exhibiting right-to-left shunt of blood, causing *cyanosis*.

cytotoxic Toxic to cellular structures.

dead space The total volume of the airways not participating in gas exchange (*see* anatomical dead space and physiological dead space).

decerebration Functional or anatomical absence of the cerebrum.

deflation Emptying of the lungs (*opposite:* inflation).

denervation Elimination of the innervation of one or more organs, by surgical or pharmacological means.

density Mass per unitary volume.

diapause Delayed implantation of the fertilized egg.

diaphragm The main inspiratory muscle in mammals, separating the abdominal and thoracic cavities.

diffusing capacity An index of the capacity of the lungs to exchange gases.

diffusion Movement of molecules as the result of a pressure (or activity) gradient.

dynamic As regards the various modes of operation of the respiratory system, characterized by the presence of airflow (*opposite:* static).

dynamic compliance Lung compliance measured during continuous breathing.

dysplasia An organ or tissue deformity, formed usually on a genetic basis.

dyspnea Difficult or labored breathing; an uncomfortable awareness of the act of breathing (also *air hunger*).

ectotherm An animal able to increase its body temperature only via external mechanisms (*opposite:* endotherm).

edema Excess fluid in bodily tissues.

efferent Sending information from the central nervous system toward a peripheral organ (*opposite:* afferent).

elastance An index of the elastic properties of the lungs, specifically a change in pressure per unitary change in volume (i.e., 1/compliance).

elastic loading An added elastance confronted during breathing.

elastic pressure A component of the total pressure needed to inflate the lungs, dependent on the degree of distention of the lung and its elastic properties, specifically the pressure determined by lung volume and compliance (also *recoil pressure*).

electrochemical gradient A force causing an ion to move across a membrane because of a difference both in concentration and in electrical charge on the two sides of the membrane.

electromyogram A record of the electrical activity of a muscle.

electron transport chain A series of multienzyme protein complexes in the inner mitochondrial membrane serving for the transport of high-energy electrons, before combining with oxygen atoms to form water (also *electron transfer chain*).

endotherm An animal able to increase its body temperature via internal mechanisms (*opposite:* ectotherm).

endotracheal tube A tube placed inside the trachea either through the mouth or through a hole in the trachea (*tracheotomy*), usually for the purpose of artificially ventilating the lungs.

energy The capacity to perform work (which see).

enzyme A protein, produced by living cells, that catalyzes (accelerates) specific chemical reactions.

epinephrine See adrenaline.

erythrocyte A hemoglobin-containing blood cell serving for the transport of oxygen and carbon dioxide (also *red cell*).

erythropoiesis The collective processes of red cell production.

estivation A state of prolonged inactivity of some animals during the summer months, as during seasonal droughts (cf. *hibernation, torpor*).

eupnea Normal spontaneous breathing.

eutherian A mammal having a chorioallantoic placenta; most mammals are eutherians.

extrafusal fiber One of the outer components of the skeletal muscle fibers forming a muscle spindle (which see).

fetal breathing movements (*FBM*) The spontaneous breathing-like movements of the fetus.

fibroblast A common cell type, found in the connective tissue, that produces collagen (which see).

flow-resistive pressure A component of the total pressure needed to inflate the lungs at a given rate, specifically the pressure dependent on lung resistance and the rate of change in lung volume (also *resistive pressure*).

fluid A state of matter, comprising both gases and liquids, characterized by a weak intermolecular connection.

force The product of mass and acceleration.

fossorial Living permanently in a burrow, as for example moles.

fractional concentration The percent of a gas within a gas mixture.

free radicals See radicals.

functional residual capacity (FRC) The amount of air remaining in the lungs at end-expiration.

γ-loop Any of the neural fibers participating in the reflex control of the muscle spindle (also *gamma efferent fibers*).

gasp A deep breath with high inspiratory flow, resulting from massive and rapid contraction of the inspiratory muscles.

gene upregulation Increased expression of a gene (*opposite:* gene downregulation).

glycolysis An anaerobic process by which glucose is phosphorylated and subsequently split to form two molecules of pyruvic acid, in the process of forming ATP.

Head's paradoxical reflex A vagal reflex, described by H. Head in the late nineteenth century, consisting of a brief facilitation of inspiratory activity that occurs when the lungs are suddenly inflated.

hematocrit The volume of blood cells, expressed as a percent of total blood volume.

hemoglobin The oxygen-binding protein in the erythrocytes (red cells).

Hering-Breuer reflexes A response of the respiratory muscles to change in lung volume: the afferent component of these reflexes is through the vagus nerves.

hibernation The state of inactivity of some animals during the winter months, usually of variable duration, often characterized by a marked drop in metabolic rate and body temperature (cf. estivation, torpor).

histamine A small molecule, derived from the amino acid histidine, producing, among its numerous effects, a contraction of the airway smooth muscle.

homeostasis The maintenance of constant conditions in the internal environment of an organism.

homeotherm An animal, typically a bird or mammal, having a tendency to maintain a constant body temperature, irrespective of changes in ambient temperature (see *homeothermy; opposite:* poikilotherm).

homeothermy The tendency of an animal to maintain a constant body temperature (*opposite:* poikilothermy).

humoral stimulus A stimulus dependent on blood transport.

hyperbaria A condition in which barometric pressure exceeds one atmosphere (*opposite:* hypobaria).

hypercapnia An increase in arterial partial pressure of carbon dioxide (*opposite:* hypocapnia).

hyperoxia *See* hypoxia.

hyperplasia An increase in numbers of cells (*opposite:* hypoplasia).

hyperpnea Increased pulmonary ventilation (*opposite:* hypopnea).

hypertension An increase in pressure, the term often applied to blood pressure (*opposite:* hypotension).

hyperthermia A condition in which body temperature is above normal (*opposite:* hypothermia).

hypertrophy An increase in cellular size (*opposite:* hypotrophy).

hyperventilation An increase in pulmonary ventilation (strictly, *alveolar ventilation*) relative to metabolic rate, irrespective of the absolute value of ventilation (*opposite:* hypoventilation).

hypobaria *See* hyperbaria.

hypocapnia *See* hypercapnia.

hypoplasia *See* hyperplasia.

hypopnea *See* hyperpnea.

hypotension *See* hypertension.

hypothermia *See* hyperthermia.

hypotrophy *See* hypertrophy.

hypoventilation *See* hyperventilation.

hypoxemia A decrease in oxygen, whether of pressure or content, in the arterial blood.

hypoxia A reduced level of oxygen, whether of concentration or pressure (*opposite:* hyperoxia); *tissue hypoxia* is a decrease in tissue oxygenation.

hysteresis The influence of the previous condition (or history) of an organ or organism on its subsequent response to a stimulus, a phenomenon lying at the basis of the difference between the lengthening and shortening phases of the force-length relationship, or between the inflation and deflation limbs of the pressure-volume relationship.

impedance Change in pressure per unitary flow (also *mechanical impedance*).

inflation The filling of the lungs (*opposite:* deflation).

inflation reflex Any of the reflex ventilatory responses to lung inflation.

interstitial edema Excess fluid in the interstitial space.

intrafusal fiber Any of a group of skeletal muscle fibers forming the inner components of a muscle spindle (which see).

intubation The positioning of a tube in the trachea, usually for the purpose of artificial ventilation.

in vitro Occurring in an artificial condition, outside the organism.

in vivo Occurring in the intact organism.

irritant receptor A rapidly adapting pulmonary receptor responding to a variety of mechanical and/or chemical stimuli.

isocapnia Constant pressure of carbon dioxide.

isoform Any of multiple forms of the same protein having small differences in amino acid sequence.

isometric force Force generated by a muscle without shortening.

joey The young of a marsupial.

juxtacapillary Of any structure lying in close proximity to a pulmonary capillary, usually referred to a group of nerve endings.

larynx An organ connected to the central portion of the trachea, and containing the vocal cords.

load compensation Any of various mechanisms tending to protect pulmonary ventilation in the event a mechanical load is imposed on the respiratory system.

loading See elastic loading, resistive loading.

lower airways All the air-conductive passages from the trachea to the alveoli.

lower critical temperature The lower temperature value of thermoneutrality (which see), below which (in homeotherms) thermogenic mechanisms increase.

lung recoil A term defining the tendency of the lung to collapse, equivalent to the elastic pressure generated by the lung at any volume.

mammal Any of various vertebrate animals characterized by the production of milk by mammary glands, the presence of hair for body covering, and a muscle (the *diaphragm*) separating the abdominal and thoracic cavities.

marsupial A pouched mammal of the subclass Metatheria (e.g., opossum, wallaby, dunnart).

mechanical impedance See impedance.

mechanoreceptor A receptor sensitive to mechanical stimuli, usually pressure or tension.

membrane oxygenation Blood oxygenation effected (by a machine) by diffusion through a membrane.

metabolic rate Average speed of the metabolic processes (*see* metabolism).

metabolism All the chemical transformations occurring within a living organism: *anabolism* is the range of synthetic (usually energy-requiring) transformations by which large molecules are made from smaller ones; *catabolism* is the range of destructive (usually energy-yielding) transformations by which larger molecules are broken down into smaller ones.

mitochondrion (pl. *mitochondria*) Any of various cellular organelles, each about the size of a bacterium, serving to transform potential energy into ATP, via oxidative phosphorylation.

mixed venous blood A pool of venous blood from different body regions, e.g., the venous blood in the right atrium.

mole A quantity, in grams, of a substance equal to its molecular weight.

muscle spindle *See* spindle.

myelin A membrane of fatty material wrapping the axon of a neural fiber and increasing its conduction velocity.

myoglobin An oxygen-binding protein, typical of muscles.

myosin One of the main molecules responsible for muscle contraction.

myosin heavy chains The two high-molecular-weight polypeptide chains forming the myosin molecule.

negative end-expiratory pressure (NEEP) A subatmospheric air pressure applied to the airway opening (cf. *positive end-expiratory pressure*).

nonshivering thermogenesis Production of heat without shivering, the term often restricted to indicate heat production by organs specialized in heat production, such as the brown adipose tissue.

normobaria Air pressure equal to that generated by one atmosphere at sea level.

normocapnia The normal level of carbon dioxide.

normoxia The normal level of oxygen.

oncotic pressure Osmotic pressure generated by molecules of high molecular weight, e.g., proteins, in which case it is also called *colloid osmotic pressure*.

ontogenesis *See* ontogeny.

ontogeny The process of growth and development of an organism; also *ontogenesis*.

orthopnea Dyspnea experienced only in the recumbent, usually supine, position, and relieved by sitting or standing.

osmolality The number of osmoles per kilogram of solvent (*see* osmole).

osmolarity The number of osmoles per liter of solution (*see* osmole).

osmole The molecular weight of a substance in grams divided by the number of freely moving particles that each molecule liberates in solution.

osmosis The movement of water molecules across a semipermeable membrane because of a difference in concentration of the solutes on the two sides of the membrane.

osmotic pressure The pressure generated by osmosis.

oxidation Loss of one or more electrons from an atom, typically during the addition of oxygen to a molecule (*opposite:* reduction).

oxidative phosphorylation The formation of high-energy phosphate compounds during the sequence of oxidative reactions at the level of the electron transport chain; *see also* phosphorylation.

oxygen consumption The amount of oxygen used by living matter per unit time.

oxygen delivery The product of cardiac output and arterial oxygen content.

oxygen uptake The amount of oxygen removed from the environment by a living organism, equal in the steady state to oxygen consumption.

panting A very rapid and shallow breathing pattern, like that adopted by some mammals in some conditions to dissipate heat (*thermal panting*).

paradoxical reflex *See* Head's paradoxical reflex.

paradoxical sleep *See* rapid-eye-movement sleep.

passive In a mode characterized by muscle relaxation (*opposite:* active).

passive inflation Lung inflation not determined by muscle contraction.

perfusion The amount of blood circulating per unit time, termed *cardiac output* when referred to the whole body.

periodic breathing An abnormal breathing pattern characterized by a series of cycles separated by pauses.

peripheral chemoreceptors The receptors, placed outside the central nervous system, that sense changes in chemical stimuli; often the term refers to the carotid and aortic bodies.

pH The negative decimal log of the concentration of hydrogen ions.

phosphorylation A reaction in which a phosphate group is incorporated into a molecule; *see also* oxidative phosphorylation.

phrenicotomy Sectioning of the phrenic nerves.

phylogeny The evolutionary pathway of an organism or group of organisms.

physiological dead space The total volume of the airways not participating in gas exchange, including all those airways that do not have alveoli (i.e., the *anatomical dead space*) plus all those airways that for functional reasons (for example, absence of blood perfusion) do not gas exchange (*see* dead space).

plethysmograph An airtight chamber wherein a subject sits while breathing through a mouthpiece and tubing connected to the outside (also *body plethysmograph,* or *body box*); many variations have been made, including some for use on animals, for the purpose of monitoring functional residual capacity and spirometry, and for the calculation of other respiratory parameters.

pleural pressure Pressure measured within the pleural space.

pneumotachograph A mechanical device for measuring airflow.

pneumothorax Collapse of the lung, or a portion of it, brought on by a presence of gas (usually air) in the pleural space.

poikilotherm An animal having more or less constantly varying body temperature, the temperature depending on environmental temperature (*see* poikilothermy; *opposite:* homeotherm).

poikilothermy A condition, common in all animals but birds and mammals, characterized by frequent and significant variation in body temperature (*see* poikilotherm; *opposite:* homeothermy).

polypnea An irregular, usually rapid, breathing pattern.

positive end-expiratory pressure (PEEP) A greater-than-atmospheric air pressure applied to the airway opening (cf. *negative end-expiratory pressure*).

pouch young The neonate of a marsupial.

power Energy produced over a period of time.

precocial Said of an animal species born at a late stage of development, therefore more mature at birth (*opposite:* altricial).

prematurity Birth at an earlier-than-normal gestational age.

pressure Force applied to a unitary surface area.

proprioceptive *See* somatic stimuli.

prostacyclin Any of various molecules derived from *prostaglandins.*

prostaglandin Any of various macromolecules, produced by most organs of the body, that are characterized by 20-carbon unsaturated fatty acids with a cyclopentane ring.

pulmonary Related to, functioning like, or associated with the lungs.

pulmonary surfactant The mixture of lipids and proteins, produced by specialized cells of the lung alveoli, that serves to lower surface tension.

pyrogen A substance inducing fever.

Q_{10} *factor* A numerical index of the changes of a variable as a function of temperature, specifically the Q_{10} (Arrhenius) factor expressing the change in reaction velocity for a 10°C change in temperature.

quadriplegic *See* tetraplegic.

radicals Molecules that are highly reactive because of having incomplete numbers of electrons.

rapid-eye-movement sleep (REM sleep) A phase of sleep characterized by rapid movements of the eyes, marked electrical activity of the brain, major depression of muscle tone, and irregular heart and breathing rates, typically associated with dreaming (also *paradoxical sleep*).

rapidly adapting receptor *See* irritant receptor.

rebreathing The inhaling of expired air.

receptor A sensory nervous terminal translating a physical stimulus into neural activity (cf. slowly adapting receptor, chemoreceptor, spindle, thermoreceptor).

recoil pressure *See* elastic pressure.

red cell *See* erythrocyte.

reduction *See* oxidation.

regulated conformism Controlled changes in metabolic rate, effected via regulatory mechanisms, whenever the external conditions vary.

REM sleep *See* rapid-eye-movement sleep.

residual mass index The exponent resulting from a mathematical operation between allometric equations.

resistance An index of the difficulty encountered by a fluid during motion, specifically the change in pressure required to generate a unitary change in flow (i.e., 1/conductance).

resistive loading Breathing through an added resistance.

resistive pressure *See* flow-resistive pressure.

respiratory controller The set of centers and neural networks responsible for the central generation and control of the respiratory pattern.

respiratory exchange ratio The ratio between carbon dioxide production and oxygen consumption, both gases measured at the airway opening, corresponding in conditions of perfect stability to the *respiratory quotient.*

respiratory quotient The ratio between carbon dioxide production and oxygen consumption of cells, tissues, organs, or the whole organism (*see* respiratory exchange ratio).

respiratory pump The set of anatomical structures and muscles responsible for the convection of air in and out the lungs.

respiratory system The combination of lungs and chest wall, taken collectively.

resting volume The amount of air present in the lungs when no forces are applied to the respiratory system.

Reynolds' number A nondimensional quantity predicting the nature of a gas flow in a tube, dependent on the velocity, density, and viscosity of the gas and the diameter of the tube.

rhythmogenesis The set of mechanisms responsible for originating rhythmic respiratory activity.

ruminant Any of various mammals that chew the cud, which is food brought back into the mouth from the first stomach, to be chewed again.

semifossorial Living intermittently in a burrow, as for example gophers and wombats.

septation *See* alveolar septation.

sham Mock, often used to indicate a control intervention; in a sham denervation, for example, the control animal is anesthetized and operated on as the experimental one would be, but the nerve is left intact.

sham operation *See* sham.

shivering thermogenesis Production of heat by more or less sustained muscle contraction.

sigh A deep, and usually audible, breath.

slowly adapting receptor A receptor responding mainly to mechanical stimuli, such as changes in lung volume or pressure, with little adaptation to a sustained stimulus (also *stretch receptor* or *slowly adapting stretch receptor*).

slow-oxidative fiber Any fiber of the skeletal muscles having a rather slow response and high oxidative capacity, thus suited for long contractions (also *Type I fiber,* or *red muscle*).

slow-twitch fiber Any fiber of the skeletal muscles having a fast response, high glycolytic capacity, and rather low oxidative capacity, thus suited for fine and brief muscle contractions (also *Type II fiber,* or *white muscle*).

sodium pump A transmembrane carrier protein in the cellular membrane, using ATP to pump sodium out of the cell and potassium into it.

somatic stimuli Sensory information from various organs of the body, as opposed to the special senses (e.g., vision, hearing, smell, taste, equilibrium), including proprioceptive (from joints and muscles), visceral, and skin stimuli.

spindle A receptor, within a skeletal muscle, that senses length and its rate of change.

spirogram A record of changes in lung volume as a function of time.

spirometer An instrument, consisting of a drum inverted over a chamber of water, serving to measure changes in lung volume.

spirometry The measuring of changes in lung volume as a function of time.

spontaneous breathing The normal act of breathing, as opposed to artificial ventilation or artificial breathing, in either of which lung ventilation is achieved by external means.

Starling equilibrium About a century ago E. H. Starling indicated that the amount of fluid filtering out from some capillaries is almost identical to the amount absorbed by others; the forces chiefly responsible for this balance are the oncotic pressures of proteins in both plasma and interstitium, the capillary pressures, and the interstitial pressures.

static As regards the various modes of operation of the respiratory system, characterized by the absence of airflow (*opposite:* dynamic).

stretch receptor See slowly adapting receptor.

superfusion The bathing of an in vitro preparation with an artificial liquid of desired composition.

surface force The force generated at the interface (meniscus) between two media, such as the air and water in the alveoli.

surface tension See *surface force.*

surfactant See pulmonary surfactant.

tachycardia See bradycardia.

tachypnea See bradypnea.

tension Force applied over a linear dimension.

tetanic force Muscle force that occurs when successive contractions occur at such a high rate that they fuse together.

tetraplegic A patient who, because of a lesion in the cervical portion of the spinal cord, has lost the voluntary control of all the muscles from the neck down, including the muscles of the four limbs (also *quadriplegic*).

thermal panting See panting.

thermodispersion Heat dissipation.

thermogenesis Heat production (also *thermoproduction*).

thermogenin A protein characteristic of the brown adipose tissue (*see* uncoupling protein).

thermoneutrality The range in ambient temperature, under normoxia, over which body temperature is maintained at the normal level with minimal consumption of oxygen.

thermoproduction *See* thermogenesis.

thermoreceptor A receptor sensing changes in temperature.

thermotaxis Movement in response to a temperature stimulus.

tidal volume The volume of air inhaled with each breath.

tissue hypoxia *See* hypoxia.

torpor A state of inactivity in an animal (occurring during some hours of the day) that is often characterized by a drop in both metabolic rate and body temperature (cf. *estivation, hibernation*).

total lung capacity The total volume of air in the lungs at maximal inspiration.

tracheobronchial tree All the airways from the trachea toward the alveoli.

tracheostomy Surgical connection of the trachea to the skin of the neck.

tracheotomy *See* endotracheal tube.

transmural pressure A pressure across a structure or an organ, by convention the difference between the inside and outside pressures (*see* transpulmonary pressure or transrespiratory system pressure).

transpulmonary pressure Pressure across the lungs, i.e., the pressure difference between the airways and the pleural space.

transrespiratory system pressure Pressure across the respiratory system, i.e., the pressure difference between the airways and the body surface.

trigeminal area The anatomical region innervated by the trigeminal nerve (or V-cranial nerve).

uncoupling protein A protein that uncouples oxidative metabolism from phosphorylation at the level of the inner mitochondrial membrane.

upper airways All the air-conductive passages from the nose and oral cavity to the larynx.

vagal feedback The afferent information carried from the lungs by the vagal fibers.

vagal fibers The axons of the vagus nerves (or X-cranial nerve).

vagal reflexes *See* Hering-Breuer reflexes.

vagotomy Sectioning of the vagus nerve.

vascular expansion An increase in the total vascular bed.

vascular resistance The flow resistance of a vascular district.

vasoactive Acting on the smooth muscle of a vessel, usually a small artery.

vasoconstriction The narrowing of a vessel or group of vessels, usually small arteries (*opposite:* vasodilatation).

vasodilatation *See* vasoconstriction.

ventilation The amount of air entering or leaving the lungs per unit time.

ventilator A machine capable of expanding the lungs for the purpose of gas exchange: some are positive-pressure ventilators, connected to the airways; others are negative-pressure ventilators, which inflate the lungs by decreasing the pressure around the chest.

ventilatory efficiency *See* ventilatory equivalent.

ventilatory equivalent Ventilation per unitary change in oxygen consumption ($\dot{V}E/\dot{V}O_2$); the reciprocal ($\dot{V}O_2/\dot{V}E$) reflects the amount of oxygen used per unit of ventilation (*ventilatory efficiency for oxygen*).

vertebrate Any of various animals whose nerve cord is enclosed in a backbone of bony segments called *vertebrae;* the main classes of vertebrates are fishes, amphibians, reptiles, birds, and mammals.

vital capacity The maximal amount of air that can be exhaled after a full inspiration.

work Energy produced by a force acting through a distance (F · d); hence the product of pressure (F/surface area) and volume (d · surface area) also has the physical dimensions of work.

Young-Laplace relationship A mathematical relationship among pressure, surface tension, and the radius of curvature of an interface (meniscus) between two media; in its most general form, tension = pressure · radius.

References Cited

Adams, F. H., T. Fujiwara, and G. Rowshan. The nature and origin of the fluid in the fetal lamb lung. *J. Pediatr.* 63:881–888, 1963.

Adams, F. H., M. Yanagisawa, D. Kuzela, and H. Martinek. The disappearance of fetal lung fluid following birth. *J. Pediatr.* 78:837–843, 1971.

Adams, R., F. B. Garry, B. M. Aldridge, M. D. Holland, and K. G. Odde. Physiologic differences between twin and single born beef calves in the first two days of life. *Cornell Vet.* 83:13–29, 1993.

Adamson, S. L., B. S. Richardson, and J. Homan. Initiation of pulmonary gas exchange by fetal sheep in utero. *J. Appl. Physiol.* 62:989–998, 1987.

Adamson, S. L., I. M. Kuipers, and D. M. Olson. Umbilical cord occlusion stimulates breathing independent of blood gases and pH. *J. Appl. Physiol.* 70:1796–1809, 1991.

Adamsons, K., E. Blumberg, and I. Joelsson. The effect of ambient temperature upon post-natal changes in oxygen consumption of the guinea pig. *J. Physiol. (London)* 202:261–269, 1969.

Adler, S. M., B. T. Thach, and I. D. Frantz III. Maturational changes of effective elastance in the first 10 days of life. *J. Appl. Physiol.* 40:539–542, 1976.

Adolph, E. F. Regulations during survival without oxygen in infant mammals. *Respir. Physiol.* 7:356–368, 1969.

———. Physiological adaptations in hypoxia in infant mammals. *Am. Zool.* 13:469–473, 1973.

Adolph, E. F., and P. A. Hoy. Ventilation of lungs in infant and adult rats and its responses to hypoxia. *J. Appl. Physiol.* 15:1075–1086, 1960.

Adrian, G. M., G. S. Dawes, M. M. L. Prichard, S. R. M. Reynolds, and D. G. Wyatt. The effect of ventilation of the foetal lungs upon the pulmonary circulation. *J. Physiol. (London)* 118:12–22, 1952.

Agata, Y., S. Hiraishi, H. Misawa, J. H. Han, K. Oguchi, Y. Horiguchi, N. Fujino, N. Takeda, and J. F. Padbury. Hemodynamic adaptations at birth and neonates delivered vaginally and by cesarean section. *Biol. Neonate* 68:404–411, 1995.

Agostoni, E. Volume-pressure relationships of the thorax and lung in the newborn. *J. Appl. Physiol.* 14:909–913, 1959.

Agostoni, E., and E. D'Angelo. Comparative features of the transpulmonary pressure. *Respir. Physiol.* 11:76–83, 1970/71.

———. Pleural liquid pressure. *J. Appl. Physiol.* 71:393–403, 1991.

Agostoni, E., A. Taglietti, A. F. Agostoni, and I. Setnikar. Mechanical aspects of the first breath. *J. Appl. Physiol.* 13:344–348, 1958.

Aherne, W., and M. J. R. Dawkins. The removal of fluid from the pulmonary airways after birth in the rabbit, and the effect on this of prematurity and prenatal hypoxia. *Biol. Neonate* 7:214–229, 1964.

Aherne, W., D. R. Ayyar, P. A. Clarke, and J. N. Walton. Muscle fibre size in normal infants, children and adolescents. An autopsy study. *J. Neurol. Sci.* 14:171–182, 1971.

Aiton, N. R., G. F. Fox, J. Alexander, D. M. Ingram, and A. D. Milner. The influence of sleeping position on functional residual capacity and effective pulmonary blood flow in healthy neonates. *Pediatr. Pulmonol.* 22:342–347, 1996.

Alberts, J. R. Huddling by rat pups: multisensory control of contact behavior. *J. Comp. Physiol. Psychol.* 92:220–230, 1978a.

———. Huddling by rat pups: group behavioral mechanisms of temperature regulation and energy conservation. *J. Comp. Physiol. Psychol.* 92:231–245, 1978b.

Alcorn, D., T. M. Adamson, T. F. Lambert, J. E. Maloney, B. C. Ritchie, and P. M. Robinson. Morphological effects of chronic tracheal ligation and drainage in the fetal lamb lung. *J. Anat.* 123:649–660, 1977.

Alcorn, D., T. M. Adamson, J. E. Maloney, and P. M. Robinson. Morphological effects of chronic bilateral phrenectomy or vagotomy in the fetal lamb lung. *J. Anat.* 130:683–695, 1980.

Alexander, G. Body temperature control in mammalian young. *Br. Med. Bull.* 31:62–68, 1975.

Alexander, G., and D. Williams. Summit metabolism and cardiovascular function in young lambs during hyperoxia and hypoxia. *J. Physiol. (London)* 208:85–97, 1970.

Allen, J. L., J. S. Greenspan, K. S. Deoras, E. Keklikian, M. R. Wolfson, and T. Shaffer. Interaction between chest wall motion and lung mechanics in normal infants and infants with bronchopulmonary dysplasia. *Pediatr. Pulmonol.* 11:37–43, 1991.

Allen, K. M., and S. G. Haworth. Impaired adaptation of pulmonary circulation to extrauterine life in newborn pigs exposed to hypoxia: an ultrastructural study. *J. Pathol.* 150:250–212, 1986.

Al-Shway, S. F., and J. P. Mortola. Respiratory effects of airflow through the upper airways in newborn kittens and puppies. *J. Appl. Physiol.* 53:805–814, 1982.

Amy, R. W. M., D. Bowes, P. H. Burri, J. Haines, and W. M. Thurlbeck. Postnatal growth of the mouse lung. *J. Anat.* 124:131–151, 1977.

Anderson, G. C. Current knowledge about skin-to-skin (kangaroo) care for preterm infants. *J. Perinat.* 11:216–226, 1991.

Anderson, J. W., and J. T. Fisher. Capsaicin-induced reflex bronchoconstriction in the newborn. *Respir. Physiol.* 93:13–27, 1993.

Anderson, D. F., C. M. Parks, and J. J. Faber. Fetal O_2 consumption in sheep during controlled long-term reductions in umbilical blood flow. *Am. J. Physiol.* 250:H1037–H1042, 1986.

Andersson, D., G. Gennser, and P. Johnson. Phase characteristics of breathing movements in healthy newborns. *J. Develop. Physiol.* 5:289–298, 1983.

Andrews, D. C., L. Fedorko, P. Johnson, and J. C. Wollner. The maturation of the ambient thermal stimulus to breathing during sleep in lambs. In: *The Physiological Development of the Fetus and the Newborn*, edited by C. T. Jones and P. W. Nathanielsz. London: Academic Press, pp. 821–825, 1985.

Andrews, D. C., M. E. Symonds, and P. Johnson. Thermoregulation and the control of breathing during non-REM sleep in the developing lamb. *J. Develop. Physiol.* 16:27–36, 1991.

Arieli, R. Can the rat detect hypoxia in inspired air? *Respir. Physiol.* 79:243–254, 1990.

Arora, N. S., and D. F. Rochester. Effect of body weight and muscularity on human diaphragm muscle mass, thickness, and area. *J. Appl. Physiol.* 52:64–70, 1982.

Asher, M. I., A. L. Coates, J. M. Collinge, and J. Milic-Emili. Measurement of pleural pressure in neonates. *J. Appl. Physiol.* 52:491–494, 1982.

Assali, N. S., J. A. Morris, R. W. Smith, and W. A. Manson. Studies on ductus arteriosus circulation. *Circ. Res.* 13:478–489, 1963.

Avery, M. E., and C. D. Cook. Volume-pressure relationships of lungs and thorax in fetal, newborn, and adult goats. *J. Appl. Physiol.* 16:1034–1038, 1961.

Avery, M. E., and J. Mead. Surface properties in relation to atelectasis and hyaline membrane disease. *Am. J. Dis. Child.* 97:517–523, 1959.

Avery, M. E., N. R. Frank, and I. Gribetz. The inflation force produced by pulmonary vascular distention in excised lungs. The possible relation of this force to that needed to inflate the lungs at birth. *J. Clin. Invest.* 38:456–462, 1959.

Avery, M. E., O. B. Gatewood, and G. Brumley. Transient tachypnea of newborn. Possible delayed resorption of fluid at birth. *Am. J. Dis. Child.* 111:380–385, 1966.

Baier, R. J., S. U. Hasan, D. B. Cates, D. Hooper, B. Nowaczyk, and H. Rigatto.

Effects of various concentrations of O_2 and umbilical cord occlusion on fetal breathing and behavior. *J. Appl. Physiol.* 68:1597–1604, 1990.

Bairam, A., B. Hannhart, C. Choné, and F. Marchal. Effects of dopamine on the carotid chemosensory response to hypoxia in newborn kittens. *Respir. Physiol.* 94:297–307, 1993.

Bamford, O. S., L. M. Sterni, M. J. Wasicko, M. H. Montrose, and J. L. Carroll. Postnatal maturation of carotid body and type I cell chemoreception in the rat. *Am. J. Physiol.* 276:L875–L884, 1999.

Banovcin, P., J. Seidenberg, and H. Von der Hardt. Assessment of tidal breathing patterns for monitoring of bronchial obstruction in infants. *Pediatr. Res.* 38:218–220, 1995.

Barron, D. H. A history of fetal respiration: from Harvey's question (1651) to Zweifel's answer (1876). In: *Fetal and Newborn Cardiovascular Physiology,* edited by L. D. Longo and D. D. Reneau. New York: Garland, vol. 1, ch. 1, pp. 1–32, 1976.

Bartels, H., R. Bartels, A. M. Rathschlag-Schaefer, H. Röbbel, and S. Lüdders. Acclimatization of newborn rats and guinea pigs to 3000 and 5000 m simulated altitudes. *Respir. Physiol.* 36:375–389, 1979.

Bartlett, D., Jr. Upper airway motor systems. In *Handbook of Physiology. The Respiratory System,* vol. 2, *Control of Breathing,* edited by N. S. Cherniack and J. G. Widdicombe. Bethesda, Md.: Am. Physiol. Soc., part 1, ch. 8, pp. 223–245, 1986.

———. Respiratory functions of the larynx. *Physiol. Rev.* 69:33–57, 1989.

Bartlett, D., Jr., and J. G. Areson. Quantitative lung morphology in newborn mammals. *Respir. Physiol.* 29:193–200, 1977.

Bartlett, D., Jr., and G. F. Birchard. Effects of hypoxia on lung volume in the garter snake. *Respir. Physiol.* 53:63–70, 1983.

Bartlett, D., Jr., and J. E. Remmers. Effects of high altitude exposure on the lungs of young rats. *Respir. Physiol.* 13:116–125, 1971.

Bartlett, D., Jr., J. P. Mortola, and E. J. Doll. Respiratory mechanics and control of the ventilatory cycle in the garter snake. *Respir. Physiol.* 64:13–27, 1986.

Battaglia, F. C., M. McGaughey, E. L. Makowski, and G. Meschia. Postnatal changes in oxygen affinity of sheep red cells: a dual role of 2,3-diphosphoglyceric acid. *Am. J. Physiol.* 219:217–221, 1970.

Baudinette, R. V., B. J. Gannon, R. G. Ryall, and P. B. Frappell. Changes in metabolic rates and blood respiratory characteristics during pouch development of a marsupial, *Macropus eugenii. Respir. Physiol.* 72:219–228, 1988.

Baydur, A., P. K. Behrakis, W. A. Zin, M. Jaeger, and J. Milic-Emili. A simple method for assessing the validity of the esophageal balloon technique. *Am. Rev. Respir. Dis.* 126:788–791, 1982.

Bazzy, A. R. Effect of hypoxia on neuromuscular transmission in the developing diaphragm. *J. Appl. Physiol.* 76:708–713, 1994.

Bazzy, A. R., and D. F. Donnelly. Failure to generate action potentials in newborn diaphragms following nerve stimulation. *Brain Res.* 600:349–352, 1993.

Bennett, F. M., and S. M. Tenney. Comparative mechanics of mammalian respiratory system. *Respir. Physiol.* 49:131–140, 1982.

Berger, P. J., A. M. Walker, R. Horne, V. Brodecky, M. H. Wilkinson, F. Wilson, and J. E. Maloney. Phasic respiratory activity in the fetal lamb during late gestation and labour. *Respir. Physiol.* 65:55–68, 1986.

Berger, P. J., R. S. C. Horne, and A. M. Walker. Cardio-respiratory responses to cool ambient temperature differ with sleep state in neonatal lambs. *J. Physiol. (London)* 412:351–363, 1989.

Berger, P. J., R. S. C. Horne, M. Soust, A. M. Walker, and J. E. Maloney. Breathing at birth and the associated blood gas and pH changes in the lamb. *Respir. Physiol.* 82:251–266, 1990.

Berger, P. J., M. Soust, J. J. Smolich, and A. M. Walker. Respiratory muscle blood flow in the fetal lamb during apnoea and breathing. *Respir. Physiol.* 97:111–121, 1994.

Berger, P. J., J. J. Smolich, C. A. Ramsden, and A. M. Walker. Effect of lung liquid volume on respiratory performance after Caesarean section delivery in the lamb. *J. Physiol. (London)* 492:905–912, 1996.

Berman, L. S., W. W. Fox, R. C. Raphaely, and J. D. Downes, Jr. Optimum levels of CPAP for tracheal extubation of newborn infants. *J. Pediatr.* 89:109–112, 1976.

Bernard, S. L., D. L. Luchtel, R. W. Glenny, and S. Lakshminarayan. Bronchial circulation in the marsupial opossum, *Didelphis marsupialis. Respir. Physiol.* 105:77–83, 1996.

Berterottière, D., A. M. D'Allest, M. Dehan, and C. Gaultier. Effects of increase in body temperature on the breathing pattern in premature infants. *J. Develop. Physiol.* 13:303–308, 1990.

Bertin, R., F. De Marco, I. Mouroux, and R. Portet. Postnatal development of nonshivering thermogenesis in rats: effects of rearing temperature. *J. Develop. Physiol.* 19:9–15, 1993.

Bhutani, V. H., S. D. Rubenstein, and T. H. Shaffer. Pressure-volume relationships of the tracheae in fetal newborn and adult rabbits. *Respir. Physiol.* 43:221–231, 1981.

Bhutani, V. H., R. J. Koslo, and T. H. Shaffer. The effect of tracheal smooth muscle tone on neonatal airway collapsibility. *Pediatr. Res.* 20:492–495, 1986.

Bigos, D., and J. J. Pérez Fontán. Contribution of viscoelastic stress to the rate-dependence of pulmonary dynamic elastance. *Respir. Physiol.* 98:53–67, 1994.

Biscoe, T. J., and M. J. Purves. Carotid body chemoreceptor activity in the newborn lamb. *J. Physiol. (London)*, 190:443–454, 1967.

Biscoe, T. J., M. J. Purves, and S. R. Sampson. Types of nervous activity which may be recorded from the carotid sinus nerve of the sheep foetus. *J. Physiol. (London)* 202:1–23, 1969.

Bisgard, G. E., A. V. Ruiz, R. F. Grover, and J. A. Will. Ventilatory acclimatization to 3400 meters altitude in the Hereford calf. *Respir. Physiol.* 21:271–296, 1974.

Blackmore, D. Body size and metabolic rate in newborn lambs of two breeds. *J. App. Physiol.* 27:241–245,1969.

Blanco, C. E. Maturation of fetal breathing activity. *Biol. Neonate* 65:182–188, 1994.

Blanco, C. E., G. S. Dawes, M. A. Hanson, and H. B. McCooke. The response to hypoxia of arterial chemoreceptors in fetal sheep and new-born lambs. *J. Physiol. (London)* 351:25–37, 1984a.

Blanco, C. E., M. A. Hanson, P. Johnson, and H. Rigatto. Breathing pattern of kittens during hypoxia. *J. Appl. Physiol.* 56:12–17, 1984b.

Blanco, C. E., C. B. Martin, Jr., M. A. Hanson, and H. B. McCooke. Breathing activity in fetal sheep during mechanical ventilation of the lungs *in utero. Eur. J. Obst. Gynecol. Reprod. Biol.* 26:175–182, 1987a.

———. Determinants of the onset of continuous air breathing at birth. *Eur. J. Obst. Gynecol. Reprod. Biol.* 26:183–192, 1987b.

Blanco , C. E., M. A. Hanson, and H. B. McCooke. Effects on carotid chemoreceptor resetting of pulmonary ventilation in the fetal lamb in utero. *J. Develop. Physiol.* 10:167–174, 1988.

Bland, R. D., M. A. Bressack, and D. D. McMillan. Labor decreases the lung water content of newborn rabbits. *Am. J. Obstet. Gynecol.* 135:364–367, 1979.

Bland, R. D., D. D. McMillan, M. A. Bressack, and L. Dong. Clearance of liquid from lungs of newborn rabbits. *J. Appl. Physiol.* 49:171–177, 1980.

Bland, R. D., T. N. Hansen, C. M. Haberken, M. A. Bressack, T. A. Hazinski, J. U. Raj, and R. B. Goldberg. Lung fluid balance in lambs before and after birth. *J. Appl. Physiol.* 53:992–1004, 1982.

Blatteis, C. M. Hypoxia and the metabolic response to cold in new-born rabbit. *J. Physiol. (London)* 172:358–368, 1964.

Blix, A. S., and J. W. Lentfer. Modes of thermal protection in polar bear cubs at birth and on emergence from the den. *Am. J. Physiol.* 236:R67–R74, 1979.

Blix, A. S., and J. B. Steen. Temperature regulation in newborn polar homeotherms. *Physiol. Rev.* 59:285–304, 1979.

Bocking, A. D., R. Gagnon, K. M. Milne, and S. E. White. Behavioral activity during prolonged hypoxemia in fetal sheep. *J. Appl. Physiol.* 65:2420–2426, 1988.

Boggs, D. F., and D. Bartlett, Jr. Chemical specificity of a laryngeal apneic reflex in puppies. *J. Appl. Physiol.* 53:455–462, 1982.

Boggs, D. F., D. L. Kilgore, Jr., and G. F. Birchard. Respiratory physiology of burrowing mammals and birds. *Comp. Biochem. Physiol.* 77A:1–7, 1984.

Bonora, M., and H. Gautier. Maturational changes in body temperature and ventilation during hypoxia in kittens. *Respir. Physiol.* 68:359–370, 1987.

Bonora, M., D. Marlot, H. Gautier, and B. Duron. Effects of hypoxia on ventilation during postnatal development in conscious kittens. *J. Appl. Physiol.* 56:1464–1471, 1984.

Bonora, M., M. Boule, and H. Gautier. Diaphragmatic and ventilatory responses to alveolar hypoxia and hypercapnia in conscious kittens. *J. Appl. Physiol.* 72:203–210, 1992.

Boon, A. W., A. D. Milner, and I. E. Hopkin. Lung expansion, tidal exchange, and formation of the functional residual capacity during resuscitation of asphyxiated neonates. *J. Pediatr.* 95:1031–1036, 1979.

Borell, U., and I. Fernstrom. The shape of the foetal chest during its passage through the birth canal. A radiographic study. *Acta Obstet. Gynecol. Scand.* 41:213–222, 1962.

Bortolotti, A., D. Castelli, and M. Bonati. Hematologic and serum chemistry values of adult, pregnant and newborn New Zealand rabbits (*Oryctolagus cuniculus*). *Lab. Anim. Sci.* 39:437–439, 1989.

Bosma, J. F. Functional anatomy of the upper airway during development. In: *Respiratory Function of the Upper Airways*, Lung Biology in Health and Disease Series, vol. 35, edited by O. P. Mathew and G. Sant'Ambrogio. New York: Marcel Dekker, ch. 3, pp. 47–86, 1988.

Bosma, J. F., and J. Lind. Roentgenologic observations of motions of the upper airway associated with establishment of respiration in the newborn infant. *Acta Paediatr. Suppl.* 123:18–55, 1960.

Bosma, J. F., J. Lind, and N. Gentz. Motions of the pharynx associated with initial aeration of the lungs of the newborn infant. *Acta Paediatr. Suppl.* 117:117–122, 1959.

Bowden, D. H., and R. A. Goyer. The size of muscle fibers in infants and children. *Arch. Pathol.* 69:188–189, 1960.

Bradley, K. H., S. D. McConnell, and R. G. Crystal. Lung collagen composition and synthesis. Characterisation and changes with age. *J. Biol. Chem.* 249:2674–2683, 1974.

Brady, J. P., and E. Ceruti. Chemoreceptor reflexes in the new-born infant: effects of varying degrees of hypoxia on heart rate and ventilation in a warm environment. *J. Physiol. (London)* 184:631–645, 1966.

Brady, J. P., and P. M. Dunn. Chemoreceptor reflexes in the newborn infant: effect of CO_2 on the ventilatory response to hypoxia. *Pediatrics* 45:206–214, 1970.

Brady, J. P., E. C. Cotton, and W. H. Tooley. Chemoreflexes in the new-born infant: effects of 100% oxygen on heart rate and ventilation. *J. Physiol. (London)* 172:332–341, 1964.

Brans, Y. W., D. S. Andrew, C. A. Schwartz, and K. D. Carey. Serial estimates of body water content and distribution during the first postnatal month in baboons. *J. Med. Primatol.* 20:75–81, 1991.

Brody, A. W. Mechanical compliance and resistance of the lung-thorax calculated from the flow recorded during passive expiration. *Am. J. Physiol.* 178:189–196, 1954.

Brody, S., and R. C. Procter. Relation between basal metabolism and mature body weight in different species of mammals and birds. *Univ. Missouri Agric. Exp. Sta. Res. Bull.* 166:89,1932, quoted at p. 345 in Brody, *Bioenergetics and Growth.* New York: Reinhold, 1945 (a reprint of this 1945 edition is available from Hafner, New York, 1974).

Brody, J. S., and W. M. Thurlbeck. Development, growth, and aging of the lung. In: *Handbook of Physiology: The Respiratory System, vol. III, Mechanics of Breathing,* edited by P. T. Macklem and J. Mead. Bethesda, Md.: Am. Physiol. Soc., part 1, ch. 22, pp. 355–386, 1986.

Brouillette, R. T., B. T. Thach, Y. K. Abu-Osba, and S. L. Wilson. Hiccups in infants: characteristics and effects on ventilation. *J. Pediatr.* 96:219–225, 1980.

Brozanski, B. S., M. J. Daood, J. F. Watchko, W. A. LaFramboise, and R. D. Guthrie. Postnatal expression of myosin isoforms in the genioglossus and diaphragm muscles. *Pediatr. Pulmonol.* 15:212–219, 1993.

Brück, K. Neonatal thermal regulation. In: *Fetal and Neonatal Physiology, 2nd ed.,* edited by R. A. Polin and W. W. Fox. Philadelphia: Saunders, vol. 1, ch. 68, pp. 676–702, 1998.

Brumley, G. W., V. Chernick, W. A. Hodson, C. Normand, A. Fenner, and M. E. Avery. Correlations of mechanical stability, morphology, pulmonary surfactant, and phospholipid content in the developing lamb lung. *J. Clin. Invest.* 46:863–873, 1967.

Bryan, A. C., G. Bowes, and J. E. Maloney. Control of breathing in the fetus and the newborn. In: *Handbook of Physiology: The Respiratory System. vol. II, Control of breathing,* edited by N. S. Cherniack and J. G. Widdicombe. Bethesda, Md.: Am. Physiol. Soc., part 2, ch. 18, pp. 621–647, 1986.

Bureau, M. A., and R. Bégin. Postnatal maturation of the respiratory response to O_2 in awake newborn lambs. *J. Appl. Physiol.* 52:428–433, 1982.

Bureau, M. A., R. Bégin, and Y. Berthiaume. Central chemical regulation of respiration in term newborn. *J. Appl. Physiol.* 47:1212–1217, 1979.

Bureau, M. A., R. Zinman, P. Foulon, and R. Begin. Diphasic ventilatory response to hypoxia in newborn lambs. *J. Appl. Physiol.* 56:84–94, 1984.

Bureau, M. A., J. Lamarche, P. Foulon, and D. Dalle. Postnatal maturation of respiration in intact and carotid body-chemodenervated lambs. *J. Appl. Physiol.* 59:869–874, 1985.

Burggren, W. W. Pulmonary blood plasma filtration in reptiles: a "wet" vertebrate lung? *Science* 215:77–78, 1982.

Burke, B. E. Glossopharyngeal breathing and its use in the treatment of respiratory poliomyelitis patients, with some notes on chest respirators. *Aust. J. Physioth.* 3:5–12, 1957.

Burri, P. H. The postnatal growth of the rat lung. III. Morphology. *Anat. Rec.* 180:77–98, 1974.

———. Structural development of the lung in the fetus and neonate. In: *Fetus and Neonate: Physiology and Clinical Applications,* edited by M. A. Hanson, J. A. D. Spencer, C. H. Rodeck, and D. Walters. Cambridge: Cambridge Univ. Press, p. 12, 1994.

Cabanac, M. Thermoregulatory behaviour. In: *Essays on Temperature Regulation,* edited by J. Bligh and R. Moore. New York: Elsevier, pp. 19–36, 1972.

Cameron, Y. L., D. Merazzi, and J. P. Mortola. Variability of the breathing pattern in newborn rats: effects of ambient temperature in normoxia or hypoxia. Pediatr. Res., 47:813–818, 2000.

Campbell, A. G. M., F. Cockburn, G. S. Dawes, and J. E. Milligan. Pulmonary vasoconstriction in asphyxia during cross-circulation between twin foetal lambs. *J. Physiol. (London)* 192:111–121, 1967.

Carroll, J. L., and M. A. Bureau. Peripheral chemoreceptor CO_2 response during hyperoxia in the 14-day-old awake lamb. *Respir. Physiol.* 73:339–350, 1988.

Carroll, J. L., O. S. Bamford, and R. S. Fitzgerald. Postnatal maturation of carotid chemoreceptor responses to O_2 and CO_2 in the cat. *J. Appl. Physiol.* 75:2383–2391, 1993.

Cassin, S., G. S. Dawes, J. C. Mott, B. B. Ross, and L. B. Strang. The vascular resistance of the foetal and newly ventilated lung of the lamb. *J. Physiol. (London)* 171:61–79, 1964.

Chan, V., and A. Greenough. Lung function and the Hering Breuer reflex in the neonatal period. *Early Hum. Develop.* 28:111–118, 1992.

Chang, J. H. T., J. C. Rutledge, D. Stoops, and R. Abbe. Hypobaric hypoxia-induced intrauterine growth retardation. *Biol. Neonate* 46:10–13, 1984.

Chapple, R. S., A. W. English, R. C. Mulley, and E. E. Lepherd. Haematology and serum biochemistry of captive unsedated chital deer (*Axis axis*) in Australia. *J. Wildlife Dis.* 27:396–406, 1991.

Chelucci, G. L., S. Boncinelli, M. Marsili, P. Lorenzi, A. Allegra, M. Linden, A. Chelucci, V. Merciai, F. Cresci, C. Rostagno, et al. Aspirin effect on early and late changes in acute lung injury in sheep. *Intens. Care Med.* 19:13–21, 1993.

Cherniack, N. S., and M. D. Altose. Central chemoreceptors. In: *The Lung: Sci-*

entific Foundation, 2nd edition, edited by R. G. Crystal, J. B. West, P. J. Barnes, and E. R. Weibel. Philadelphia: Lippincott-Raven, chap. 131, pp. 1767–1776, 1997.

Cheung, E., N. Wong, and J. P. Mortola. Compliance of the respiratory system in newborn and adult rats after gestation in hypoxia. *J. Comp. Physiol.,* 170:193–199, 2000.

Christensson, K., C. Siles, T. Cabrera, A. Belaustequi, P. de la Fuente, H. Lagercrantz, P. Puyol, and J. Winberg. Lower body temperatures in infants delivered by caesarean section than in vaginally delivered infants. *Acta Paediatr.* 82:128–31, 1993.

Clark, D. J., and J. E. Fewell. Decreased body-core temperature during acute hypoxemia in guinea pigs during postnatal maturation: a regulated thermoregulatory response. *Can. J. Physiol. Pharmacol.* 74:331–336, 1996.

Clarke, W. R., G. Gause, B. E. Marshall, and S. Cassin. The role of lung perfusate PO_2 in the control of the pulmonary vascular resistance of exteriorized fetal lambs. *Respir. Physiol.* 79:19–32, 1990.

Clement, M. G., J. P. Mortola, M. Albertini, and G. Aguggini. Effects of vagotomy on respiratory mechanics in newborn and adult pigs. *J. Appl. Physiol.* 60:1992–1999, 1986.

Clements, J. A. Surface tension of lung extracts. *Proc. Soc. Exp. Biol. Med.* 95:170–172, 1957.

Clements, J. A., R. F. Hustead, R. R. Johnson, and I. Gribetz. Pulmonary surface tension and alveolar stability. *J. Appl. Physiol.* 16:444–450, 1961.

Clerch, L. B., and D. Massaro. Rat lung antioxidant enzymes: differences in perinatal gene expression and regulation. *Am. J. Physiol.* L466–L470, 1992.

Clerici, C., A. Harf, C. Gaultier, and F. Roudot. Cholinergic component of histamine-induced bronchoconstriction in newborn guinea pigs. *J. Appl. Physiol.* 66:2145–2149, 1989.

Clewlow, F., G. S. Dawes, B. M. Johnston, and D. W. Walker. Changes in breathing, electrocortical and muscle activity in unanaesthetized fetal lambs with age. *J. Physiol. (London)* 341:463–476, 1983.

Coates, A. L., and J. Stocks. Esophageal pressure manometry in human infants. *Pediatr. Pulmonol.* 11:350–360, 1991.

Cobb, M. A., W. A. Schutt, Jr., J. L. Petrie, and J. W. Hermanson. Neonatal development of the diaphragm in the horse, *Equus caballus. Anat. Rec.* 238:311–316, 1994.

Cohen, G., G. Malcolm, and D. Henderson-Smart. Ventilatory response of the newborn infant to mild hypoxia. *Pediatr. Pulmonol.* 24:163–172, 1997.

Cohn, R. Factors affecting the postnatal growth of the lung. *Anat. Rec.* 75:195–205, 1939.

Colebatch, H. J. H., I. A. Greaves, and C. K. Y. Ng. Exponential analysis of elastic recoil and aging in healthy males and females. *J. Appl. Physiol.* 47:683–691, 1979.

Coleridge, H. M., and J. C. G. Coleridge. Reflexes evoked from tracheobronchial tree and lungs. In: *Handbook of Physiology: The Respiratory System, vol. II, Control of Breathing,* edited by N. S. Cherniack and J. G. Widdicombe. Bethesda, Md.: Am. Physiol. Soc., part 1, ch. 12, pp. 395–429, 1986.

Collins, M. H., J. Kleinerman, A. C. Moessinger, A. H. Collins, L. S. James, and W. A. Blanc. Morphometric analysis of the growth of the normal fetal guinea pig lung. *Anat. Rec.* 216:381–391, 1986.

Comroe, J. H., Jr. Premature science and immature lung. 1. Some premature discoveries. *Am. Rev. Respir. Dis.* 116:127–135, 1977a.

———. Premature science and immature lung. 2. Chemical warfare and the newly born. *Am. Rev. Respir. Dis.* 116:311–323, 1977b.

———. Premature science and immature lung. 3. The attack on immature lungs. *Am. Rev. Respir. Dis.* 116:497–518, 1977c.

Connors, G., C. Hunse, L. Carmichael, R. Natale, and B. Richardson. The role of carbon dioxide in the generation of human fetal breathing movements. *Am. J. Obstet. Gynecol.* 158:322–327, 1988.

Cook, C. D., R. B. Cherry, D. O'Brien, P. Karlberg, and C. A. Smith. Studies of respiratory physiology in the newborn infant. I. Observations on normal premature and full-term infants. *J. Clin. Invest.* 34:975–982, 1955.

Cook, C. D., J. M. Sutherland, S. Segal, R. B. Cherry, J. Mead, M. B. McIlroy, and C. A. Smith. Studies of respiratory physiology in the newborn infant. III. Measurements of mechanics of respiration. *J. Clin. Invest.* 36:440–448, 1957.

Cook, C. D., P. J. Helliesen, and S. Agathon. Relation between mechanics of respiration, lung size and body size from birth to young adulthood. *J. Appl. Physiol.* 13:349–352, 1958.

Cook, C. D., J. Mead, G. L. Schreiner, N. R. Frank, and J. M. Craig. Pulmonary mechanics during induced pulmonary edema in anesthetized dogs. *J. Appl. Physiol.* 14:177–186, 1959.

Coombs, H. C., and F. H. Pike. The nervous control of respiration in kittens. *Am. J. Physiol.* 95:681–693, 1930.

Coppoletta, J. M., and S. B. Wolbach. Body length and organ weight of infants and children. *Am. J. Pathol.* 9:55–70, 1933.

Côté, A., and H. Porras. Respiratory, cardiovascular, and metabolic adjustments to hypoxemia during sleep in piglets. *Can. J. Physiol. Pharmacol.* 76:747–755, 1998.

Côté, A., K. Yunis, P. W. Blanchard, J. P. Mortola, and M. A. Bureau. Dynamics of breathing in the hypoxic awake lamb. *J. Appl. Physiol.* 64:354–359, 1988.

Côté, A., P. W. Blanchard, and B. Meehan. Metabolic and cardiorespiratory effects of doxapram and theophylline in sleeping newborn piglets. *J. Appl. Physiol.* 72:410–415, 1992.

Côté, A., H. Porras, and B. Meehan. Age-dependent vulnerability to carotid chemodenervation in piglets. *J. Appl. Physiol.* 80:323–331, 1996.

Cotton, E. K., and M. M. Grunstein. Effects of hypoxia on respiratory control in neonates at high altitude. *J. Appl. Physiol.* 48:587–595, 1980.

Crelin, E. S. *Functional Anatomy of the Newborn.* New Haven, Conn.: Yale University Press, p. 38, 1973.

Crighton, G. W., and R. Pownall. The homeothermic status of the neonatal dog. *Nature* 251:142–144, 1974.

Crosfill, M. L., and J. G. Widdicombe. Physical characteristics of the chest and lungs and the work of breathing in different mammalian species. *J. Physiol. (London)* 158:1–14, 1961.

Cross, K. W. Head's paradoxical reflex. *Brain* 84:529–534, 1961.

Cross, K. W., and P. Warner. The effect of inhalation of high and low oxygen concentrations on the respiration of the newborn infant. *J. Physiol. (London)* 114:283–295, 1951.

Cross, K. W., J. M. D. Hooper, and J. M. Lord. Anoxic depression of the medulla in the new-born infant. *J. Physiol. (London)* 125:628–640, 1954.

Cross, K. W., J. P. M. Tizard, and D. A. H. Trythall. The gaseous metabolism of the new-born infant breathing 15% oxygen. *Acta Paediatr.* 46:265–285, 1958.

Cross, K. W., M. Klaus, W. H. Tooley, and K. Weisser. The response of the newborn baby to inflation of the lungs. *J. Physiol. (London)* 151:551–565, 1960.

Croteau, J. R., and C. D. Cook. Volume-pressure and length-tension measurements in human tracheal and bronchial segments. *J. Appl. Physiol.* 16:170–172, 1961.

Cummings, J. J., D. P. Carlton, F. R. Poulain, J. U. Raj, and R. D. Bland. Hypoproteinemia slows lung liquid clearance in young lambs. *J. Appl. Physiol.* 74:153–160, 1993.

Cunningham, E. L., J. S. Brody, and B. P. Jain. Lung growth induced by hypoxia. *J. Appl. Physiol.* 37:362–366, 1974.

Curzi-Dascalova, L. Phase relationships between thoracic and abdominal respiratory movement during sleep in 31–38 weeks CA normal infants. Comparison with full-term (39–41 weeks) newborns. *Neuropediatrics* 13, Suppl.: 15–20, 1982.

Dail, C. W., J. E. Affeldt, and C. R. Collier. Clinical aspects of glossopharyngeal breathing. Report of use by one hundred postpoliomyelitic patients. *J. Am. Med. Assoc.* 158:445–449, 1955.

Daily, W. J. R., M. Klaus, and H. B. P. Meyer. Apnea in premature infants: monitoring, incidence, heart rate changes, and an effect of environmental temperature. *Pediatrics* 43:510–518, 1969.

D'Albis, A., C. Janmot, and R. Couteaux. Species- and muscle-type dependence of perinatal isomyosin transition. *Int. J. Dev. Biol.* 35:53–56, 1991.

Danon, J., W. S. Druz, N. B. Goldberg, and J. T. Sharp. Function of the isolated paced diaphragm and the cervical accessory muscles in C1 quadriplegics. *Am. Rev. Respir. Dis.* 119:909–919, 1979.

Daubenspeck, J. A. Mechanical aspects of loaded breathing. In: *The Thorax. Second Revision, Revised and Expanded,* Lung Biology in Health and Disease Series, vol. 85, edited by C. Roussos. New York: Marcel Dekker, ch. 32, Part A: Physiology, pp. 953–985, 1995.

Davey, M. G., D. P. Johns, and R. Harding. Postnatal development of respiratory function in lambs studied serially between birth and 8 weeks. *Respir. Physiol.* 113:83–93, 1998.

Davey, M. G., S. B. Hooper, M. L. Tester, D. P. Johns, and R. Harding. Respiratory function in lambs after in utero treatment of lung hypoplasia by tracheal obstruction. *J. Appl. Physiol.* 87:2296–2304, 1999.

Davi, M., K. Sankaran, M. MacCallum, D. Cates, and H. Rigatto. Effect of sleep state on chest distortion and on the ventilatory response to CO_2 in neonates. *Pediatr. Res.* 13:982–986, 1979.

Davi, M., K. Sankaran, and H. Rigatto. Effect of inhaling 100% O_2 on ventilation and acid-base balance in cerebrospinal fluid of neonates. *Biol. Neonate* 38:85–89, 1980.

Davidson, M. B. The relationship between diaphragm and body weight in the rat. *Growth* 32:221–223, 1968.

Davies, A., M. Dixon, D. Callanan, A. Huszczuk, J. G. Widdicombe, and J. C. M. Wise. Lung reflexes in rabbits during pulmonary stretch receptor block by sulphur dioxide. *Respir. Physiol.* 34:83–101, 1978.

Davies, A. M., J. S. Koenig, and B. T. Thach. Characteristics of upper airway chemoreflex prolonged apnea in human infants. *Am. Rev. Respir. Dis.* 139:668–673, 1989.

Davis, F. C., and R. A. Gorski. Development of hamster circadian rhythms: role of the maternal suprachiasmatic nucleus. *J. Comp. Physiol.* A:162:601–610, 1988.

Davis, G. M., M. A. Bureau, and C. Gaultier. The sustained ventilatory response to hypoxic challenge in the awake newborn piglet with an intact upper airway. *Respir. Physiol.* 71:307–314, 1988a.

Davis, G. M., A. L. Coates, D. Dalle, and M. A. Bureau. Measurement of pulmonary mechanics in the newborn lamb: a comparison of three techniques. *J. Appl. Physiol.* 64:972–981, 1988b.

Davis, G. M., A. L. Coates, A. Papageorgiou, and M. A. Bureau. Direct measurement of static chest wall compliance in animal and human neonates. *J. Appl. Physiol.* 65:1093–1098, 1988c.

Dawes, G. S., and J. C. Mott. The increase in oxygen consumption of the lamb after birth. *J. Physiol. (London)* 146:295–315, 1959.

Dawes, G. S., J. C. Mott, J. G. Widdicombe, and D. G. Wyatt. Changes in the lungs of the new-born lamb. *J. Physiol. (London)* 121:141–162, 1953.

Dawes, G. S., S. L. B. Duncan, B. V. Lewis, C. L. Merlet, J. B. Owen-Thomas, and J. T. Reeves. Hypoxaemia and aortic chemoreceptor function in foetal lamb. *J. Physiol. (London)* 201:105–116, 1969.

Dawes, G. S., H. E. Fox, B. M. Leduc, G. C. Liggins, and R. T. Richards. Respiratory movements and rapid eye movement sleep in the fetal lamb. *J. Physiol. (London)* 220:119–143, 1972.

Dejours, P. Intérêt méthodologique de l'étude d'un organisme vivant à la phase initiale de rupture d'un équilibre physiologique. *C. R. Acad. Sci. (Paris)* 245:1946–1948, 1957.

Delacourt, C., E. Canet, J.-P. Praud, and M. A. Bureau. Influence of vagal afferents on diphasic ventilatory response to hypoxia in newborn lambs. *Respir. Physiol.* 99:29–39, 1995.

Delacourt, C., E. Canet, and M. A. Bureau. Predominant role of peripheral chemoreceptors in the termination of apnea in maturing newborn lambs. *J. Appl. Physiol.* 80:892–898, 1996.

Delivoria-Papadopoulos, M., and J. E. McGowan. Oxygen transport and delivery. In *Fetal and Neonatal Physiology,* 2nd ed., edited by R. A. Polin and W. W. Fox. Philadelphia: Saunders, vol. 1, ch. 102, pp. 1105–1117, 1998.

Delivoria-Papadopoulos, M., R. J. Martens, R. E. Forster, II, and F. A. Oski. Postnatal changes in oxygen-hemoglobin affinity and erythrocyte 2,3-diphosphoglycerate in piglets. *Pediatr. Res.* 8:64–66, 1974.

DeNeef, K. J., J. R. C. Jansen, and A. Versprille. Developmental morphometry and physiology of the rabbit vagus nerve. *Dev. Brain Res.* 4:265–274, 1982.

De Troyer, A., and M. Estenne. Coordination between rib cage muscles and diaphragm during quiet breathing in humans. *J. Appl. Physiol.* 57:899–906, 1984.

Devlieger, H., H. Daniels, G. Marchal, P. Moerman, P. Casaer, and E. Eggermont. The diaphragm of the newborn infant: anatomical and ultrasonographic studies. *J. Dev. Physiol.* 16:321–329, 1991.

Dhindsa, D. S., A. S. Hoversland, and J. W. Templeton. Postnatal changes in oxygen affinity and concentrations of 2,3-diphosphoglycerate in dog blood. *Biol. Neonate* 20:226–235, 1972.

Dickson, K. A., J. E. Maloney, and P. J. Berger. State-related changes in lung liquid secretion and tracheal flow rate in fetal lambs. *J. Appl. Physiol.* 62:34–38, 1987.

Dinwiddie, R., and G. Russel. Relationship of intraesophageal pressure to intrapleural pressure in the newborn. *J. Appl. Physiol.* 33:415–417, 1972.

Dolfin, T., P. Duffy, D. Wilkes, S. England, and H. Bryan. Effects of a face mask and pneumotachograph on breathing in sleeping infants. *Am. Rev. Respir. Dis.* 128:977–979, 1983.

Donawick, W. J., and A. E. Baue. Blood gases, acid-base balance, and alveolar-arterial oxygen gradient in calves. *Am. J. Vet. Res.* 29:561–567, 1968.

Donnelly, D. F., and G. G. Haddad. Prolonged apnea and impaired survival in piglets after sinus and aortic nerve section. *J. Appl. Physiol.* 68:1048–1052, 1990.

Dornan, J. C., J. W. K. Ritchie, and C. Meban. Fetal breathing movements and lung maturation in the congenitally abnormal human fetus. *J. Dev. Physiol.* 6:367–375, 1984.

Dotta, A., and J. P. Mortola. Effects of hyperoxia on the metabolic response to cold of the newborn rat. *J. Dev. Physiol.* 17:247–250, 1992a.

———. Postnatal development of the denervated lung in normoxia, hypoxia, or hyperoxia. *J. Appl. Physiol.* 73:1461–1466, 1992b.

Downing, S. E., and S. C. Lee. Laryngeal chemosensitivity: a possible mechanism for sudden infant death. *Pediatrics* 55:640–649, 1975.

Dreier, T., P. H. Wolff, E. E. Cross, and W. D. Cochran. Patterns of breath intervals during non-nutritive sucking in full-term and 'at risk' preterm infants with normal neurological examinations. *Early Hum. Develop.* 3:187–199, 1979.

Dreshaj, I. A., R. J. Martin, M. J. Miller, and M. A. Haxhiu. Responses of lung parenchyma and airways to tachykinin peptides in piglets. *J. Appl. Physiol.* 77:147–151, 1994.

Dreshaj, I. A., M. A. Haxhiu, C. F. Potter, F. H. Agani, and R. J. Martin. Maturational changes in response of tissue and airway resistance to histamine. *J. Appl. Physiol.* 81:1785–1791, 1996.

Dripps, R. D., and J. H. Comroe, Jr. The effect of inhalation of high and low oxygen concentrations on respiration, pulse rate, ballistocardiogram and arterial oxygen saturation (oximeter) of normal individuals. *Am. J. Physiol.* 149:277–291, 1947.

Drorbaugh, J. E. Pulmonary function in different animals. *J. Appl. Physiol.* 15:1069–1072, 1960.

Duara, S., G. S. Neto, T. Gerhardt, C. Suguihara, and E. Bancalari. Metabolic and respiratory effects of flow-resistive loading in preterm infants. *J. Appl. Physiol.* 70:895–899, 1991.

Duara, S., C. Suguihara, T. Gerhardt, and E. Bancalari. Metabolic, hemodynamic, and ventilatory responses to respiratory load in sedated neonatal piglets. *J. Appl. Physiol.* 75:181–184, 1993.

Ducros, G., and T. Trippenbach. Respiratory effects of lactic acid injected into the jugular vein of newborn rabbits. *Pediatr. Res.* 29:548–552, 1991.

Duffy, T. E., and R. C. Vannucci. Metabolic aspects of cerebral anoxia in the fetus and newborn. In: *Brain, Fetal and Infant,* edited by S. R. Berenberg. The Hague: Martinus Nijhoff Medical Division, pp. 316–323, 1977.

Duffy, T. E., S. J. Kohle, and R. C. Vannucci. Carbohydrate and energy metabolism in perinatal rat brain: relation to survival in anoxia. *J. Neurochem.* 24:271–276, 1975.

Dunnil, M. S. Postnatal growth of the lung. *Thorax* 17:329–333, 1962.

Dupré, R. K., A. M. Romero, and S. C. Wood. Thermoregulation and metabolism in hypoxic animals. In: *Oxygen Transfer from Atmosphere to Tissues,* edited by N. C. Gonzalez and M. R. Fedde. New York: Plenum, pp. 347–351, 1988.

Durmowicz, A. G., E. C. Orton, and K. R. Stenmark. Progressive loss of vasodilator responsive component of pulmonary hypertension in neonatal calves exposed to 4,570 m. *Am. J. Physiol.* 265:H2175–H2183, 1993.

Economos, A. C. Gravity, metabolic rate and body size of mammals. *Physiologist* 22 (Suppl.):S71–S72, 1979.

Edelman, N. H., S. Lahiri, L. Braudo, N. S. Cherniack, and A. P. Fishman. The blunted ventilatory response to hypoxia in cyanotic congenital heart disease. *New Engl. J. Med.* 282:405–411, 1970.

Edelman, N. H., J. E. Melton, and J. A. Neubauer. Central adaptation to hypoxia. In: *Response and Adaptation to Hypoxia,* edited by S. Lahiri, N. S. Cherniack, and R. S. Fitzgerald, Am. Physiol. Soc. New York: Oxford University Press, ch. 22, pp. 235–244, 1991.

———. Effects of CNS hypoxia on breathing. In: *The Lung: Scientific Foundation,* 2nd edition, edited by R. G. Crystal, J. B. West, P. J. Barnes, and E. R. Weibel. Philadelphia: Lippincott-Raven, ch. 130, pp. 1757–1765, 1997.

Edelstone, D. I. The fetus: Responses to reduced O_2 delivery. In: *Hypoxia, The Tolerable Limits,* edited by J. R. Sutton, C. S. Houston, and G. Coates. Indianapolis: Benchmark, pp. 251–261, 1988.

Eden, G. J., and M. A. Hanson. Effects of chronic hypoxia from birth on the ventilatory response to acute hypoxia in the newborn rat. *J. Physiol. (London)* 392:11–19, 1987.

Egan, E. A., R. E. Olver, and L. B. Strang. Changes in non-electrolyte permeability of alveoli and the absorption of lung liquid at the start of breathing in the lamb. *J. Physiol. (London)* 244:161–179, 1975.

Eisenberg, J. F. *Mammals of the Neotropics. The Northern Neotropics.* Chicago: University of Chicago Press, 1989.

Ekholm, J. Postnatal changes in cutaneous reflexes and in the discharge pattern of cutaneous and articular sense organs. A morphological and physiological study of the cat. *Acta Physiol. Scand. Suppl.* 297:85–113, 1967.

Eldridge, F. L., and D. E. Millhorn. Oscillation, gating, and memory in the respiratory control system. In: *Handbook of Physiology, Sect. 3, Respiration, vol.*

II, Control of Breathing, edited by N. S. Cherniack and J. G. Widdicombe. Bethesda, Md.: Am. Physiol. Soc., part 1, ch. 3, pp. 93–114, 1986.

El-Khodor, B. F., and P. Boksa. Long-term reciprocal changes in dopamine levels in prefrontal cortex verus nucleus accumbens in rats born by Caesarean section compared to vaginal birth. *Exp. Neurol.* 145: 118–129, 1997.

Else, P. L. Oxygen consumption and sodium pump thermogenesis in a developing mammal. *Am. J. Physiol.* 261:R1575–R1578, 1991.

Else, P. L., and A. J. Hulbert. Comparison of the "mammal machine" and the "reptile machine": energy production. *Am. J. Physiol.* 240:R3–R9, 1981.

————. An allometric comparison of the mitochondria of mammalian and reptilian tissues: the implications for the evolution of endothermy. *J. Comp. Physiol. B* 156:3–11, 1985a.

————. Mammals: an allometric study of metabolism at tissue and mitochondrial level. *Am. J. Physiol.* 248:R415–R421, 1985b.

————. Evolution of mammalian endothermic metabolism: "leaky" membranes as a source of heat. *Am. J. Physiol.* 253:R1–R7, 1987.

Emmanouilides, G. C., A. J. Moss, E. R. Duffie, Jr., and F. H. Adams. Pulmonary arterial pressure changes in human newborn infants from birth to 3 days of age. *J. Pediatr.* 65:327–333, 1964.

Engel, S. The structure of the respiratory tissue in the newly-born. *Acta Anat.* 19:353–365, 1953.

Erickson, J. T., C. Mayer, A. Jawa, L. Ling, E. B. Olson, Jr., E. H. Vidruk, G. S. Mitchell, and D. M. Katz. Chemoafferent degeneration and carotid body hypoplasia following chronic hyperoxia in newborn rats. *J. Physiol. (London)* 509:519–526, 1998.

Estol, P. C., H. Piriz, S. Basalo, F. Simini, and C. Grela. Oro-naso-pharyngeal suction at birth: effects on respiratory adaptation of normal term vaginally born infants. *J. Perinat. Med.* 20:297–305, 1992.

Euler, C., von. Brain stem mechanisms for generation and control of breathing pattern. In: *Handbook of Physiology. The Respiratory System, vol. II, Control of Breathing,* edited by N. S. Cherniack and J. G. Widdicombe. Bethesda, Md.: Am. Physiol. Soc., part 1, ch. 1, pp. 1–67, 1986.

Fagan, D. G. Post-mortem studies of the semistatic volume-pressure characteristics of infants' lungs. *Thorax* 31:534–543,1976.

————. Shape changes in static V-P loops from children's lungs related to growth. *Thorax* 32:198–202,1977.

Fagenholz, S. A., J. C. Lee, and S. E. Downing. Laryngeal reflex apnea in the chemodenervated newborn piglet. *Am. J. Physiol.* 237:R10–R14, 1979.

Fahey, J. T., and G. Lister. Response to low cardiac output: developmental differences in metabolism during oxygen deficit and recovery in lambs. *Pediatr. Res.* 26:180–187, 1989.

Farber, J. P. Development of pulmonary reflexes and pattern of breathing in the Virginia opossum. *Respir. Physiol.* 14:278–286, 1972.

———. Laryngeal effects and respiration in the suckling opossum. *Respir. Physiol.* 35:189–201, 1978.

———. Expiratory motor responses in the suckling opossum. *J. Appl. Physiol.* 54:919–925, 1983.

———. Development of pulmonary and chest wall reflexes influencing breathing. In *Developmental Neurobiology of Breathing,* Lung Biology in Health and Disease Series, vol. 53, edited by G. G. Haddad and J. P. Farber. New York: Marcel Dekker, ch. 8, pp. 245–269, 1991.

Farber, J. P., and E. E. Lawson. Neurophysiological organization of respiratory neurons in early life. In: *Developmental Neurobiology of Breathing,* Lung Biology in Health and Disease Series, vol. 53, edited by G. G. Haddad and J. P. Farber. New York: Marcel Dekker, ch. 6, pp. 199–215, 1991.

Farber, J. P., and T. A. Marlow. An obstructive apnea in the suckling opossum. *Respir. Physiol.* 34:295–305, 1978.

Farber, J. P., H. N. Hultgren, and S. M. Tenney. Development of the chemical control of breathing in the Virginia opossum. *Respir. Physiol.* 14:267–277, 1972.

Farber, J. P., J. T. Fisher, and G. Sant'Ambrogio. Airway receptor activity in the developing opossum. *Am. J. Physiol.* 246:R753–R758, 1984.

Faridy, E. E. Effect of distension on release of surfactant in excised dogs' lungs. *Respir. Physiol.* 27:99–114, 1976.

———. Instinctive resuscitation of the newborn rat. *Respir. Physiol.* 51:1–19, 1983.

———. Air opening pressure in fetal lungs. *Respir. Physiol.* 68:293–300, 1987.

Faridy, E. E., and S. Permutt. Surface forces and airway obstruction. *J. Appl. Physiol.* 30:319–321, 1971.

Faridy, E. E., M. R. Sanii, and J. A. Thliveris. Fetal lung growth: influence of maternal hypoxia and hyperoxia in rats. *Respir. Physiol.* 73:225–242, 1988.

Fauré-Fremiet, E., and J. Dragoiu. Le développement du poumon foetal chez le mouton. *Arch. Anat. Microsc. Morphol. Exp.* 19:411–474, 1923.

Fawcitt, J., J. Lind, and C. Wegelius. The first breath. A preliminary communication describing some methods of investigation of the first breath of a baby and the results obtained from them. *Acta Paediatr. Suppl.* 123:5–17, 1960.

Faxelius, G., K. Bremme, and H. Lagercrantz. An old problem revisited—Hyaline membrane disease and cesarean section. *Europ. J. Pediatr.* 139:121–124, 1982.

Fedorko, L., E. N. Kelly, and S. J. England. Importance of vagal afferents in determining ventilation in newborn rats. *J. Appl. Physiol.* 65:1033–1039, 1988.

Fedullo, A. J., Y. Jung-Legg, G. L. Snider, and J. B. Karlinsky. Hysteresis ratio: a measure of the mechanical efficiency of fibrotic and emphysematous hamster lung tissue. *Am. Rev. Respir. Dis.* 122:47–52, 1980.

Feldman, H. A., and T. A. McMahon. The 3/4 exponent for energy metabolism is not a statistical artifact. *Respir. Physiol.* 52:149–163, 1983.

Feldman, J. D., A. R. Bazzy, T. R. Cummins, and G. G. Haddad. Developmental changes in neuromuscular transmission in the rat diaphragm. *J. Appl. Physiol.* 71:280–286, 1991.

Fewell, J. E. Fever in young lambs: carotid denervation alters the febrile response to a small dose of bacterial pyrogen. *Pediatr. Res.* 31:107–111, 1992.

Fewell, J. E., C. S. Kondo, V. Dascalu, and S. C. Filyk. Influence of carotid denervation on the arousal and cardiopulmonary response to rapidly developing hypoxemia in lambs. *Pediatr. Res.* 25:473–477, 1989a.

———. Influence of carotid-denervation on the arousal and cardiopulmonary responses to alveolar hypercapnia in lambs. *J. Develop. Physiol.* 12:193–199, 1989b.

Fewell, J. E., B. J. Taylor, C. S. Kondo, V. Dascalu, and S. C. Filyk. Influence of carotid denervation on the arousal and cardiopulmonary responses to upper airway obstruction in lambs. *Pediatr. Res.* 28:374–378, 1990a.

Fewell, J. E., C. S. Kondo, and V. Dascalu. Influence of sleep on the cardiovascular and metabolic responses to a decrease in ambient temperature in lambs. *J. Develop. Physiol.* 13:223–230, 1990b.

Fewell, J. E., M. Kang, and H. L. Eliason. Autonomic and behavioral thermoregulation in guinea pigs during postnatal maturation. *J. Appl. Physiol.* 83:830–836, 1997.

Fidone, S. J., and C. Gonzalez. Initiation and control of chemoreceptor activity in the carotid body. In *Handbook of Physiology, Sect. 3, Respiration, Vol. II, Control of Breathing,* edited by N. S. Cherniack and J. G. Widdicombe. Bethesda, Md.: Am. Physiol. Soc., part 1, ch. 9, pp. 247–312, 1986.

Fike, C. D., and M. R. Kaplowitz. Chronic hypoxia alters nitric oxide-dependent pulmonary vascular responses in lungs of newborn pigs. *J. Appl. Physiol.* 81:2078–2087, 1996.

Fike, C. D., and S. J. Lai-Fook. Effect of airway and left atrial pressures on microcirculation of newborn lungs. *J. Appl. Physiol.* 69:1063–1072, 1990.

Finkel, M. L. The character of respiration in mature neonates during diurnal sleep (transl. from Russian). *Vopr. Okhr. Materin. Det.* 1:47–52, 1972.

Finkelstein, D. I., P. Andrianakis, A. R. Luff, and D. W. Walker. Developmental changes in hindlimb muscles and diaphragm of sheep. *Am. J. Physiol.* 263:R900–R908, 1992.

Fisher, A. B. Intracellular production of oxygen-derived free radicals. In: *Oxygen Radicals and Tissue Injury,* Upjohn Symposium edited by B. Halliwell, Bethesda, Md.: Fed. Am. Soc. Exp. Biol., pp.34–39, 1988.

Fisher, J. T., and J. P. Mortola. Statics of the respiratory system in newborn mammals. *Respir. Physiol.* 41:155–172, 1980a.

———. Immediate ventilatory response to inspiratory elastic loads in newborn and adult rabbits. *Acc. Naz. Lincei* 68:449–454, 1980b.

———. Statics of the respiratory system and growth: an experimental and allometric approach. *Am. J. Physiol.* 241:R336–R341, 1981.

Fisher, J. T., and G. Sant'Ambrogio. Location and discharge properties of respiratory vagal afferents in the newborn dog. *Respir. Physiol.* 50:209–220, 1982.

———. Airway and lung receptors and their reflex effects in the newborn. *Pediatr. Pulmonol.* 1:112–126, 1985.

Fisher, J. T., J. P. Mortola, J. B. Smith, G. S. Fox, and S. Weeks. Respiration in newborns. Development of the control of breathing. *Am. Rev. Respir. Dis.* 125:650–657, 1982.

———. Neonatal pattern of breathing following cesarean section: epidural versus general anesthesia. *Anesthesiology* 59:385–389, 1983a.

Fisher, J. T., F. B. Sant'Ambrogio, and G. Sant'Ambrogio. Stimulation of tracheal slowly adapting stretch receptors by hypercapnia and hypoxia. *Respir. Physiol.* 53:325–339, 1983b.

Fisher, J. T., O. P. Mathew, F. B. Sant'Ambrogio, and G. Sant'Ambrogio. Reflex effects and receptor responses to upper airway pressure and flow stimuli in developing puppies. *J. Appl. Physiol.* 58:258–264, 1985.

Fisher, J. T., M. A. Waldron, and C. J. Armstrong. Effects of hypoxia on lung mechanics in the newborn cat. *Can. J. Physiol. Pharmacol.* 65:1234–1238, 1987.

Fisher, J. T., K. L. Brundage, M. A. Waldron, and B. J. Connelly. Vagal cholinergic innervation of the airways in newborn cat and dog. *J. Appl. Physiol.* 69:1525–1531, 1990.

Fisher, J. T., O. P. Mathew, and G. Sant'Ambrogio. Morphological and neurophysiological aspects of airway and pulmonary receptors. In: *Developmental Neurobiology of Breathing*, Lung Biology in Health and Disease Series, vol. 53, edited by G. G. Haddad and J. P. Farber. New York: Marcel Dekker, ch. 7, pp. 219–244, 1991.

Fisher, J. T., M. A. Haxhiu, and R. J. Martin. Regulation of lower airway function. In: *Fetal and Neonatal Physiology, 2nd ed.*, edited by R. A. Polin and W. W. Fox, Philadelphia: Saunders, vol. 1, ch. 98, pp. 1060–1070, 1998.

Fitzgerald, L. R. The oxygen consumption of neonatal mice. *J. Exp. Zool.* 124:415–425, 1953.

Fleming, P. J., A. C. Bryan, and M. H. Bryan. Functional immaturity of pulmonary irritant receptors and apnea in newborn preterm infants. *Pediatrics* 61:515–518, 1978.

Fleming, P. J., N. L. Muller, M. H. Bryan, and A. C. Bryan. The effects of abdominal loading on rib cage distortion in premature infants. *Pediatrics* 64:425–428, 1979.

Fleming, P. J., M. T. Levine, and A. L. Goncalves. Changes in respiratory pattern resulting from the use of a facemask to record respiration in newborn infants. *Pediatr. Res.* 16:1031–1034, 1982.

Fletcher, B. D., B. F. Sachs, and R. V. Kotas. Radiologic demonstration of postnatal liquid in the lungs of newborn lambs. *Pediatrics* 46:252–258, 1970.

Forrest, J. B. Lung tissue plasticity: morphometric analysis of anisotropic strain in liquid filled lungs. *Respir. Physiol.* 27:223–239, 1976.

Forster, H. V., J. A. Dempsey, M. L. Birnbaum, W. G. Reddan, J. Thoden, R. F. Grover, and J. Rankin. Effect of chronic exposure to hypoxia on ventilatory response to CO_2 and hypoxia. *J. Appl. Physiol.* 31:586–592, 1971.

Forsyth, A. *Mammals of the Canadian Wild.* Camden, Ont., Canada: Camden House, 1985.

Fournier, M., M. Alula, and G. C. Sieck. Neuromuscular transmission failure during postnatal development. *Neurosci. Letters* 125:34–36, 1991.

Fowler, S. J., and C. Kellogg. Ontogeny of thermoregulatory mechanisms in the rat. *J. Comp. Physiol. Psychol.* 89:738–746, 1975.

Fox, R. E., P. C. Kosch, H. A. Feldman, and A. R. Stark. Control of inspiratory duration in premature infants. *J. Appl. Physiol.* 64:2597–2604, 1988.

Frank, L., and E. E. Groseclose. Preparation for birth into an O_2-rich environment: the antioxidant enzymes in the developing rabbit lung. *Pediatr. Res.* 18:240–244, 1984.

Frank, L., and I. R. S. Sosenko. Prenatal development of lung antioxidant enzymes in four species. *J. Pediatr.* 110:106–110, 1987.

———. Failure of premature rabbits to increase antioxidant enzymes during hyperoxic exposure: increased susceptibility to pulmonary oxygen toxicity compared with term rabbits. *Pediatr. Res.* 29:292–296, 1991.

Frank, L., J. R. Bucher, and R. J. Roberts. Oxygen toxicity in neonatal and adult animals of various species. *J. Appl. Physiol.* 45:699–704, 1978.

Frank, L., I. R. S. Sosenko, and J. Gerdes. Pathophysiology of lung injury and repair: special features of the immature lung. In *Fetal and Neonatal Physiology,* edited by R. A. Polin and W. W. Fox. Philadelphia: Saunders, ch. 107, pp. 1175–1188, 1998.

Frantz, I. D., III, and J. Milic-Emili. The progressive response of the newborn infant to added respiratory loads. *Respir. Physiol.* 24:233–239, 1975.

Frappell, P. B., and R. V. Baudinette. Scaling of respiratory variables and the breathing pattern in adult marsupials. *Respir. Physiol.* 100:83–90, 1995.

Frappell, P. B., and J. P. Mortola. Respiratory mechanics in small newborn mammals. *Respir. Physiol.* 76:25–36, 1989.

————. Hamsters vs. rat: metabolic and ventilatory response to development in chronic hypoxia. *J. Appl. Physiol.* 77:2748–2752, 1994.

————. Passive body movement and gas exchange in the frilled lizard (*Chlamydosaurus kingii*) and goanna (*Varanus gouldii*). *J. Exp. Biol.* 201:2307–2311, 1998.

————. Respiratory function in a newborn marsupial with skin gas exchange. *Respir. Physiol.* 120:35–45, 2000.

Frappell, P., C. Saiki, and J. P. Mortola. Metabolism during normoxia, hypoxia and recovery in the newborn kitten. *Respir. Physiol.* 86:115–124, 1991.

Frappell, P. B., A. Dotta, and J. P. Mortola. Metabolism during normoxia, hyperoxia and recovery in newborn rats. *Can. J. Physiol. Pharmacol.* 70:408–411, 1992a.

Frappell P., C. Lanthier, R. V. Baudinette, and J. P. Mortola. Metabolism and ventilation in acute hypoxia: a comparative analysis in small mammalian species. *Am. J. Physiol.* 262:R1040–R1046, 1992b.

Frappell, P. B., F. León-Velarde, L. Aguero, and J. P. Mortola. Response to cooling temperature in infants born at an altitude of 4,330 m. *Am. J. Respir. Crit. Care Med.* 158:1751–1756, 1998.

Frigo, L., and P. A. Woolley. Growth and development of pouch young of the stripe-faced dunnart, *Sminthopsis macroura* (Marsupialia: Dasyuridae), in captivity. *Austr. J. Zool.* 45:157–170, 1997.

Gallivan, G. J., W. N. McDonell, and J. B. Forrest. Comparative pulmonary mechanics in the horse and the cow. *Res. Vet. Sci.* 46:322–330, 1989.

Gaultier, C., and F. Girard. Croissance pulmonaire normale et pathologique: relations structure-fonction. *Bull. Europ. Physiopath. Resp.* 16:791–842, 1980.

Gaultier, C., and J. P. Mortola. Hering-Breuer reflex in young and adult mammals. *Can. J. Physiol. Pharmacol.* 59:1017–1021, 1981.

Gaultier, C., and R. Zinman. Maximal static pressures in healthy children. *Respir. Physiol.* 51:45–61, 1983.

Gaultier, C., A. Harf, A. M. Lorino, and G. Atlan. Lung mechanics in growing guinea pigs. *Respir. Physiol.* 56:217–228, 1984.

Gaultier, C., J. P. Praud, E. Canet, M. F. Delaperche, and A. M. D'Allest. Paradoxical inward rib cage motion during rapid eye movement sleep in infants and young children. *J. Dev. Physiol.* 9:391–397, 1987.

Gautier, H. Pattern of breathing during hypoxia or hypercapnia of the awake or anesthetized cat. *Respir. Physiol.* 27:193–206, 1976.

————. Interactions among metabolic rate, hypoxia, and control of breathing. *J. Appl. Physiol.* 81:521–527, 1996.

Gautier, H., M. Bonora, and J. H. Gaudy. Breuer-Hering inflation reflex and breathing pattern in anesthetized humans and cats. *J. Appl. Physiol.* 51:1162–1168, 1981.

Gemmell, R. T. Lung development in the marsupial bandicoot, *Isoodon macrourus. J. Anat.* 148:193–204, 1986.

Gerdin, E., O. Tydén, and U. J. Eriksson. The development of antioxidant enzymatic defense in perinatal rat lung: activities of superoxide dismutase, glutathione peroxidase, and catalase. *Pediatr. Res.* 19:687–691, 1985.

Gerhardt, T., and E. Bancalari. Chest wall compliance in full-term and premature infants. *Acta Pediatr. Scand.* 69:359–364, 1980.

———. Maturational changes of reflexes influencing inspiratory timing in newborns. *J. Appl. Physiol.* 50:1282–1285, 1981.

Geubelle, F., P. Karlberg, G. Koch, J. Lind, G. Wallgren, and G. Wegelius. L'aération du poumon chez le nouveau-né. *Biol. Neonat.* 1:169–210, 1959.

Gibson, G. J., N. B. Pride, J. Davis, and R. C. Schroter. Exponential description of the static pressure-volume curve of normal and diseased lungs. *Am. Rev. Respir. Dis.* 120:799–811, 1979.

Gilbert, R. D., L. A. Cummings, M. R. Juchau, and L. D. Longo. Placental diffusing capacity and fetal development in exercising or hypoxic guinea pig. *J. Appl. Physiol.* 46:828–834, 1979.

Gillespie, J. R. Postnatal lung growth and function in the foal. *J. Reprod. Fert. Suppl.* 23:667–671, 1975.

———. Mechanisms that determine functional residual capacity in different mammalian species. *Am. Rev. Respir. Dis.* 128, Suppl. S74–S77, 1983.

Gingell, R. L., D. R. Pieroni, and M. G. Hornung. Growth problems associated with congenital heart disease in infancy. In: *Textbook of Gastroenterology and Nutrition in Infancy,* vol. 2, edited by E. Lebenthal. New York: Raven, pp. 853–860, 1981.

Glass, H. G., F. F. Snyder, and E. Webster. The rate of decline in resistance to anoxia of rabbits, dogs and guinea pigs from the onset of viability to adult life. *Am. J. Physiol.* 140:609–615, 1944.

Glebovskii, V. D. Contractile properties of respiratory muscles in fully grown and neonate animals. *Sechenov Physiol. J. USSR* (English transl. from *Fiziol. Zh. SSSR Im. I. M. Sechenova*) 47:427–435, 1961.

Gleed, R. D., and J. P. Mortola. Ventilation in newborn rats after gestation at simulated high altitude. *J. Appl. Physiol.* 70:1146–1151, 1991.

Gluck, L., M. V. Kulovich, R. C. Borer, P. H. Brenner, G. C. Anderson, and W. N. Spellacy. Diagnosis of the respiratory distress syndrome by amniocentesis. *Am. J. Obstet. Gynecol.* 109:440–445, 1971.

Gluckman, P. D., and B. M. Johnston. Lesions in the upper lateral pons abolish the hypoxic depression of breathing in unanaesthetized fetal lambs in utero. *J. Physiol. (London)* 382:373–383, 1987.

Gluckman, P. D., T. R. Gunn, and B. M. Johnston. The effect of cooling on breathing and shivering in unanaesthetized fetal lambs in utero. *J. Physiol. (London)* 343:495–506, 1983.

Goldenberg, V. E., S. Buckingham, and S. C. Sommers. Pulmonary alveolar lesions in vagotomized rats. *Lab. Invest.* 16:693–705, 1967.

Goldman, M. D., and J. Mead. Mechanical interaction between the diaphragm and rib cage. *J. Appl. Physiol.* 35:197–204, 1973.

Goldstein, J. D., and L. M. Reid. Pulmonary hypoplasia resulting from phrenic nerve agenesis and diaphragmatic amyoplasia. *J. Pediatr.* 97:282–287, 1980.

Gonzalez-Crussi, F., and R. W. Boston. The absorptive function of the neonatal lung. Ultrastructural study of horseradish peroxidase uptake at the onset of ventilation. *Lab. Invest.* 6:114–121, 1972.

Gordon, C. J., and L. Fogelson. Comparative effects of hypoxia on behavioral thermoregulation in rats, hamsters, and mice. *Am. J. Physiol.* 260:R120–R125, 1991.

Gosselin, L. E., B. D. Johnson, and G. C. Sieck. Age-related changes in diaphragm muscle contractile properties and myosin heavy chain isoforms. *Am. J. Respir. Crit. Care Med.* 150:174–178, 1994.

Grauw, T. J. de, R. E. Myers, and W. J. Scott. Fetal growth retardation in rats from different levels of hypoxia. *Biol. Neonate* 49:85–89, 1986.

Greenough, A., C. J. Morley, and J. A. Davis. Provoked augmented inspirations in ventilated premature babies. *Early Hum. Develop.* 9:111–117, 1984.

Greer, J. J., D. W. Allan, M. Martin-Caraballo, and R. P. Lemke. An overview of phrenic and diaphragm muscle development in the perinatal rat. *J. Appl. Physiol.* 86:779–786, 1999.

Gregory, J. E., and U. Proske. Responses of muscle receptors in the kitten. *J. Physiol. (London)* 366:27–45, 1985.

Gregory, G. A., J. A. Kitterman, R. H. Phibbs, W. H. Tooley, and W. K. Hamilton. Treatment of the idiopathic respiratory-distress syndrome with continuous positive airway pressure. *N. Engl. J. Med.* 284:1333–1340, 1971.

Gribetz, I., N. R. Frank, and M. E. Avery. Static volume-pressure relations of excised lungs of infants with hyaline membrane disease, newborn and stillborn infants. *J. Clin. Invest.* 38:2168–2175, 1959.

Griffiths, G. B., A. Noworaj, and J. P. Mortola. End-expiratory level and breathing pattern in the newborn. *J. Appl. Physiol.* 55:243–249, 1983.

Griffiths, R. I., and P. J. Berger. Functional development of the sheep diaphragmatic ligament. *J. Physiol. (London)* 492:913–919, 1996.

Griffiths, R. I., J. Baldwin, and P. J. Berger. Metabolic development of the sheep diaphragm during foetal and newborn life. *Respir. Physiol.* 95:337–347, 1994.

Grogaard, J., E. Kreuger, D. Lindstrom, and H. Sundell. Effects of carotid body maturation and terbutaline on the laryngeal chemoreflex in newborn lambs. *Pediatr. Res.* 20:724–729, 1986.

Gruenwald, P. Surface tension as a factor in the resistance of neonatal lungs to aeration. *Am. J. Obstet. Gynecol.* 53:996–1007, 1947.

————. Normal and abnormal expansion of the lungs of newborn infants obtained at autopsy. II. Opening pressure, maximal volume and stability of expansion. *Lab. Invest.* 12:563–576, 1963.

Gryboski, J. D. Suck and swallow in the premature infant. *Pediatrics* 43:96–102, 1969.

Günther, B., and B. L. De La Barra. Comparative physiometry of the mechanics of breathing in mammals. *Respir. Physiol.* 2:129–134, 1967.

Guntheroth, W. G. Primary apnea, hypoxic apnea and gasping. In: *SIDS*, edited by R. R. Robinson. Toronto: Canadian Foundation for the Study of Infant Deaths, pp. 243–247, 1974.

Guntheroth, W. G., and I. Kawabori. Hypoxic apnea and gasping. *J. Clin. Invest.* 56:1371–1377, 1975.

Guslits, B. G., S. E. Gaston, M. H. Bryan, S. J. England, and A. C. Bryan. Diaphragmatic work of breathing in premature human infants. *J. Appl. Physiol.* 62:1410–1415, 1987.

Guthrie, R. D., W. A. LaFramboise, T. A. Standaert, G. Van Belle, and D. E. Woodrum. Ventilatory interaction between oxygen and carbon dioxide in the preterm primate. *Pediatr. Res.* 19:528–533, 1985.

Haddad, G. G., H. L. Leistner, R. A. Epstein, M. A. F. Epstein, W. K. Grodin, and R. B. Mellins. CO_2-induced changes in ventilation and ventilatory pattern in normal sleeping infants. *J. Appl. Physiol.* 48:684–688, 1980.

Haddad, G. G., M. R. Gandhi, and R. B. Mellins. Maturation of ventilatory responses to hypoxia in puppies during sleep. *J. Appl. Physiol.* 52:309–314, 1982.

Hagan, R., A. C. Bryan, M. H. Bryan, and G. Gulston. Neonatal chest wall afferents and regulation of respiration. *J. Appl. Physiol.* 42:362–367, 1977.

Hakim, T. S., and J. P. Mortola. Pulmonary vascular resistance in adult rats exposed to hypoxia in the neonatal period. *Can. J. Physiol. Pharmacol.* 68:419–424, 1990.

Hales, K. A., M. A. Morgan, and G. R. Thurnau. Influence of labor and route of delivery on the frequency of respiratory morbidity in term neonates. *Int. J. Gynecol. Obst.* 43:35–40, 1993.

Haltenorth, T., and H. Diller (translated by R. W. Hayman). *Mammals of Africa.* Lexington, Mass.: Stephen Green Press, 1988.

Hanson, M. A., G. J. Eden, J. G. Nijhuis, and P. J. Moore. Peripheral chemoreceptors and other oxygen sensors in the fetus and newborn. In: *Chemoreceptors and Reflexes in Breathing: Cellular and Molecular Aspects,* edited by S. Lahiri, R. E. Forster II, R. O. Davies, and A. I. Pack. New York: Oxford University Press, pp. 113–120, 1989a.

Hanson, M. A., P. Kumar, and B. A. Williams. The effect of chronic hypoxia upon the development of respiratory chemoreflexes in the newborn kitten. *J. Physiol. (London)* 411:563–574, 1989b.

Harding, R. State-related developmental changes in laryngeal function. *Sleep* 3:307–322, 1980.

―――. Fetal breathing: Relation to postnatal breathing and lung development. In: *Fetus and Neonate–Physiology and Clinical Applications,* edited by M. A. Hanson, J. A. D. Spencer, C. H. Rodeck, and D. Walters. Cambridge: Cambridge University Press., ch. 4, pp. 63–84, 1994.

Harding, R., and S. B. Hooper. Regulation of lung expansion and lung growth before birth. *J. Appl. Physiol.* 81:209–224, 1996.

Harding, R., and D. A. Titchen. Oesophageal and diaphragmatic activity during sucking in lambs. *J. Physiol. (London)* 321:317–329, 1981.

Harding, R., P. Johnson, M. E. McClelland, C. N. McLeod, P. L. Whyte, and A. R. Wilkinson. Respiratory and cardiovascular responses to feeding in lambs (abstract). *J. Physiol. (London)* 275:40P–41P, 1978a.

Harding, R., P. Johnson, and M. E. McClelland. Liquid-sensitive laryngeal receptors in the developing sheep, cat and monkey. *J. Physiol. (London)* 277:409–422, 1978b.

―――. Respiratory function of the larynx in developing sheep and the influence of sleep state. *Respir. Physiol.* 40:165–179, 1980.

Harned, H. S., Jr., R. T. Herrington, and J. I. Ferreiro. The effects of immersion and temperature on respiration in newborn lambs. *Pediatrics* 45:598–605, 1970.

Harned, H. S., Jr., J. Myracle, and J. Ferreiro. Respiratory suppression and swallowing from introduction of fluids into the laryngeal region of the lamb. *Pediatr. Res.* 12:1003–1009, 1978.

Harpin, V. A., G. Chellappah, and N. Utter. Responses of the newborn infant to overheating. *Biol. Neonate* 44:65–75, 1983.

Harris, W. H., V. D. Jansen, and S. Yamashiro. Plasma and erythrocyte volume of newborn rabbits delivered after 30 days of gestation. *Lab. Anim.* 17:294–297, 1983.

Harrison, V. C., H. de V. Heese, and M. Klein. The significance of grunting in hyaline membrane disease. *Pediatrics* 41:549–559, 1968.

Hart, J. S., O. Heroux, W. H. Cottle, and C. A. Mills. The influence of climate on metabolic and thermal responses of infant caribou. *Can. J. Zool.* 39:845–856, 1961.

Harvey, J. W., R. L. Asquith, P. K. McNulthy, J. Kivipelto, and J. E. Bauer. Haematology of foals up to one year old. *Equine Vet. J.* 16:347–353, 1984.

Hasan, S. U., and A. Rigaux. The effects of lung distension, oxygenation and gestational age on fetal behavior and breathing movements in sheep. *Pediatr. Res.* 30:193–201, 1991.

―――. Arterial oxygen tension threshold range for the onset of arousal and breathing in fetal sheep. *Pediatr. Res.* 32:342–349, 1992a.

———. Effect of bilateral vagotomy on oxygenation, arousal, and breathing movements in fetal sheep. *J. Appl. Physiol.* 73:1402–1412, 1992b.

Hasan, S. U., H. B. Sarnat, and R. N. Auer. Vagal nerve maturation in the fetal lamb: an ultrastructural and morphometric study. *Anat. Rec.* 237:527–537, 1993.

Haxhiu-Poskurica, B., W. A. Carlo, M. J. Miller, J. M. DiFiore, M. A. Haxhiu, and R. J. Martin. Maturation of respiratory reflex responses in the piglet. *J. Appl. Physiol.* 70:608–616, 1991.

Head, H. On the regulation of respiration. *J. Physiol. (London)* 10:1–70, 1889.

Heldt, G. P. Development of stability of the respiratory system in preterm infants. *J. Appl. Physiol.* 65:441–444, 1988.

Hellbrügge, T. The development of circadian rhythms in infants. *Cold Spring Harbor Symp. Quant. Biol.* 25:311–323, 1960.

Helms, P., C. S. Beardsmore, and J. Stocks. Absolute intraesophageal pressure at functional residual capacity in infancy *J. Appl. Physiol.* 51:279–275, 1981.

Hemmingsen, A. M. Energy metabolism as related to body size and respiratory surfaces, and its evolution. *Reports of the Steno Memorial Hospital and Nordinsk Insulin Laboratorium* 9:6–110, 1960 (quoted and elaborated upon in R. H. Peters, 1983).

Henderson-Smart, D. J., and D. J. C. Read. Depression of intercostal and abdominal muscle activity and vulnerability to asphyxia during active sleep in the newborn. In: *Sleep Apnea Syndromes,* edited by C. Guilleminault and W. C. Dement. New York: Liss, pp. 93–117, 1078.

Herget, J., and V. Hampl. Pulmonary vasculature of adult rats is influenced by perinatal experience of hypoxia. In: *Pulmonary Blood Vessels in Lung Disease,* edited by J. Widimský and J. Herget. Basel: Karger, vol. 26, pp. 70–76, 1990.

Herrington, R. T., H. S. Harned, Jr., J. I. Ferreiro, and C. A. Griffin, III. The role of the central nervous system in perinatal respiration: studies of chemoregulatory mechanisms in the term lamb. *Pediatrics* 47:857–864, 1971.

Hershenson, M. B., A. R. Stark, and J. Mead. Action of the inspiratory muscles of the rib cage during breathing in newborns. *Am. Rev. Respir. Dis.* 139:1207–1212, 1989.

Hertzberg, T., and H. Lagercrantz. Postnatal sensitivity of the peripheral chemoreceptors in newborn infants. *Arch. Dis. Child.* 62:1238–1241, 1987.

Hertzberg, T., S. Hellström, H. Lagercrantz, and J. M. Pequignot. Development of the arterial chemoreflex and turnover of carotid body catecholamines in the newborn rat. *J. Physiol. (London)* 425:211–225, 1990.

Heusner, A. A. Energy metabolism and body size. I. Is the 0.75 mass exponent of Kleiber's equation a statistical artifact? *Respir. Physiol.* 48:1–12, 1982.

Hey, E. N. The relation between environmental temperature and oxygen consumption in the new-born baby. *J. Physiol (London)* 200:589–603, 1969.

Hilaire, G., and B. Duron. Maturation of the mammalian respiratory system. *Physiol. Rev.* 79:325–360, 1999.

Hildebran, J. N., J. Goerke, and J. A. Clements. Surfactant release in excised rat lung is stimulated by air inflation. *J. Appl. Physiol.* 51:905–910, 1981.

Hill, J. R. The oxygen consumption of newborn and adult mammals. Its dependence on the oxygen tension in the inspired air and on the environmental temperature. *J. Physiol. (London)* 149:346–373, 1959.

Hissa, R. Postnatal development of thermoregulation in the Norwegian lemming and the golden hamster. *Ann. Zool. Finn.* 5:345–383, 1968.

Hissa, R., and K. Lagerspetz. The postnatal development of homoiothermy in the golden hamster. *Ann. Med. Exp. Finn.* 42:43–45, 1964.

Hoch, B., M. Bernhard, and A. Hinsch. Different patterns of sighs in neonates and young infants. *Biol. Neonate* 74:16–21, 1998.

Hodson, W. A., A. Fenner, G. Brumley, V. Chernick, and M. E. Avery. Cerebrospinal fluid and blood acid-base relationships in fetal and neonatal lambs and pregnant ewes. *Respir. Physiol.* 4:322–332, 1968.

Hodson, W. A., S. Palmer, G. A. Blakely, J. H. Murphy, D. E. Woodrum, and T. E. Morgan. Lung development in the fetal primate *Macaca nemestrina*. I. Growth and compositional changes. *Pediatr. Res.*11:1009–1014 (actually printed at pp. 1051–1056), 1977.

Hofer, M. A. Lethal respiratory disturbance in neonatal rats after arterial chemoreceptor denervation. *Life Sci.* 34:489–496, 1984.

———. Role of carotid sinus and aortic nerves in respiratory control of infant rats. *Am. J. Physiol.* 251:R811–R817, 1986.

Hohimer, A. R., J. M. Bissonnette, B. S. Richardson, and C. M. Machida. Central chemical regulation of breathing movements in fetal lambs. *Respir. Physiol.* 52:99–111, 1983.

Holland, R. A. B., S. J. Calvert, R. M. Hope, and C. M. Chesson. Blood O_2 transport in newborn and adult of a very small marsupial (*Sminthopsis crassicaudata*). *Respir. Physiol.* 98:69–81, 1994.

Holloway, D. A., and A. G. Heath. Ventilatory changes in the golden hamster, *Mesocricetus auratus*, compared to the laboratory rat, *Rattus norvegicus*, during hypercapnia and/or hypoxia. *Comp. Biochem. Physiol.* 77A:267–273, 1984.

Hooper, S. B., and R. Harding. Changes in lung liquid dynamics induced by prolonged fetal hypoxemia. *J. Appl. Physiol.* 69:127–135, 1990.

Hooper, S. B., K. A. Dickson, and R. Harding. Fetal lung liquid secretion, flow and volume in response to reduced uterine blood flow in fetal sheep. *J. Dev. Physiol.* 10:473–485, 1988.

Hooper, S. B., V. K. M. Han, and R. Harding. Changes in lung expansion alter pulmonary DNA synthesis and IGF-II gene expression in fetal sheep. *Am. J. Physiol.* 265:L403–L409, 1993.

Hope, P. J., D. Pyle, C. B. Daniels, I. Chapman, M. Horowitz, J. E. Morley, P. Trayhurn, J. Kumaratilake, and G. Wittert. Identification of brown fat and mechanisms for energy balance in the marsupial, *Sminthopsis crassicaudata. Am. J. Physiol.* 273:R161–R167, 1997.

Horne, R. S. C., P. J. Berger, G. Bowes, and A. M. Walker. Effect of sinoaortic denervation on arousal responses to hypotension in newborn lambs. *Am. J. Physiol.* 256:H434–H440, 1989.

Howatt, W. F., and G. R. Demuth. Configuration of the chest. *Pediatrics* 35:177–184, 1965.

Howatt, W. F., M. E. Avery, P. W. Humphreys, I. C. S. Normand, L. Reid, and L. B. Strang. Factors affecting pulmonary surface properties in the foetal lamb. *Clin. Sci.* 29:239–248, 1965.

Hughes, D. T. D., H. R. Parker, and J. V. Williams. The response of foetal sheep and lambs to pulmonary inflation. *J. Physiol. (London)* 189:177–187, 1967.

Huisman, T. H. J., J. P. Lewis, M. H. Blunt, H. R. Adams, A. Miller, A. M. Dozy, and E. M. Boyd. Hemoglobin C in newborn sheep and goats: a possible explanation for its function and biosynthesis. *Pediatr. Res.* 3:189–198, 1969.

Hulbert, A. J. Metabolism and the Development of Endothermy. In: *The Developing Marsupial. Models for Biomedical Research,* edited by C. H. Tyndale-Biscoe and P. A. Janssens. Berlin: Springer-Verlag, ch.11, pp. 148–161, 1988.

Hulbert, A. J., and P. L. Else. The cellular basis of endothermic metabolism: a role for "leaky" membranes? *News in Physiol. Sci.* 5:25–28, 1990.

Hulbert, A. J., W. Mantaj, and P. A. Janssens. Development of mammalian endothermic metabolism: quantitative changes in tissue mitochondria. *Am. J. Physiol.* 261:R561–R568, 1991.

Hull, D. Oxygen consumption and body temperature of new-born rabbits and kittens exposed to cold. *J. Physiol. (London)* 177:192–202, 1965.

———. Lung expansion and ventilation during resuscitation of asphyxiated newborn infants. *J. Pediatr.* 75: 47–58, 1969.

Hull, D., and M. M. Segall. The contribution of brown adipose tissue to heat production in the new-born rabbit. *J. Physiol. (London)* 181:449–457, 1965.

Hull, D., and O. R. C. Smales. Heat production in the newborn. In: *Temperature Regulation and Energy Metabolism in the Newborn,* edited by J. C. Sinclair. New York: Grune & Stratton, pp. 129–156, 1978.

Hull, D., J. Hull, and J. Vinter. The preferred environmental temperature of newborn rabbits. *Biol. Neonate* 50:323–330, 1986.

Humphreys, P. W., and L. B. Strang. Effects of gestation and prenatal asphyxia on pulmonary surface properties of the foetal rabbit. *J. Physiol. (London)* 192:53–62, 1967.

Humphreys, P. W., I. C. S. Normand, E. O. R. Reynolds, and L. B. Strang. Pulmonary lymph flow and the uptake of liquid from the lungs of the lamb at the start of breathing. *J. Physiol. (London)* 193:1–29, 1967.

Hutchison, A. A., K. R. Ross, and G. Russell. The effect of posture on ventilation and lung mechanics in preterm and light-for-date infants. *Pediatrics* 64:429–432,1979.

Ianuzzo, C. D., and D. A. Hood. Cellular and molecular adaptations of the respiratory muscles. In: *The Thorax, second edition, revised and expanded. Part A : Physiology,* Lung Biology in Health and Disease Series, vol.85, edited by C. Roussos. New York: Marcel Dekker, ch. 11, pp. 313–348, 1995.

Ikegami, M., F. H. Adams, B. Towers, and A. B. Osher. The quantity of natural surfactant necessary to prevent the respiratory distress syndrome in premature lambs. *Pediatr. Res.* 14:1082–1085, 1980.

Inscore, S. C., K. R. Stenmark, C. Orton, and C. G. Irvin. Neonatal calves develop airflow limitation due to chronic hypobaric hypoxia. *J. Appl. Physiol.* 70:384–390, 1990.

Ioffe, S., A. H. Jansen, and V. Chernick. Maturation of spontaneous fetal diaphragmatic activity and fetal response to hypercapnia and hypoxemia. *J. Appl. Physiol.* 62:609–622, 1987.

Iscoe, S. D. Central control of the upper airways. In: *Respiratory Function of the Upper Airways,* edited by O. P. Mathew and G. Sant'Ambrogio, Lung Biology in Health and Disease Series, vol. 35. New York: Marcel Dekker, ch. 5, pp. 125–192, 1988.

Issa, F. G., and J. E. Remmers. Identification of a subsurface area in the ventral medulla sensitive to local changes in PCO_2. *J. Appl. Physiol.* 72:439–446, 1992.

Iversen, J. A., and J. Krog. Heat production and body surface area in seals and sea otters. *Norw. J. Zool.* 21:51–54,1973.

Jacobs, R., J. S. Robinson, J. A. Owens, J. Falconer, and M. E. D. Webster. The effect of prolonged hypobaric hypoxia on growth of fetal sheep. *J. Dev. Physiol.* 10:97–112, 1988.

Jansen, A. H., and V. Chernick. Development of respiratory control. *Physiol. Rev.* 63:437–483, 1983.

———. Fetal breathing and development of control of breathing. *J. Appl. Physiol.* 70:1431–1446, 1991.

Jansen, A. H., S. Ioffe, B. J. Russell, and V. Chernick. Effect of carotid chemoreceptor denervation on breathing in utero and after birth. *J. Appl. Physiol.* 51:630–633, 1981.

———. Influence of sleep state on the response to hypercapnia in fetal lambs. *Respir. Physiol.* 48:125–142, 1982.

Janský, L. Non-shivering thermogenesis and its thermoregulatory significance. *Biol. Rev.* 48:85–132, 1973.

Jaykka, S. Capillary erection and the structural appearance of fetal and neonatal lungs. *Acta Paediatr.* 47:484–500, 1958.

Johanson, C. E., J. Allen, and C. D. Withrow. Regulation of pH and HCO₃ in brain and CSF of the developing mammalian central nervous system. *Dev. Brain Res.* 38:255–264, 1988.

Johnson, P., D. M. Salisbury, and A. T. Storey. Apnoea induced by stimulation of sensory receptors in the larynx. In: *Development of Upper Respiratory Anatomy and Function,* edited by J. F. Bosma and J. Showacre. Rockville, Md.: U.S. Dept. HEW, ch. 11, pp. 160–178, 1975.

Johnson, B. D., L. E. Wilson, W. Z. Zhan, J. F. Watchko, M. J. Daood, and G. C. Sieck. Contractile properties of the developing diaphragm correlate with myosin heavy chain phenotype. *J. Appl. Physiol.* 77:481–487, 1994.

Johnston, B. M., and P. D. Gluckman. Lateral pontine lesions affect central chemosensitivity in unanesthetized fetal lambs. *J. Appl. Physiol.* 67:1113–1118, 1989.

Jost, A., and A. Policard. Contribution expérimentale a l'étude du développement prénatal du poumon chez le lapin. *Arch. Anat. Microsc. Morphol. Exp.* 37:325–332, 1948.

Jouvet-Mounier, D., L. Astic, and D. Lacote. Ontogenesis of the states of sleep in rat, cat and guinea pig during the first postnatal month. *Dev. Psychobiol.* 2:216–239, 1970.

Kalia, M. Visceral and somatic reflexes produced by J pulmonary receptors in newborn kittens. *J. Appl. Physiol.* 41:1–6, 1976.

Karlberg, P. The adaptive changes in the immediate postnatal period, with particular reference to respiration. *J. Pediatr.* 56:585–604, 1960.

Karlberg, P., F. H. Adams, F. Geubelle, and G. Wallgren. Alterations of the infant's thorax during vaginal delivery. *Acta Ostet. Gynecol. Scand.* 41:223–229, 1962a.

Karlberg, P., R. B. Cherry, F. E. Escardo, and G. Koch. Respiratory studies in newborn infants. II. Pulmonary ventilation and mechanics of breathing in the first minutes of life, including the onset of respiration. *Acta Paediatr.* 51:121–136, 1962b.

Kattwinkel, J., H. S. Nearman, A. A. Fanaroff, P. G. Katona, and M. H. Klaus. Apnea of prematurity. *J. Pediatr.* 86:588–592, 1975.

Keeley, F. W., D. G. Fagan, and S. I. Webster. Quantity and character of elastin in developing human lung parenchymal tissues of normal infants and infants with respiratory distress syndrome. *J. Lab. Clin. Med.* 90:981–989, 1977.

Keens, T. G., A. C. Bryan, H. Levison, and C. D. Ianuzzo. Developmental pattern of muscle fiber types in human ventilatory muscles. *J. Appl. Physiol.* 44:909–913,1978.

Khater-Boidin, J., F. Wallois, P. Toussaint, and B. Duron. Nonvagal reflex apnea in the newborn kitten and during the early postnatal period. *Biol. Neonate* 65:41–50, 1994.

Kholwadwala, D., and D. F. Donnelly. Maturation of carotid body chemoreceptor sensitivity to hypoxia: in vitro studies. *J. Physiol. (London)* 453:461–473, 1992.

Kida, K., and W. M. Thurlbeck. The effects of β-aminoproprionitrile on the growing rat lung. *Am. J. Pathol.* 101:693–710, 1980.

Kikkawa, Y., M. Kaibara, E. K. Motoyama, M. M. Orzalesi, and C. D. Cook. Morphologic development of fetal rabbit lung and its acceleration with cortisol. *Am. J. Pathol.* 64:423–442, 1971.

Kiorpes, A. L., G. E. Bisgard, and M. Manohar. Pulmonary function value in healthy Holstein-Friesian calves. *Am. J. Vet. Res.* 39:773–778, 1978.

Kitterman, J. A. Arachidonic acid metabolites and control of breathing in the fetus and newborn. *Semin. Perinatol.* 11:43–52, 1987.

Klaus, M., W. H. Tooley, K. H. Weaver, and J. A. Clements. Lung volume in the newborn infant. *Pediatrics* 30:111–116, 1962a.

Klaus, M., O. K. Reiss, W. H. Tooley, C. Piel, and J. A. Clements. Alveolar epithelial cell mitochondria as source of the surface-active lung lining. *Science* 137:750–751, 1962b.

Kleiber, M. Body size and metabolism. *Hilgardia* 6:315–353, 1932.

———. *The Fire of Life: An Introduction to Animal Energetics.* New York: Wiley, p. 454, 1961.

Klingenberg, M. Mechanism and evolution of the uncoupling protein of brown adipose tissue. *Trends in Biochem. Sci.* 15:108–112, 1990.

Knill, R., and A. C. Bryan. An intercostal-phrenic inhibitory reflex in human newborn infants. *J. Appl. Physiol.* 40:352–356, 1976.

Knill, R., W. Andrews, A. C. Bryan, and M. H. Bryan. Respiratory load compensation in infants. *J. Appl. Physiol.* 40:357–361, 1976.

Kochi, T., S. Okubo, W. A. Zin, and J. Milic-Emili. Chest wall and respiratory system mechanics in cats: effects of flow and volume. *J. Appl. Physiol.* 64:2636–2646, 1988.

Koepke, G. H., A. J. Murphy, J. W. Rae, Jr., and D. G. Dickinson. An electromyographic study of some of the muscles used in respiration. *Arch. Phys. Med. Rehabil.* 36:217–222, 1955.

Konno, K., and J. Mead. Measurement of the separate volume changes of rib cage and abdomen during breathing. *J. Appl. Physiol.* 22:407–422, 1967.

Koos, B. J., and H. Sameshima. Effects of hypoxaemia and hypercapnia on breathing movements and sleep state in sinoaortic-denervated fetal sheep. *J. Develop. Physiol.* 10:131–144, 1988.

Koos, B. J., T. Kitanaka, K. Matsuda, R. D. Gilbert, and L. D. Longo. Fetal breathing adaptation to prolonged hypoxaemia in sheep. *J. Develop. Physiol.* 10:161–166, 1988.

Korpas, J., and G. Kalocsayova. Mechanoreception of the cat respiratory tract on the first days of postnatal life. *Physiol. Bohemoslov.* 22:365–373, 1973.

Kosch, P. C., and A. R. Stark. Dynamic maintenance of end-expiratory lung volume in full-term infants. *J. Appl. Physiol.* 57:1126–1133, 1984.

Kosch, P. C., P. W. Davenport, J. A. Wozniak, and A. R. Stark. Reflex control of inspiratory duration in newborn infants. *J. Appl. Physiol.* 60:2007–2014, 1986.

Kotas, R. V., and M. E. Avery. Accelerated appearance of pulmonary surfactant in the fetal rabbit. *J. Appl. Physiol.* 30:358–361, 1971.

Kotas, R. V., B. D. Fletcher, J. Torday, and M. E. Avery. Evidence of independent regulators of organ maturation in fetal rabbits. *Pediatrics* 47:57–64, 1971.

Kotas, R. V., P. M. Farrel, R. E. Ulane, and R. A. Chez. Foetal rhesus monkey lung development: lobar differences and discordances between stability and distensibility. *J. Appl. Physiol.* 43:92–98, 1977.

Koterba, A. M., and P. C. Kosch. Respiratory mechanics and breathing pattern in the neonatal foal. *J. Reprod. Fert. Suppl.* 35:575–586, 1987.

Koterba, A. M., J. A. Wozniak, and P. C. Kosch. Respiratory mechanics of the horse during the first year of life. *Respir. Physiol.* 95:21–41, 1994.

Kovar, I., U. Selstam, W. Z. Catterton, M. T. Stahlman, and H. W. Sundell. Laryngeal chemoreflex in newborn lambs: respiratory and swallowing response to salts, acids, and sugars. *Pediatr. Res.* 13:1144–1149, 1979.

Kozuma, S., N. Hidenori, N. Unno, H. Kagawa, A. Kikuchi, T. Fujii, K. Baba, T. Okai, Y. Kuwabara, and Y. Taketani. Goat fetuses disconnected from the placenta, but reconnected to an artificial placenta, display intermittent breathing movements. *Biol. Neonate* 75:388–397, 1999.

Krauss, A. N., D. W. Thibeault, and P. A. M. Auld. Acid-base in cerebrospinal fluid of newborn infants. *Biol. Neonate* 21:25–34, 1972.

Krebs, H. A. Body size and tissue respiration. *Biochim. Biophys. Acta* 4:249–269, 1950.

Krous, H. F., J. Jordan, J. Wen, and J. P. Farber. Developmental morphometry of the vagus nerve in the opossum. *Dev. Brain Res.* 20:155–159, 1985.

Kugelman, A., T. G. Keens, R. deLemos, and M. Durand. Comparison of dynamic and passive measurements of respiratory mechanics in ventilated newborn infants. *Pediatr. Pulmonol.* 20:258–264, 1995.

Kuhnen, G. O_2 and CO_2 concentrations in burrows of euthermic and hibernating golden hamsters. *Comp. Biochem. Physiol.* 84A:517–522, 1986.

Kuipers, I. M., W. J. Maertzdorf, H. Keunen, D. S. De Jong, M. A. Hanson, and C. E. Blanco. Fetal breathing is not initiated after cord occlusion in the unanaesthetized fetal lamb in utero. *J. Develop. Physiol.* 17:233–240, 1992.

Kuipers, I. M., W. J. Maertzdorf, D. S. De Jong, M. A. Hanson, and C. E. Blanco.

Effect of mild hypocapnia on fetal breathing and behavior in unanesthetized normoxic fetal lambs. *J. Appl. Physiol.* 76:1476–1480, 1994.

———. The effect of hypercapnia and hypercapnia associated with central cooling on breathing in unanesthetized fetal lamb. *Pediatr. Res.* 41:90–95, 1997a.

———. Initiation and maintenance of continuous breathing at birth. *Pediatr. Res.* 42:163–168, 1997b.

Kumar, P., and M. A. Hanson. Re-setting of the hypoxic sensitivity of aortic chemoreceptors in the new-born lamb. *J. Develop. Physiol.* 11:199–206, 1989.

Lachmann, B., G. Grossmann, R. Nilsson, and B. Robertson. Lung mechanics during spontaneous ventilation in premature and fullterm rabbit neonates. *Respir. Physiol.* 38:283–302, 1979.

LaFramboise, W. A., R. D. Guthrie, T. A. Standaert, and D. E. Woodrum. Pulmonary mechanics during the ventilatory response to hypoxemia in the newborn monkey. *J. Appl. Physiol.* 55:1008–1014, 1983.

LaFramboise, W. A., D. E. Woodrum, and R. D. Guthrie. Influence of vagal activity on the neonatal ventilatory response to hypoxemia. *Pediatr. Res.* 19:903–907, 1985.

LaFramboise, W. A., T. A. Standaert, R. D. Guthrie, and D. E. Woodrum. Developmental changes in the ventilatory response of the newborn to added airway resistance. *Am. Rev. Respir. Dis.* 136:1075–1083, 1987.

Lagercrantz, H. Stress, arousal, ane gene activation at birth. *News Physiol. Sci.* 11:214–218, 1996.

Lagercrantz, H., and T. A. Slotkin. The "stress" of being born. *Sci. Am.* 254:100–107, 1986.

Lahiri, S. Blood oxygen affinity and alveolar ventilation in relation to body weight in mammals. *Am. J. Physiol.* 229:529–536, 1975.

———. Respiratory control in Andean and Himalayan high-altitude natives. In: *High Altitude and Man,* edited by J. B. West and S. Lahiri. Bethesda, Md.: Am. Physiol. Soc., ch. 3, pp. 147–162, 1984.

———. Physiological responses: peripheral chemoreceptors and chemoreflexes. In: *The Lung: Scientific Foundation,* 2nd edition, edited by R. G. Crystal, J. B. West, P. J. Barnes, and E. R. Weibel. Philadelphia: Lippincott-Raven, ch. 129, pp. 1747–1756, 1997.

Lahiri, S., J. S. Brody, E. K. Motoyama, and T. M. Velasquez. Regulation of breathing in newborns at high altitude. *J. Appl. Physiol.* 44:673–678, 1978.

Laing, I. A., R. L. Teele, and A. R. Stark. Diaphragmatic movement in newborn infants. *J. Pediatr.* 112:638–643, 1988.

Langman, J. *Medical Embryology,* 3rd edition. Baltimore: Williams & Wilkins, pp. 305–307, 1977.

Langston, C., K. Kida, M. Reed, and W. M. Thurlbeck. Human lung growth in late gestation and in the neonate. *Am. Rev. Respir. Dis.* 129:607–613, 1984.

Lanier, B., M. A. Richardson, and C. Cummings. Effect of hypoxia on laryngeal reflex apnea. Implications for sudden infant death. *Otolaryngol. Head Neck Surg.* 91:597–604, 1983.

Larson, J. E., and W. M. Thurlbeck. The effect of experimental maternal hypoxia on fetal lung growth. *Pediatr. Res.* 24:156–159, 1988.

Lavoie, J. P., J. E. Madigan, J. S. Cullor, and E. E. Powell. Haemodynamic, pathological, haematological and behavioural changes during endotoxin infusion in equine neonates. *Equine Vet. J.* 22:23–29, 1990.

Lawson, E. E., and W. A. Long. Central origin of biphasic breathing pattern during hypoxia in newborns. *J. Appl. Physiol.* 55:483–488, 1983.

Lawson, E. E., R. L. Birdwell, P. S. Huang, and H. W. Taeusch. Augmentation of pulmonary surfactant secretion by lung expansion at birth. *Pediatr. Res.* 13:611–614, 1979.

LeBlanc, P. H., J. C. Baker, P. R. Gray, N. E. Robinson, and F. J. Derksen. Effects of bovine respiratory syncytial virus on airway function in neonatal calves. *Am. J. Vet. Res.* 52:1401–1406, 1991.

Lechner, A. J., and N. Banchero. Lung morphometry in guinea pigs acclimated to cold during growth. *J. Appl. Physiol.* 48:886–891, 1980.

———. Advanced pulmonary development in newborn guinea pigs (*Cavia porcellus*). *Am. J. Anat.* 163:235–246, 1982.

Lechner, A. J., D. C. Winston, and J. E. Bauman. Lung mechanics, cellularity, and surfactant after prenatal starvation in guinea pigs. *J. Appl. Physiol.* 60:1610–1614, 1986.

Lechner, A. J., C. I. Blake, and N. Banchero. Pulmonary development in growing guinea pigs exposed to chronic hypercapnia. *Respiration* 52:108–114, 1987.

Lee, T. M., and I. Zucker. Vole infant development is influenced perinatally by maternal photoperiodic history. *Am. J. Physiol.* 255:R831–R838, 1988.

Leffler, C. W., T. L. Tyler, and S. Cassin. Effect of indomethacin on pulmonary vascular response to ventilation in fetal goats. *Am. J. Physiol.* 234:H346–H351, 1978.

Leffler, C. W., J. R. Hessler, and R. S. Green. The onset of breathing at birth stimulates pulmonary vascular prostacyclin synthesis. *Pediatr. Res.* 18:938–942, 1984a.

———. Mechanisms of stimulation of pulmonary prostacyclin synthesis at birth. *Prostaglandins* 28:877–887, 1984b.

Leiter, J. C., J. P. Mortola, and S. M. Tenney. A comparative analysis of contractile characteristics of the diaphragm and of respiratory system mechanics. *Respir. Physiol.* 64:267–276, 1986.

Leith, D. E. Comparative mammalian respiratory mechanics. *Physiologist* 19:485–510, 1976.

———. Comparative mammalian respiratory mechanics. *Am. Rev. Respir. Dis.*, Suppl. 128:S77–S82, 1983.

Leith, D. E., and J. R. Gillespie. Respiratory mechanics of normal horses and one with chronic obstructive lung disease (abstract). *Feder. Proc.* 30:556Abs, 1971.

Lekeux, P., R. Hajer, and H. J. Breukink. Effect of somatic growth on pulmonary function values in healthy Friesian cattle. *Am. J. Vet. Res.* 45:2003–2007, 1984.

Leonard, C. M. Thermotaxis in golden hamster pups. *J. Comp. Physiol. Psychol.* 86:458–469, 1974.

LeSouëf, P. N., J. M. Lopes, S. J. England, M. H. Bryan, and A. C. Bryan. Influence of chest wall distortion on esophageal pressure. *J. Appl. Physiol.* 55:353–358, 1983.

LeSouef, P. N., S. J. England, and A. C. Bryan. Passive respiratory mechanics in newborns and children. *Am. Rev. Respir. Dis.* 129:552–556, 1984.

LeSouëf, P. N., S. J. England, H. A. F. Stogryn, and A. C. Bryan. Comparison of diaphragmatic fatigue in newborn and older rabbits. *J. Appl. Physiol.* 65:1040–1044, 1988.

Lieberman, D. A., L. C. Maxwell, and J. A. Faulkner. Adaptation of guinea pig diaphragm muscle to aging and endurance training. *Am. J. Physiol.* 222:556–560, 1972.

Lindroth, M., B. Johnson, H. Ahlstrom, and N. W. Svenningsen. Pulmonary mechanics in early infancy. Subclinical grunting in low-birth-weight infants. *Pediatr. Res.* 15:979–984, 1981.

Lindstedt, S. L., and W. A. Calder, III. Body size, physiological time, and longevity of homeothermic animals. *Quart. Rev. Biol.* 56:1–16, 1981.

Ling, L., E. B. Olson, Jr., E. H. Vidruk, and G. S. Mitchell. Attenuation of the hypoxic ventilatory response in adult rats following one month of perinatal hyperoxia. *J. Physiol. (London)* 495:561–571, 1996.

———. Developmental plasticity of the hypoxic ventilatory response. *Respir. Physiol.* 110:261–268, 1997.

Litmanovitz, I., I. Dreshaj, M. J. Miller, M. A. Haxhiu, and R. J. Martin. Cental chemosensitivity affects respiratory muscle responses to laryngeal stimulation in the piglet. *J. Appl. Physiol.* 76:403–408, 1994.

Littlejohn, A. Aspects of respiration in anaesthetized newborn foals. *J. Reprod. Fert. Suppl.* 23:681–684, 1975.

Lodrup Carlsen, K. C., P. Magnus, and K.-H. Carlsen. Lung function by tidal breathing in awake healthy newborn infants. *Europ. Respir. J.* 7:1660–1668, 1994.

Lopes, J. M., N. L. Muller, M. H. Bryan, and A. C. Bryan. Synergistic behavior of inspiratory muscles after diaphragmatic fatigue in the newborn. *J. Appl. Physiol.* 51:547–551, 1981.

Lorino, H., G. Moriette, C. Mariette, A.-M. Lorino, A. Harf, and P.-H. Jarreau. Inspiratory work of breathing in ventilated preterm infants. *Pediatr. Pulmonol.* 21:323–327, 1996.

Lowry, T. F., H. V. Forster, L. G. Pan, M. A. Korducki, J. Probst, R. A. Franciosi, and M. Forster. Effect of carotid body denervation on breathing in neonatal goats. *J. Appl. Physiol.* 87:1026–1034, 1999a.

Lowry, T. F., H. V. Forster, L. G. Pan, A. Serra, J. Wenninger, R. Nash, D. Sheridan, and R. A. Franciosi. Effects on breathing of carotid body denervation in neonatal piglets. *J. Appl. Physiol.* 87:2128–2135, 1999b.

Lucier, G. E., A. T. Storey, and B. J. Sessle. Effects of upper respiratory tract stimuli on neonatal respiration: reflex and single neuron analyses in the kitten. *Biol. Neonate* 35:82–89, 1979.

Ludwig, M. S., I. Dreshaj, J. Solway, A. Munoz, and R. H. Ingram, Jr. Partitioning of pulmonary resistance during constriction in the dog: effects of volume history. *J. Appl. Physiol.* 62:807–815, 1987.

Lyrene, R. K., K. A. Welch, G. Godoy, and J. B. Philips, III. Alkalosis attenuates hypoxic pulmonary vasoconstriction in neonatal lambs. *Pediatr. Res.* 19:1268–1271, 1985.

Mahaffey, L. W., and P. D. Rossdale. A convulsive syndrome in newborn foals resembling pulmonary syndrome in the newborn infant. *Lancet* (1):1223–1225, 1959.

Marchal, F., A. Bairam, P. Haouzi, J. P. Crance, C. Di Giulio, P. Vert, and S. Lahiri. Carotid chemoreceptor response to natural stimuli in the newborn kitten. *Respir. Physiol.* 87:183–193, 1992.

Marlot, D., and B. Duron. Postnatal development of vagal control of breathing in the kitten. *J. Physiol. (Paris)* 75:891–900, 1979a.

———. Postnatal maturation of phrenic, vagus, and intercostal nerves in the kitten. *Biol. Neonate* 36:264–272, 1979b.

Marlot, D., and J. P. Mortola. Positive- and negative-pressure breathing in newborn rat before and after anesthesia. *J. Appl. Physiol.* 57:1454–1461, 1984.

Marlot, D., J. P. Mortola, and B. Duron. Functional localization of pulmonary stretch receptors in the tracheobronchial tree of the kitten. *Can. J. Physiol. Pharmacol.* 60:1073–1077, 1982.

Marsland, D. W., B. J. Callahan, and D. C. Shannon. The afferent vagus and regulation of breathing in response to inhaled CO_2 in awake newborn lambs. *Biol. Neonate* 27:102–107, 1975.

Martin, C. J., S. Chihara, and D. B. Chang. A comparative study of the mechanical properties in aging alveolar wall. *Am. Rev. Respir. Dis.* 115:981–998, 1977.

Martin, R. J., A. Okken, P. G. Katona, and M. H. Klaus. Effect of lung volume on expiratory time in the newborn infant. *J. Appl. Physiol.* 45:18–23, 1978.

Martin, R. J., W. A. Carlo, S. S. Robertson, W. R. Day, and E. N. Bruce. Biphasic response of respiratory frequency to hypercapnea in preterm infants. *Pediatr. Res.* 19:791–796, 1985.

Martin, R. J., I. A. Dreshaj, M. J. Miller, and M. A. Haxhiu. Neurochemical control of tissue resistance in piglets. *J. Appl. Physiol.* 79:812–817, 1995.

Martin-Body, R. L., and B. M. Johnston. Central origin of the hypoxic depression of breathing in the newborn. *Respir. Physiol.* 71:25–32, 1988.

Martin-Caraballo, M., P. A. Campagnaro, Y. Gao, and J. J. Greer. Contractile and fatigue properties of the rat diaphragm musculature during the perinatal period. *J. Appl. Physiol.* 88:573–580, 2000.

Massaro, G. D., and D. Massaro. Morphologic evidence that large inflations of the lung stimulate secretion of surfactant. *Am. Rev. Respir. Dis.* 127:235–236, 1983.

Massaro, G. D., J. Olivier, and D. Massaro. Short term perinatal 10% O_2 alters postnatal development of lung alveoli. *Am. J. Physiol.* 257:L221–L225, 1989.

Massaro, G. D., J. Olivier, C. Dzikowski, and D. Massaro. Postnatal development of lung alvcoli: suppression by 13% O_2 and a critical period. *Am. J. Physiol.* 258:L321–L327, 1990.

Mathew, O. P. Regulation of breathing pattern during feeding. Role of suck, swallow, and nutrients. In: *Respiratory Function of the Upper Airways,* Lung Biology in Health and Disease Series, vol. 35, edited by O. P. Mathew and G. Sant'Ambrogio. New York: Marcel Dekker, ch. 15, pp. 535–560, 1988.

Mathew, O. P., M. L. Clark, and M. H. Pronske. Breathing pattern of neonates during nonnutritive sucking. *Pediatr. Pulmonol.* 1:204–206, 1985.

Mathew, O. P., J. W. Anderson, G. P. Orani, F. B. Sant'Ambrogio, and G. Sant'Ambrogio. Cooling mediates the ventilatory depression associated with airflow through the larynx. *Respir. Physiol.* 82:359–368, 1990.

Matoth, Y., R. Zaizov, and I. Varsano. Postnatal changes in some red cell parameters. *Acta Paediatr. Scand.* 60:317–323, 1971.

Matsuoka, T., and J. P. Mortola. Effects of hypoxia and hypercapnia on the Hering-Breuer reflex of the conscious newborn rat. *J. Appl. Physiol.* 78:5–11, 1995.

Matsuoka, T., T. Yoda, S. Ushikubo, S. Matsuzawa, T. Sasano, and A. Komiyama. Repeated acute hypoxia temporarily attenuates the ventilatory response to hypoxia in conscious newborn rats. *Pediatr. Res.* 46:120–125, 1999.

Maxwell, L. C., R. J. M. McCarter, T. J. Kuehl, and J. L. Robotham. Development of histochemical and functional properties of baboon respiratory muscles. *J. Appl. Physiol.* 54:551–561, 1983.

Maxwell, L. C., T. J. Kuehl, R. J. M. McCarter, and J. L. Robotham. Regional distribution of fiber types in developing baboon diaphragm muscles. *Anat. Rec.* 224:66–78, 1989.

May, P. L'action immédiate de l'oxygène sur la ventilation chez l'homme normal. *Helv. Physiol. Acta* 15:230–240, 1957.

Mayock, D. E., J. Hall, J. F. Watchko, T. A. Standaert, and D. E. Woodrum. Diaphragmatic muscle fiber type development in swine. *Pediatr. Res.* 22:449–454, 1987.

Mayock, D. E., M. J. Dooad, D. E. Woodrum, and J. F. Watchko. Diaphragmatic myosin heavy chain expression in the subhuman primate. *Macaca nemestrina* (abstract). *Am. J. Resp. Crit. Care Med.* 149:A275, 1994.

McCarter, R. J. M. Age-related changes in skeletal muscle function. *Aging* 2:27–38, 1990.

McIlroy, M. B., D. F. Tierney, and J. A. Nadel. A new method for measurement of compliance and resistance of lungs and thorax. *J. Appl. Physiol.* 18:424–427, 1963.

McMahon, T. Size and shape in biology. *Science* 179:1201–1204, 1973.

Mead, J. Mechanical properties of lungs. *Physiol. Rev.* 41:281–330, 1961.

Mead, J., and J. L. Whittenberger. Physical properties of human lung measured during spontaneous respiration. *J. Appl. Physiol.* 5:779–796, 1953.

Medeiros, L. O., S. Ferri, S. R. Barcelos, and O. Miguel. Hematologic standards for healthy newborn thoroughbred. *Biol. Neonate* 17:351–360, 1971.

Merazzi, D., and J. P. Mortola. Effects of changes in ambient temperature on the Hering-Breuer reflex of the conscious newborn rat. *Pediatr. Res.* 45:370–376, 1999a.

———. Hering-Breuer reflex in conscious newborn rats: effects of changes in ambient temperature during hypoxia. *J. Appl. Physiol.* 87: 1656–1661, 1999b.

Milerad, J., and H. W. Sundell. Reduced inspiratory drive following laryngeal chemoreflex apnea during hypoxia. *Respir. Physiol.* 116:35–45, 1999.

Milic-Emili, J., and W. A. Zin. Breathing responses to imposed mechanical loads. In: *Handbook of Physiology: The Respiratory System,* vol. II, Control of Breathing, edited by N. S. Cherniack and J. G. Widdicombe. Bethesda, Md.: Am. Physiol. Soc., part 2, chap. 23, pp. 751–769, 1986.

Miller, A. J., and C. R. Dunmire. Characterization of the postnatal development of superior laryngeal nerve fibers in the postnatal kitten. *J. Neurobiol.* 7:483–494, 1976.

Miller, A. J., and R. F. Loizzi. Anatomical and functional differentiation of superior laryngeal nerve fibers affecting swallowing and respiration. *Exp. Neurol.* 42:369–387, 1974.

Miller, J. A., Jr., and F. S. Miller. Physiological, biochemical, and clinical aspects of hypothermia in neonatal asphyxia. In: *Depressed Metabolism,* edited by X. J. Musacchia and J. F. Saunders. New York: Elsevier, pp. 427–452, 1969.

Miller, H. C., G. O. Proud, and F. C. Behrle. Variations in the gag, cough, and swallow reflexes and tone of the vocal cords as determined by direct laryngoscopy in newborn infants. *Yale J. Biol. Med.* 24:284{endash}291, 1952.

Miller, J. A., Jr., F. S. Miller, and B. Westin. Hypothermia in the treatment of asphyxia neonatorum. *Biol. Neonate* 6:148–163, 1964.

Miller, K., M. Rosenmann, and P. Morrison. Oxygen uptake and temperature regulation of young harbor seals (*Phoca vitulina richardi*) in water. *Comp. Biochem. Physiol. A Comp. Physiol.* 54:105–107, 1976.

Milner, A. D., and R. A. Saunders. Pressure and volume changes during the first breath of human neonates. *Arch. Dis. Child.* 52:918–924, 1977.

Milner, A. D., and H. Vyas. Lung expansion at birth. *J. Pediatr.* 101:879–886, 1982.

Milner, A. D., R. A. Saunders, and I. E. Hopkin. Effects of delivery by caesarian section on lung mechanics and lung volumes in the human neonate. *Arch. Dis. Child.* 53:545–548, 1978a.

————. Tidal pressure/volume and flow/volume respiratory loop patterns in human neonates. *Clin. Sci. Mol. Med.* 54:257–264, 1978b.

Milsom, W. K. Comparative aspects of vertebrate pulmonary mechanics. In: *Comparative Pulmonary Physiology*, Lung Biology in Health and Disease Series, vol. 39, edited by S. C. Wood. New York: Marcel Dekker, ch. 19, pp. 587–619, 1989.

————. Control of breathing in hibernating mammals. In: *Physiological Adaptations in Vertebrates. Respiration, Circulation, and Metabolism*, edited by S. C. Wood, R. E. Weber, A. R. Hargens, and R. W. Millard., New York: Marcel Dekker, ch. 7, pp. 119–148, 1992.

Miserocchi, G., and J. Milic-Emili. Effect of mechanical factors on the relation between rate and depth of breathing in cats. *J. Appl. Physiol.* 41:277–284, 1976.

Miserocchi, G., T. Nakamura, and E. Agostoni. Change pattern of pleural deformation pressure on varying lung height and volume. *Respir. Physiol.* 43:197–208, 1981.

Miserocchi, G., D. Negrini, and J. P. Mortola. Comparative features of Starling-lymphatic interaction at the pleural level in mammals. *J. Appl. Physiol.* 56:1151–1156, 1984.

Moessinger, A. C., R. Harding, T. M. Adamson, M. Singh, and G. T. Kiu. Role of lung fluid volume in growth and maturation of the fetal sheep lung. *J. Clin. Invest.* 86:1270–1277, 1990.

Monge, C., and F. León-Velarde. Physiological adaptation to high altitude: oxygen transport in mammals and birds. *Physiol. Rev.* 71:1135–1172, 1991.

Moore, R. Y. Circadian rhythms: basic neurobiology and clinical applications. *Ann. Rev. Med.* 48:253- 266, 1997.

Moore, C. R., and D. Price. A study at high altitudes of reproduction, growth, sexual maturity, and organ weights. *J. Exp. Zool.* 108:171–216, 1948.

Moore, L. G., S. S. Rounds, D. Jahnigen, R. F. Grover, and J. T. Reeves. Infant birth weight is related to maternal arterial oxygenation at high altitude. *J. Appl. Physiol.* 52:695–699, 1982.

Moore, L. G., P. Brodeur, O. Chumbe, J. D'Brot, S. Hofmeister, and C. Monge.

Maternal hypoxic ventilatory response, ventilation, and infant birth weight at 4,300 m. *J. Appl. Physiol.* 60:1401–1406, 1986.

Moore, B. J., H. A. Feldman, and M. B. Reid. Developmental changes in diaphragm contractile properties. *J. Appl. Physiol.* 75:522–526, 1993.

Morrison, J. J., J. M. Rennie, and P. J. Milton. Neonatal respiratory morbidity and mode of delivery at term: influence of timing of elective caesarean section. *Br. J. Obstetr. Gynaecol.* 102:101–106, 1995.

Mortola, J. P. Body posture and breathing frequency in newborn mammals. *Pediatr. Res.* 14:1403–1407, 1980.

———. Comparative aspects of the dynamics of breathing in newborn mammals. *J. Appl. Physiol.* 54:1229–1235, 1983.

———. Breathing pattern in newborns. *J. Appl. Physiol.* 56:1533–1540, 1984.

———. Dynamics of breathing in newborn mammals. *Physiol. Rev.* 67:187–243, 1987.

———. Hamsters versus rats: ventilatory responses in adults and newborns. *Respir. Physiol.* 85:305–317, 1991.

———. Ventilatory responses to hypoxia in mammals. In: *Tissue Oxygen Deprivation: Developmental, Molecular and Integrated Function,* Lung Biology in Health and Disease Series, vol. 95, edited by G. G. Haddad and G. Lister. New York: Marcel Dekker, ch. 15, pp. 433–477, 1996.

———. How newborn mammals cope with hypoxia. *Respir. Physiol.* 116:95–103, 1999.

Mortola, J. P., and A. Dotta. Effects of hypoxia and ambient temperature on gaseous metabolism of newborn rats. *Am. J. Physiol.* 263:R267–R272, 1992.

Mortola, J. P., and C. Feher. Hypoxia inhibits cold-induced huddling in rat pups. *Respir. Physiol.* 113:213–222, 1998.

Mortola, J. P., and J. T. Fisher. Comparative morphology of the trachea in newborn mammals. *Respir. Physiol.* 39:297–302, 1980.

———. Mouth and nose resistance in newborn kittens and puppies. *J. Appl. Physiol.* 51:641–645, 1981.

———. Upper airway reflexes in newborns. In: *Respiratory Function of the Upper Airways,* Lung Biology in Health and Disease Series, vol. 35, edited by O. P. Mathew and G. Sant'Ambrogio. New York: Marcel Dekker, ch. 9, pp. 303–357, 1988.

Mortola, J. P., and P. B. Frappell. Ventilatory responses to changes in temperature in mammals and other vertebrates. *Ann Rev. Physiol.,* 62:847–874, 2000.

Mortola, J. P., and H. Gautier. Interaction between metabolism and ventilation: effects of respiratory gases and temperature. In: *Regulation of Breathing, 2nd Edition Revised and Expanded,* Lung Biology in Health and Disease Series, vol. 79, edited by J. A. Dempsey and A. I. Pack, New York: Marcel Dekker, ch. 23, pp. 1011–1064, 1995.

Mortola, J. P., and C. Lanthier. Normoxic and hypoxic breathing pattern in newborn grey seals. *Can. J. Zool.* 67:483–487,1989.

———. The ventilatory and metabolic response to hypercapnia in newborn mammalian species. *Respir. Physiol.* 103:263–270, 1996.

Mortola, J. P., and T. Matsuoka. Interaction between CO_2 production and ventilation in the hypoxic kitten. *J. Appl. Physiol.* 74:905–910, 1993.

Mortola, J. P., and L. Naso. Electrophoretic analysis of contractile proteins of the diaphragm in chronically hypoxic rats. *Am. J. Physiol.* 269:2371–2376, 1995.

———. Brown adipose tissue and its uncoupling protein in chronically hypoxic rats. *Clin. Sci.* 93:349–354, 1997.

———. Thermogenesis in newborn rats after prenatal or postnatal hypoxia. *J. Appl. Physiol.* 85:84–90, 1998.

Mortola, J. P., and R. Rezzonico. Metabolic and ventilatory rates in newborn kittens during acute hypoxia. *Respir. Physiol.* 73:55–68, 1988.

———. Ventilation in kittens with chronic section of the superior laryngeal nerves. *Respir. Physiol.* 76:369- 382, 1989.

Mortola, J. P., and G. Sant'Ambrogio. Motion of the rib cage and the abdomen in tetraplegic patients. *Cli. Sci. Mol. Med.* 54:25–32, 1978.

———. Mechanics of the trachea and behaviour of its slowly adapting stretch receptors. *J. Physiol. (London)* 286:577–590, 1979.

Mortola, J. P., and S. M. Tenney. Effects of hyperoxia on ventilatory and metabolic rates of newborn mice. *Respir. Physiol.* 63:267–274, 1986.

Mortola, J. P., J. T. Fisher, J. B. Smith, G. S. Fox, S. Weeks, and D. Willis. Onset of respiration in infants delivered by cesarean section. *J. Appl. Physiol.* 52:716–724, 1982a.

Mortola, J. P., J. T. Fisher, B. Smith, G. Fox, and S. Weeks. Dynamics of breathing in infants. *J. Appl. Physiol.* 52:1209–1215, 1982b.

Mortola, J. P., S. Al-Shway, and A. Noworaj. Importance of upper airway airflow in the ventilatory depression of laryngeal origin. *Pediatr. Res.* 17:550–552, 1983.

Mortola, J. P., J. Milic-Emili, A. Noworaj, B. Smith, G. Fox, and S. Weeks. Muscle pressure and flow during expiration in infants. *Am. Rev. Respir. Dis.* 129:49–53, 1984a.

Mortola, J. P., A. Rossi, and L. Zocchi. Pressure-volume curve of lung and lobes in kittens. *J. Appl. Physiol.* 56:948–953, 1984b.

Mortola, J. P., J. T. Fisher, and G. Sant'Ambrogio. Vagal control of the breathing pattern and respiratory mechanics in the adult and newborn rabbit. *Pflügers Arch. - Eur. J. Physiol.* 401:281–286, 1984c.

Mortola, J. P., D. Magnante, and M. Saetta. Expiratory pattern of newborn mammals. *J. Appl. Physiol.* 58:528–533, 1985a.

Mortola, J. P., M. Saetta, G. Fox, B. Smith, and S. Weeks. Mechanical aspects of chest wall distortion. *J. Appl. Physiol.* 59:295–304, 1985b.

Mortola, J. P., A.-M. Lauzon, and B. Mott. Expiratory flow pattern and respiratory mechanics. *Can. J. Physiol. Pharmacol.* 65:1142–1145, 1987a.

Mortola, J. P., M. Saetta, and D. Bartlett, Jr. Postnatal development of the lung following denervation. *Respir. Physiol.* 67:137–145, 1987b.

Mortola, J. P., R. Rezzonico, and C. Lanthier. Ventilation and oxygen consumption during acute hypoxia in newborn mammals: a comparative analysis. *Respir. Physiol.* 78:31–43, 1989.

Mortola, J. P., R. Rezzonico, J. T. Fisher, N. Villena-Cabrera, E. Vargas, R. Gonzales, and F. Peña. Compliance of the respiratory system in infants born at high altitude. *Am. Rev. Respir. Dis.* 142:43–48, 1990a.

Mortola, J. P., L. Xu, and A.-M. Lauzon. Body growth, lung and heart weight, and DNA content in newborn rats exposed to different levels of chronic hypoxia. *Can J. Physiol. Pharmacol.* 68:1590–1594, 1990b.

Mortola, J. P., P. B. Frappell, A. Dotta, T. Matsuoka, G. Fox, S. Weeks, and D. Mayer. Ventilatory and metabolic responses to acute hyperoxia in newborns. *Am. Rev. Respir. Dis.* 146:11–15, 1992a.

Mortola, J. P., P. B. Frappell, D. E. Frappell, N. Villena-Cabrera, M. Villena-Cabrera, and F. Peña. Ventilation and gaseous metabolism in infants born at high altitude, and their responses to hyperoxia. *Am. Rev. Respir. Dis.* 146:1206–1209, 1992b.

Mortola, J. P., G. Hemmings, T. Matsuoka, C. Saiki, and G. Fox. Referencing lung volume for measurements of respiratory system compliance in infants. *Pediatr. Pulmonol.* 16:248–253, 1993.

Mortola, J. P., T. Matsuoka, C. Saiki, and L. Naso. Metabolism and ventilation in hypoxic rats: effect of body mass. *Respir. Physiol.* 97:225–234, 1994.

Mortola, J. P., T. Trippenbach, R. Rezzonico, J. T. Fisher, M. Diaz, N. Villena-Cabrera, and F. Peña. Hering-Breuer reflexes in high-altitude infants. *Clin. Sci.* 88:345–350, 1995a.

Mortola, J. P., C. Saiki, and G. Fox. Insensitivity of the Hering-Breuer reflex to spontaneous changes in metabolic rate in the infant. *Cli. Sci.* 89:101–105, 1995b.

Mortola, J. P., P. B. Frappell, and P. A. Woolley. Breathing through skin in a newborn mammal. *Nature,* 397:660, 1999a.

Mortola, J. P., D. Merazzi, and L. Naso. Blood flow to the brown adipose tissue of conscious young rabbits during hypoxia in cold and warm conditions. *Pflügers Arch. - Eur. J. Physiol.* 437:255–260, 1999b.

Mortola, J. P., P. B. Frappell, L. Aguero, and K. Armstrong. Birth weight and altitude: A study in peruvian communities. *J. Pediatr.,* 136:324–329, 2000.

Moss, T. J., and R. Harding. Ventilatory and arousal responses of sleeping lambs to respiratory challenges: influence of maternal anemia during pregnancy. *J. Appl. Physiol.* 88:641–648, 2000.

Moss, I. R., A. J. Mautone, and E. M. Scarpelli. Effect of temperature on regulation of breathing and sleep/wake state in fetal lambs. *J. Appl. Physiol.* 54:536–543, 1983.

Moss, M., G. Moreau, and G. Lister. Oxygen transport and metabolism in the conscious lamb: the effects of hypoxemia. *Pediatr. Res.* 22:177–183, 1987.

Moss, T. J., A. E. Jakubowska, G. J. McCrabb, K. Billings, and R. Harding. Ventilatory responses to progressive hypoxia and hypercapnia in developing sheep. *Respir. Physiol.* 100:33–44, 1995.

Moss, T. J., M. G. Davey, G. J. McCrabb, and R. Harding. Development of ventilatory responsiveness to progressive hypoxia and hypercapnia in low-birth-weight lambs. *J. Appl. Physiol.* 81:1555–1561, 1996.

Mott, J. C. The ability of young mammals to withstand total oxygen lack. *Brit. Med. Bull.* 17:144–148, 1961.

Mount, L. *Adaptation to Thermal Environment: Man and His Productive Animals.* Baltimore: University Park Press, p. 190, 1979.

Mount, L. E., and J. G. Rowell. Body size, body temperature and age in relation to the metabolic rate of the pig in the first five weeks after birth. *J. Physiol. (London)* 154:408–416,1960.

Mourek, J. Oxygen consumption during ontogenesis in rats in environments with a high and low oxygen content. *Physiol. Bohemoslov.* 8:106–111, 1959.

Mueggler, P. A., J. S. Peterson, R. D. Koler, J. Metcalfe, and J. A. Black. Postnatal regulation of oxygen delivery: hematologic parameters of postnatal dogs. *Am. J. Physiol.* 237:H71–H75, 1979.

Mueggler, P. A., G. Jones, J. S. Peterson, J. M. Bissonnette, R. D. Koler, J. Metcalfe, R. T. Jones, and J. A. Black. Postnatal regulation of canine oxygen delivery: erythrocyte components affecting Hb function. *Am. J. Physiol.* 238:H73–H79, 1980.

Muller, N., G. Gulston, D. Cade, J. Whittoin, A. B. Froese, M. H. Bryan, and A. C. Bryan. Diaphragmatic muscle fatigue in the newborn. *J. Appl. Physiol.* 46:688–695, 1979.

Mulligan, E. M. Discharge properties of the carotid bodies. Developmental aspects. In: *Developmental Neurobiology of Breathing,* Lung Biology in Health and Disease Series, vol. 53, edited by G. G. Haddad and J. P. Farber. New York: Marcel Dekker, ch. 11, pp. 321–340, 1991.

Mumm, B., R. Kaul, G. Heldmaier, and I. Schmidt. Endogenous 24-hour cycle of core temperature and oxygen consumption in week-old Zucker rat pups. *J. Comp. Physiol. B* 159:569–575, 1989.

Mundie, T. G., D. Easa, K. C. Finn, E. L. Stevens, G. Hashiro, and V. Balaraman. Effect of baseline lung compliance on the subsequent response to positive end-expiratory pressure in ventilated piglets with normal lungs. *Crit. Care Med.* 22:1631–1638, 1994.

Murai, D. T., C. H. Lee, L. D. Wallen, and J. A. Kitterman. Denervation of peripheral chemoreceptors decreases breathing movements in fetal sheep. *J. Appl. Physiol.* 59:575–579, 1985.

Murphy, T. M., D. W. Ray, L. E. Alger, I. J. Phillips, J. C. Roach, A. R. Leff, and J. Solway. Ontogeny of dry gas hyperpnea-induced bronchoconstriction in guinea pigs. *J. Appl. Physiol.* 76:1150–1155, 1994.

Mutoh, T., W. J. E. Lamm, L. J. Embree, J. Hildebrandt, and R. K. Albert. Abdominal distension alters regional pleural pressure and chest wall mechanics in pigs in vivo. *J. Appl. Physiol.* 70:2611–2618, 1991.

———. Volume infusion produces abdominal distension, lung compression, and chest wall stiffening in pigs. *J. Appl. Physiol.* 72:575–582, 1992.

Nagai, A., W. M. Thurlbeck, C. Deboeck, S. Ioffe, and V. Chernick. The effect of maternal CO_2 breathing on lung development of fetuses in the rabbit. Morphologic and morphometric studies. *Am. Rev. Respir. Dis.* 135:130–136, 1987.

Naifeh, K. H., S. E. Huggins, H. E. Hoff, T. W. Hugg, and R. E. Norton. Respiratory patterns in crocodilian reptiles. *Respir. Physiol.* 9:31–42, 1970.

Nardell, E. A., and J. S. Brody. Determinants of mechanical properties of rat lung during postnatal development. *J. Appl. Physiol.* 53:140–148, 1982.

Nardo, L., S. B. Hooper, and R. Harding. Stimulation of lung growth by tracheal obstruction in fetal sheep: relation to luminal pressure and lung liquid volume. *Pediatr. Res.* 43:184–190, 1998.

Nattie, E. E. Ventilation during acute HCl infusion in intact and chemodenervated conscious rabbit. *Respir. Physiol.* 54:97–107, 1983.

———. Central chemoreception. In: *Regulation of Breathing, 2nd Edition Revised and Expanded,* Lung Biology in Health and Disease Series, vol. 79, edited by J. A. Dempsey and A. I. Pack. New York: Marcel Dekker, ch. 10, pp. 473–510, 1995.

Nattie, E. E., and W. H. Edwards. The effects of acute total asphyxia and metabolic acidosis on cerebrospinal fluid bicarbonate regulation in newborn puppies. *Pediatr. Res.* 14:286–290, 1980.

Nault, M. A., S. G. Vincent, and J. T. Fisher. Mechanisms of capsaicin- and lactic acid-induced bronchoconstriction in the newborn dog. *J. Physiol. (London)* 515:567–578, 1999.

Nedergaard, J., and B. Cannon. The uncoupling protein thermogenin and mitochondrial thermogenesis. In: *Molecular Mechanisms in Bioenergetics.* Amsterdam: Elsevier Sci. Publishers, ch.17, pp. 385–420, 1992.

———. Brown adipose tissue: development and function. In: *Fetal and Neonatal Physiology,* 2nd ed., edited by R. A. Polin and W. W. Fox. Philadelphia: Saunders, vol. 1, ch. 47, pp. 478–489, 1998.

Newman, L. M., E. M. Johnson, and J. M. Roth. Lung volume and compliance in neonatal rats. *Lab. Anim. Sci.* 34:371–375,1984a.

Newman, S., J. Road, F. Bellemare, J. P. Clozel, C. M. Lavigne, and A. Grassino. Respiratory muscle length measured by sonomicrometry. *J. Appl. Physiol.* 56:753–764, 1984b.

Nichols, D. G. Respiratory muscle performance in infants and children. *J. Pediatr.* 118:493–502, 1991.

Nichols, D. G., S. Howell, J. Massik, R. C. Koehler, C. A. Gleason, J. R. Buck, R. S. Fitzgerald, R. J. Traystman, and J. L. Robotham. Relationship of diaphragmatic contractility to diaphragmatic blood flow in newborn lambs. *J. Appl. Physiol.* 66:120–127, 1989.

Nicolai, T., C. J. Lanteri, and P. D. Sly. Frequency dependence of elastance and resistance in ventilated children with and without chest opened. *Europ. Respir. J.* 6:1340–1346, 1993.

Nightingale, D. A., and C. C. Richards. Volume-pressure relations of the respiratory system of curarized infants. *Anesthesiology* 26:710–714, 1965.

Nilsson, R. Lung compliance and lung morphology following artificial ventilation in the premature and full-term rabbit neonate. *Scand. J. Respir. Dis.* 60:206–214, 1979.

Nioka, S., and B. Chance. Energy metabolism and ionic homeostasis during hypoxic stress in the developing mammalian brain. In: *Developmental Neurobiology of Breathing,* Lung Biology in Health and Disease Series, vol. 53, edited by G. G. Haddad and J. P. Farber. New York: Dekker, ch. 20, pp. 615–641, 1991.

Noback, G. J. The developmental topography of the larynx, trachea and lungs in the fetus, new-born, infant and child. *Am. J. Dis. Child.* 26:515–533, 1923.

Normand, I. C. S., R. E. Olver, E. O. R. Reynolds, and L. B. Strang. Permeability of lung capillaries and alveoli to non-electrolytes in the foetal lamb. *J. Physiol. (London)* 219:303–330, 1971.

Nuesslein, B., and I. Schmidt. Development of circadian cycle of core temperature in juvenile rats. *Am. J. Physiol.* 259:R270–R276, 1990.

Nugent, S. T., and J. P. Finley. Spectral analysis of the EMG and diaphragmatic muscle fatigue during periodic breathing in infants. *J. Appl. Physiol.* 58:830–833,1985.

Numa, A. H., and C. J. L. Newth. Anatomic dead space in infants and children. *J. Appl. Physiol.* 80:1485–1489, 1996.

Numa, A. H., J. Hammer, and C. J. L. Newth. Effect of prone and supine positions on functional residual capacity, oxygenation, and respiratory mechanics in ventilated infants and children. *Am J. Respir. Crit. Care Med.* 156:1185–1189, 1997.

O'Brien, M. J. Respiratory EMG findings in relation to periodic breathing in infants. *Early Hum. Dev.* 11:43–60, 1985.

O'Brien, M. J., L. A. van Eykern, and H. F. R. Prechtl. Diaphragmatic, intercostal and abdominal muscle tonic EMG activity in normal and hypotonic newborns. In: *Ontogenesis of the Brain. Vol.3: The Biochemical, Functional and Structural Development of the Nervous System,* edited by S. Trojan and F. Stastny. Prague: Czechoslovakia: Univ. Karlova Press, pp. 471–479, 1980.

O'Dempsey, T. J. D., B. E. Laurence, T. F. McArdle, J. E. Todd, A. C. Lamont, and B. M. Greenwood. The effect of temperature reduction on respiratory rate in febrile illnesses. *Arch Dis. Child.* 68:492–95, 1993.

Okubo, S., and J. P. Mortola. Long-term respiratory effects of neonatal hypoxia in the rat. *J. Appl. Physiol.* 64:952–958, 1988.

————. Respiratory mechanics in adult rats hypoxic in the neonatal period. *J. Appl. Physiol.* 66:1772–1778, 1989.

————. Control of ventilation in adult rats hypoxic in the neonatal period. *Am. J. Physiol.* 259:R836–R841, 1990.

Olinsky, A., M. H. Bryan, and A. C. Bryan. Influence of lung inflation on respiratory control in neonates. *J. Appl. Physiol.* 36:426–429, 1974a.

————. Response of newborn infants to added respiratory loads. *J. Appl. Physiol.* 37:190–193, 1974b.

Oliver, T. K., Jr., J. A. Demis, and G. D. Bates. Serial blood-gas tensions and acid-base balance during the first hour of life in human infants. *Acta Pediatr.* 50:346–360, 1961.

Olsen, C. R., A. E. Stevens, and M. B. McIlroy. Rigidity of tracheae and bronchi during muscular constriction. *J. Appl. Physiol.* 23:27–34, 1967.

Olver, R. E., E. E. Schneeberger, and D. V. Walters. Epithelial solute permeability, ion transport and tight junction morphology in the developing lung of the fetal lamb. *J. Physiol. (London)* 315:395–412, 1981.

Openshaw, P., S. Edwards, and P. Helms. Changes in rib cage geometry during childhood. *Thorax* 39:624–627, 1984.

Otis, A. B. Quantitative relationships in steady-state gas exchange. In: *Handbook of Physiology. Section 3, Respiration, vol. I,* edited by W. O. Fenn and H. Rahn. Washington, D.C.: Am. Physiol. Soc., ch. 27, pp. 681–698, 1964.

Otis, A. B., W. O. Fenn, and H. Rahn. Mechanics of breathing in man. *J. Appl. Physiol.* 2:592–607, 1950.

Ousey, J. C., A. J. McArthur, and P. D. Rossdale. Metabolic changes in thoroughbred pony foals during the first 24 h post partum. *J. Reprod. Fert. Suppl.* 44:561–570, 1991.

Paintal, A. S. Vagal sensory receptors and their reflex effects. *Physiol. Rev.* 53:159–227, 1973.

Papastamelos, C., H. B. Panitch, S. E. England, and J. L. Allen. Developmental changes in chest wall compliance in infancy and early childhood. *J. Appl. Physiol.* 78:179–184, 1995.

Parer, J. T., A. S. Hoversland, and J. Metcalfe. Comparative studies of the respiratory functions of mammalian blood. VI. Young lion and tiger. *Respir. Physiol.* 10:30–37, 1970.

Parks, Y. A., C. S. Beardsmore, U. M. McFadden, D. J. Fallot, P. C. Goodenough, R. Carpenter, and H. Simpson. The effect of a single breath of 100% oxygen on breathing in infants at 1, 2, and 3 months of age. *Am. Rev. Respir. Dis.* 144:141–145, 1991.

Parmeggiani, P. L. Thermoregulation during sleep in mammals. *News in Physiol. Sci.* 5:208–212, 1990.

Pascucci, R. C., M. B. Hershenson, N. F. Sethna, S. H. Loring, and A. R. Stark. Chest wall motion of infants during spinal anesthesia. *J. Appl. Physiol.* 68:2087–2091, 1990.

Patrick, J., W. Fetherston, H. Vick, and R. Voegelin. Human fetal breathing movements and gross fetal body movements at weeks 34 to 35 of gestation. *Am. J. Obst. Gynecol.* 130:693–699, 1978.

Pearson, O. P. Metabolism of small animals with remarks on the lower limit of mammalian size. *Science* 108:44–46, 1948.

Pedraz, C., and J. P. Mortola. CO_2 production, body temperature and ventilation in hypoxic newborn cats and dogs before and after body warming. *Pediatr. Res.* 30:165–169, 1991.

Pepelko, W. E. Effects of hypoxia and hypercapnia, singly and combined, on growing rats. *J. Appl. Physiol.* 28:646–651, 1970.

Pereyra, P. M., W. Zhang, M. Schmidt, and L. E. Becker. Development of myelinated and unmyelinated fibers of human vagus nerve during the first year of life. *J. Neurol. Sci.* 110:107–113, 1992.

Pérez Fontán, J. J., and L. P. Kinloch. Control of bronchomotor tone during perinatal development in sheep. *J. Appl. Physiol.* 75:1486–1496, 1993.

Pérez Fontán, J. J., A. O. Ray, and T. R. Oxland. Stress relaxation of the respiratory system in developing piglets. *J. Appl. Physiol.,* 73:1297–1309, 1992.

Perks, A. M., and S. Cassin. The effects of arginine vasopressin and epinephrine on lung liquid production in fetal goats. *Can. J. Physiol. Pharmacol.* 67:491–498, 1989.

Perlstein, P. H., N. K. Edwards, and J. M. Sutherland. Apnea in premature infants and incubator air-temperature changes. *New Eng. J. Med.* 282:461–466, 1970.

Peslin, R., F. Marchal, and C. Choné. Viscoelastic properties of rabbit lung during growth. *Respir. Physiol.* 86:189–198, 1991.

Peters, R. H. *The Ecological Implications of Body Size.* Cambridge: Cambridge University Press, p. 329, 1983.

Peters, J. L., T. O. Hess, C. S. Leach, T. Imlay, R. O. Gardner, F. M. Donovan, Jr., and K. R. Van Kampen. Hemic and spirometric profiles in calves used for cardiovascular research. *Am. J. Vet. Res.* 34:1595–1597, 1975.

Pettit, T. N., and G. C. Whittow. The initiation of pulmonary respiration in a bird embryo: blood and air cell gas tensions. *Respir. Physiol.* 48:199–208, 1982.

Piazza, T., A. M. Lauzon, and J. P. Mortola. Time course of adaptation to hypoxia in newborn rats. *Can. J. Physiol. Pharmacol.* 66:152–158, 1988.

Pillai, M., and D. James. Hiccups and breathing in human fetuses. *Arch. Dis. Child.* 65:1072–1075, 1990.

Plaxico, D. T., and G. M. Loughlin. Nasopharyngeal reflux and neonatal apnea. *Am. J. Dis. Child.* 135:793–794, 1981.

Polgar, G., and T. R. Weng. The functional development of the respiratory system. From the period of gestation to adulthood. *Am. Rev. Respir. Dis.* 120:625–695, 1979.

Ponte, J., and M. J. Purves. Types of afferent nervous activity which may be measured in the vagus nerve of the sheep foetus. *J. Physiol. (London)* 229:51–76, 1973.

Poore, E. R., and D. W. Walker. Chest wall movements during fetal breathing in the sheep. *J. Physiol. (London)* 301:307–315, 1980.

Porcellini, A., G. Lucarelli, L. Ferrari, and U. Butturini. L'eritropoiesi nella cavia durante la vita neonatale ed adulta. *Boll. Soc. It. Biol. Sper.* 42:1336–1338, 1966.

Porter, R. K., and M. D. Brand. Body mass dependence of H^+ leak in mitochondria and its relevance to metabolic rate. *Nature* 362:628–630, 1993.

Potter, E. L., and G. P. Bohlender. Intrauterine respiration in relation to development of fetal lung with report of 2 unusual anomalies of respiratory system. *Am. J. Obstet. Gynecol.* 42:14–22, 1941.

Potter, C. F., I. A. Dreshaj, M. A. Haxhiu, E. K. Stork, R. L. Chatburn, and R. J. Martin. Effect of exogenous and endogenous nitric oxide on the airway and tissue components of lung resistance in the newborn piglet. *Pediatr. Res.* 41:886–891, 1997.

Powell, J. T., and P. L. Whitney. Postnatal development of rat lung. Changes in lung lectin, elastin, acetylcholinesterase and other enzymes. *Biochem. J.* 188:1–8, 1980.

Powell, F. L., W. K. Milsom, and G. S. Mitchell. Time domains of the hypoxic ventilatory response. *Respir. Physiol.* 112:123–134, 1998.

Powers, S. K., J. Lawler, D. Criswell, S. Dodd, and H. Silverman. Age-related changes in enzyme activity in the rat diaphragm. *Respir. Physiol.* 83:1–10, 1991.

Prader, A., J. M. Tanner, and G. A. von Harnack. Catch-up growth following illness or starvation. *J. Pediatr.* 62:646–659, 1963.

Prakash, Y. S., M. Fournier, and G. C. Sieck. Effects of prenatal undernutrition on developing rat diaphragm. *J. Appl. Physiol.* 75:1044–1052, 1993.

Praud, J.-P., L. Egreteau, M. Benlabed, L. Curzi-Dascalova, H. Nedelcoux, and

C. Gaultier. Abdominal muscle activity during CO_2 rebreathing in sleeping neonates. *J. Appl. Physiol.* 70:1344–1350, 1991.

Prothero, J. W. Scaling of blood parameters in mammals. *Comp. Biochem. Physiol.* 67A:649–657, 1980.

———. Scaling of standard energy metabolism in mammals: I. Neglect of circadian rhythms. *J. Theor. Biol.* 106:1–8, 1984.

Purves, M. J. The respiratory response of the new-born lamb to inhaled CO_2 with and without accompanying hypoxia. *J. Physiol. (London)* 185:78–94, 1966a.

———. The effect of a single breath of oxygen on respiration in the newborn lamb. *Respir. Physiol.* 1:297–307, 1966b.

———. The effects of hypoxia in the new-born lamb before and after denervation of the carotid chemoreceptors. *J. Physiol. (London)* 185:60–77, 1966c.

———. Respiratory and circulatory effects of breathing 100% oxygen in the new-born lamb before and after denervation of the carotid chemoreceptors. *J. Physiol. (London)* 185:42–59, 1966d.

Rabbette, P. S., and J. Stocks. Influence of volume dependency and timing of airway occlusions on the Hering-Breuer reflex in infants. *J. Appl. Physiol.* 85:2033–2039, 1998.

Rabbette, P. S., K. L. Costeloe, and J. Stocks. Persistence of the Hering-Breuer reflex beyond the neonatal period. *J. Appl. Physiol.* 71:474–480, 1991.

Rabbette, P. S., M. E. Fletcher, C. A. Dezateux, H. Soriano-Brucher, and J. Stocks. Hering-Breuer reflex and respiratory system compliance in the first year of life: a longitudinal study. *J. Appl. Physiol.* 76:650–656, 1994.

Rabinovich, M., W. J. Gamble, O. S. Miettinen, and L. Reid. Age and sex influence on pulmonary hypertension of chronic hypoxia and on recovery. *Am. J. Physiol.* 240:H62–H72, 1981.

Radvanyi-Bouvet, M. F., M. Monset-Couchard, F. Morel-Kahn, G. Vilente, and C. Dreyfus-Brisac. Expiratory patterns during sleep in normal full-term and premature neonates. *Biol. Neonate* 41:74–84, 1982.

Rawson, R. E., G. D. DelGiudice, H. E. Dziuk, and L. D. Mech. Energy metabolism and hematology of white-tailed deer fawns. *J. Wildlife Dis.* 28:91–94, 1992.

Recabarren, S. E., M. Vergara, A. J. Llanos, and M. Serón-Ferré. Circadian variation of rectal temperature in newborn sheep. *J. Develop. Physiol.* 9:399–408, 1987.

Redford, K. H., and J. F. Eisenberg. *Mammals of the Neotropics: The Southern Cone.* Chicago: University of Chicago Press, 1992.

Redlin, U., B. Nuesslein, and I. Schmidt. Circadian changes of brown adipose tissue thermogenesis in juvenile rats. *Am. J. Physiol.* 262:R504–R508, 1992.

Reeves, R. B. The interaction of body temperature and acid-base balance in ectothermic vertebrates. *Ann. Rev. Physiol.* 39:559–586, 1977.

Reeves, J. T., and J. E. Leathers. Circulatory changes following birth of the calf and the effect of hypoxia. *Circ. Res.* 15:343–354, 1964.

———. Postnatal development of pulmonary and bronchial arterial circulations in the calf and the effects of chronic hypoxia. *Anat. Rec.* 157:641–656, 1967.

Refinetti, R., and M. Menaker. The circadian rhythm of body temperature. *Physiol. Behav.* 51:613–637, 1992.

Reid, L. The embryology of the lung. In: *Development of the Lung,* edited by A. V. S. de Reuck and R. Porter. Boston: Little, Brown, pp. 109–124 (Ciba Found. Symp.), 1967.

Reynolds, R. N., and B. E. Etsten. Mechanics of respiration in apneic anesthetized infants. *Anesthesiology* 27:13–19, 1966.

Rezzonico, R., and J. P. Mortola. Respiratory adaptation to chronic hypercapnia in newborn rats. *J. Appl. Physiol.* 67:311–315, 1989.

Rezzonico, R., R. D. Gleed, and J. P. Mortola. Respiratory mechanics in adult rats hypercapnic in the neonatal period. *J. Appl. Physiol.* 68:2274–2279, 1990.

Ricciuti, F., and J. E. Fewell. Fever in young lambs: hypoxemia alters the febrile response to a small dose of bacterial pyrogen. *J. Dev. Physiol.* 17:29–38, 1992.

Richards, C. C., and L. Bachman. Lung and chest wall compliance of apneic paralysed infants. *J. Clin. Invest.* 40:273–278, 1961.

Richardson, B., R. Natale, and J. Patrick. Human fetal breathing activity during electively induced labor at term. *Am. J. Obstet. Gynecol.* 133:247–255, 1979.

Riegel, K. von, and G. Ruhrmann. Über die atemgastransportfunktion des blutes und die erythropoese junger kaninchen. *Acta Haemat.* 32:129–145, 1964.

Riegel, K. von, P. Hilpert, and H. Bartels. Vergleichende untersuchungen der erythrocytenmorphologie, des fetalen hämoglobins und der sauerstoff-affinität des blutes von säuglingen, zicklein und lämmern. *Acta Haemat.* 25:164–183, 1961.

Rigatto, H., and J. P. Brady. Periodic breathing and apnea in preterm infants. II. Hypoxia as a primary event. *Pediatrics* 50:219–227, 1972.

Rigatto, H., R. De La Torre Verduzco, and D. B. Cates. Effects of O_2 on the ventilatory response to CO_2 in preterm infants. *J. Appl. Physiol.* 39:896–899, 1975.

Rigatto, H., Z. Kalapesi, F. N. Leahy, M. Durand, M. MacCallum, and D. Cates. Chemical control of respiratory frequency and tidal volume during sleep in preterm infants. *Respir. Physiol.* 41:117-125, 1980.

Rigatto, H., C. Wiebe, C. Rigatto, D. S. Lee, and D. Cates. Ventilatory response to hypoxia in unanesthetized newborn kittens. *J. Appl. Physiol.* 64:2544–2551, 1988a.

Rigatto, H., S. U. Hasan, A. Jansen, D. Gibson, B. Nowaczyk, and D. Cates. The effect of total peripheral chemodenervation on fetal breathing and the establishment of breathing at birth in sheep. In: *Research in Perinatal Medicine. VII. Fetal and Neonatal Development.*, edited by C. T. Jones. Ithaca, N.Y.: Perinatology Press, pp. 613–621, 1988b.

Rismiller, P. D., and R. S. Seymour. The echidna. *Sci. Am.* 96:96–103, 1991.

Ritchie, J. W. K., and K. Lakhani. Fetal breathing movements in response to maternal inhalation of 5% carbon dioxide. *Am. J. Obstet. Gynecol.* 136:386–388, 1980.

Robinson, S. L., C. A. Richardson, M. M. Willis, and G. A. Gregory. Halothane anesthesia reduces pulmonary function in the newborn lamb. *Anesthesiology*, 62:578–581, 1985.

Rohlicek, C. V., C. Saiki, T. Matsuoka, and J. P. Mortola. Cardiovascular and respiratory consequences of body warming during hypoxia in conscious newborn cats. *Pediatr. Res.* 40:1–5, 1996.

———. Oxygen transport in conscious newborn dogs during hypoxic hypometabolism. *J. Appl. Physiol.* 84:763–768, 1998.

Ronca, A. E., and J. R. Alberts. Simulated uterine contractions facilitate fetal and newborn respiratory behaviour in rats. *Physiol. Behav.* 58:1035–1041, 1995.

Rose, D. K., and A. B. Froese. Changes in respiratory pattern affect dead space/tidal volume ratio during spontaneous but not during controlled ventilation: a study in pediatric patients. *Anesth. Analg.* 59:341–349, 1980.

Rosen, C. L., W. S. Schecter, R. B. Mellins, and G. G. Haddad. Effect of acute hypoxia on metabolism and ventilation in awake piglets. *Respir. Physiol.* 91:307–319, 1993.

Rossdale, P. D. Blood gas tensions and pH values in the normal thoroughbred foal at birth and in the following 42 h. *Biol. Neonat.* 13:18–25, 1968.

Rossdale, P. D., R. E. Pattle, and L. W. Mahaffey. Respiratory distress in a newborn foal with failure to form lung lining film. *Nature* 215:1498–1499, 1967.

Rossi, A., and J. P. Mortola. Vagal influence on respiratory mechanics in newborn kittens. *Bull. Eur. Physiopathol. Respir.* 23:61–66, 1987.

Rubner, M. Ueber den einfluss der körpergrösse auf stoffund kraftwechsel. *Z. Biol. (Munich)* 19:535–562, 1883.

Rudolph, A. M. Fetal and neonatal pulmonary circulation. *Ann. Rev. Physiol.* 41:383–395, 1979.

———. Oxygenation in the fetus and neonate: A perspective. *Sem. Perinatol.* 8:158–167, 1984.

Runciman, S. I. C., B. J. Gannon, and R. V. Baudinette. Central cardiovascular shunts in the perinatal marsupial. *Anat. Rec.* 243:71–83, 1995.

Ru-Yung, S., and Z. Jinxiang. Postnatal development of thermoregulation in the root vole (*Microtus oeconomus*) and the quantitative index of homeothermic ability. *J. Therm. Biol.* 12:267–272, 1987.

Sachis, P. N., D. L. Armstrong, L. E. Becker, and A. C. Bryan. Myelination of the human vagus nerve from 24 weeks postconceptional age to adolescence. *J. Neuropathol. Exp. Neurol.* 41:466–472, 1982.

Saetta, M., and J. P. Mortola. Exponential analysis of the lung pressure-volume curve in newborn mammals. *Pediatr. Pulmonol.* 1:193–197, 1985a.

———. Breathing pattern and CO_2 response in newborn rats before and during anesthesia. *J. Appl. Physiol.* 58:1988–1996, 1985b.

———. Interaction of hypoxic and hypercapnic stimuli on breathing pattern in the newborn rat. *J. Appl. Physiol.* 62:506–512, 1987.

Sahebjami, H., and C. L. Vassallo. Static volume-pressure relationship in normal rats at various stages of growth. *Respir. Physiol.* 36:131–142, 1979.

Saiki, C., and S. Matsumoto. Effect of neonatal anoxia on the ventilatory response to hypoxia in developing rats. *Pediatr. Pulmonol.* 28:313–320, 1999.

Saiki, C., and J. P. Mortola. Ventilatory control in infant rats after daily episodes of anoxia. *Pediatr. Res.* 35:490–493, 1994.

———. Hypoxia abolishes the morning-night differences of metabolism and ventilation in 6-day-old rats. *Can. J. Physiol. Pharmacol.* 73:159–164, 1995.

———. Effect of CO_2 on the metabolic and ventilatory responses to ambient temperature in conscious adult and newborn rats. *J. Physiol. (London)* 491:261–269, 1996.

Sankaran, K., F. N. Leahy, D. Cates, M. MacCallum, and H. Rigatto. Effect of lung inflation on ventilation and various phases of the respiratory cycle in preterm infants. *Biol. Neonate* 40:160- 166, 1981.

Sant'Ambrogio, G., and J. E. Remmers. Reflex influences acting on the respiratory muscles of the chest wall. In: *The Thorax, Part A,* Lung Biology in Health and Disease Series, vol. 29, edited by C. Roussos and P. T. Macklem. New York: Marcel Dekker, ch. 17, pp. 531–594, 1985.

Sant'Ambrogio, G., D. T. Frazier, M. F. Wilson, and E. Agostoni. Motor innervation and pattern of activity of cat diaphragm. *J. Appl. Physiol.* 18:43–46, 1963.

Sarnat, H. B. Ontogenesis of striated muscle. In *Fetal and Neonatal Physiology,* 2nd edition, edited by R. A. Polin and W. W. Fox. Philadelphia: Saunders, ch. 199, pp. 2226–2247, 1998.

Sarrus et Rameaux. Rapport sur un mémoire adressé a l'Académie royale de Médicine. *Bull. Acad. Nat. Roy. (Paris)* 3:1094–1100, 1838–39.

Saunders, R. A., and A. D. Milner. Pulmonary pressure/volume relationships during the last phase of delivery and the first postnatal breaths in human subjects. *J. Pediatr.* 93:667–673, 1978.

Sawczenko, A., and P. J. Fleming. Thermal stress, sleeping position, and the Sudden Infant Death Syndrome. *Sleep* 19:S267–S270, 1996.

Scarpelli, E. M. Intrapulmonary foam at birth: an adaptational phenomenon. *Pediatr. Res.* 12:1070–1076, 1978.

Scarpelli, E. M., E. J. Agasso, and Y. Kikkawa. Demonstration of the significance of pulmonary surfactants at birth. *Respir. Physiol.* 12:110–122, 1971.

Scarpelli, E. M., B. C. Clutario, and D. Traver. Failure of immature lungs to produce foam and retain air at birth. *Pediatr. Res.* 13:1285–1289, 1979.

Scarpelli, E. M., A. Kumar, C. Doyle, and B. C. Clutario. Functional anatomy and volume-pressure characteristics of immature lungs. *Respir. Physiol.* 45:25–41, 1981.

Scheffer, V. B. Exploring the lives of whales. *Nat. Geogr. Mag.* 150:752–767, 1976.

Schleman, M., N. Gootman, and P. M. Gootman. Cardiovascular and respiratory responses to right atrial injections of phenyl diguanide in pentobarbital-anesthetized newborn piglets. Pediatr. Res. 13:1271–1274, 1979.

Schmidt-Nielsen, K. Energy metabolism, body size, and problems of scaling. *Fed. Proc.* 29:1524–1532, 1970.

———. *Scaling. Why Is Animal Size So Important?* Cambridge: Cambridge University Press, p. 241, 1984.

Schmidt-Nielsen, K., and J. L. Larimer. Oxygen dissociation curves of mammalian blood in relation to body size. *Am. J. Physiol.* 195:424–428, 1958.

Schreiber, M. D., M. A. Heymann, and S. J. Soifer. Increased arterial pH, not decreased $PaCO_2$, attenuates hypoxia-induced pulmonary vasoconstriction in newborn lambs. *Pediatr. Res.* 20:113–117, 1986.

Schroter, R. C. Quantitative comparisons of mammalian lung pressure volume curves. *Respir. Physiol.* 42:101–107, 1980.

Schulte, F. J. Developmental neurophysiology. In: *Scientific Foundations of Paediatrics,* 2nd ed., edited by J. A. Davis and J. Dobbing. London: Heinemann, ch. 41, pp. 785–829, 1981.

Schumacher, G. H., E. Wolff, and E. Jutzi. Quantitative Untersuchungen über das postnale Organwachstum des Goldhamster (*Mesocricetus auratus WTRH*). II. Lunge-Leber-Milz. *Morphol. Jb.* 108:18–40, 1965.

Schwieler, G. H. Respiratory regulation during postnatal development in cats and rabbits and some of its morphological substrate. *Acta Physiol. Scand. Suppl.* 304:3–123, 1968.

Seddon, P. C., G. M. Davis, and A. L. Coates. Do tidal expiratory flow patterns reflect lung mechanics in infants? *Am. J. Respir. Crit. Care Med.* 153:1248–1252, 1996.

Segal, A. N. Postnatal growth, metabolism and thermoregulation in the stoat. *Sov. J. Ecol.* (English transl. from *Ekologiya*) 6:28–32, 1975.

Sekhon, H. S., and W. M. Thurlbeck. Lung growth in hypobaric normoxia, normobaric hypoxia, and hypobaric hypoxia in growing rats. I. Biochemistry. *J. Appl. Physiol.* 78:124–131, 1995.

———. Lung morphometric changes after exposure to hypobaria and/or hypoxia and undernutrition. *Respir. Physiol.* 106:99–107, 1996.

Sekhon, H. S., J. L. Wright, and W. M. Thurlbeck. Pulmonary function alterations after 3 wk of exposure to hypobaria and/or hypoxia in growing rats. *J. Appl. Physiol.* 78:1787–1792, 1995.

Sellick, H., and J. G. Widdicombe. The activity of lung irritant receptors during pneumothorax, hyperpnoea, and pulmonary vascular congestion. *J. Physiol. (London)* 203:359–381, 1969.

Selwood, L., and P. A. Woolley. A timetable of embryonic development, and ovarian and uterine changes during pregnancy, in the stripe-faced dunnart, *Sminthopsis macroura* (Marsupialia: Dasyuridae). *J. Reprod. Fert.* 91:213–227, 1991.

Serwer, G. A. Postnatal circulatory adjustments. In: *Fetal and Neonatal Physiology,* edited by R. A. Polin and W. W. Fox. Philadelphia: Saunders, vol. 1, ch. 70, pp. 710–721, 1992.

Setnikar, I. Origine e significato delle proprietà meccaniche del polmone. *Arch. Fisiol.* 55:349–374, 1955.

Setnikar, I., E. Agostoni, and A. Taglietti. Entità, caratteristiche e origine della depressione pleurica. *Arch. Sci. Biol.* 41:312–325, 1957.

———. The fetal lung, a source of amniotic fluid. *Proc. Soc. Exp. Biol. Med.* 101:842–845, 1959.

Seymour, R. S. Diving Physiology. Reptiles. In: *Comparative Pulmonary Physiology,* Lung Biology in Health and Disease Series, vol. 39, edited by S. C. Wood, New York: Marcel Dekker, ch. 23, pp. 697–720, 1989.

Shaffer, T. H., M. Delivoria-Papadopoulos, E. Arcinue, P. Paez, and A. B. Dubois. Pulmonary function in premature lambs during the first few hours of life. *Respir. Physiol.* 28:179–188, 1976.

Shaffer, T. H., P. A. Koen, G. D. Muskowitz, J. D. Ferguson, and M. Delivoria-Papadopoulos. Positive end expiratory pressure: effects on lung mechanics of premature lambs. *Biol. Neonate* 34:1–10, 1978.

Shaffer, T. H., M. R. Wolfson, and D. Forman. Developmental alterations in pulmonary function of the lamb. *Respiration* 47:129–137, 1985.

Shannon, R. Reflexes from respiratory muscles and costovertebral joints. In: *Handbook of Physiology: The Respiratory System, vol. II, Control of Breathing,* edited by N. S. Cherniack and J. G. Widdicombe. Bethesda, Md.: Am. Physiol. Soc., part 1, ch. 13, pp. 431–447, 1986.

Shardonofsky, F. R., D. Perez-Chada, E. Carmuega, and J. Milic-Emili. Airway pressures during crying in healthy infants. *Pediatr. Pulmonol.* 6:14–18, 1989.

Sharp, J. T., W. S. Druz, R. C. Balagot, V. R. Bandelin, and J. Danon. Total respiratory compliance in infants and children. *J. Appl. Physiol.* 29:775–779, 1970.

Sharp, J. T., N. B. Goldberg, W. S. Druz, and J. Danon. Relative contributions of rib cage and abdomen to breathing in normal subjects. *J. Appl. Physiol.* 39:608–618, 1975.

Shivpuri, C. R., R. J. Martin, W. A. Carlo, and A. A. Fanaroff. Decreased ventilation in preterm infants during oral feeding. *J. Pediatr.* 103:285–289, 1983.

Sidi, D., J. R. G. Kuipers, M. A. Heymann, and A. M. Rudolph. Effects of ambient temperature on oxygen consumption and the circulation in newborn lambs at rest and during hypoxemia. *Pediatr. Res.* 17:254–258, 1983.

Sieck, G. C., and M. Fournier. Developmental aspects of diaphragm muscle cells. Structural and functional organization. In: *Developmental Neurobiology of Breathing*, Lung Biology in Health and Disease Series, vol. 53, edited by G. G. Haddad and J. P. Farber. New York: Marcel Dekker, ch. 13, pp. 375–428, 1991.

Sieck, G. C., M. Fournier, and C. E. Blanco. Diaphragm muscle fatigue resistance during postnatal development. *J. Appl. Physiol.* 71:458–464, 1991.

Sime, F., N. Banchero, D. Peñaloza, R. Gamboa, J. Cruz, and E. Marticorena. Pulmonary hypertension in children born and living at high altitudes. *Am. J. Cardiol.* 11:143–149, 1963.

Sladek, M., R. A. Parker, J. B. Grögaard, and H. W. Sundell. Long-lasting effect of prolonged hypoxemia after birth on the immediate ventilatory response to changes in arterial partial pressure of oxygen in young lambs. *Pediatr. Res.* 34:821–828, 1993a.

Sladek, M., J. B. Grögaard, R. A. Parker, and H. W. Sundell. Prolonged hypoxemia enhances and acute hypoxemia attenuates laryngeal reflex apnea in young lambs. *Pediatr. Res.* 34:813–820, 1993b.

Slocombe, R. F., N. E. Robinson, and F. J. Derksen. Effect of vagotomy on respiratory mechanics and gas exchange in the neonatal calf. *Am. J. Vet. Res.* 43:1168–1171, 1982.

Sly, P. D., and C. J. Lanteri. Differential responses of the airways and pulmonary tissues to inhaled histamine in young dogs. *J. Appl. Physiol.* 68:1562–1567, 1990.

Smejkal, V., F. Palacek, and M. Frydrychova. Développement postnatal du réflexe de Breuer-Hering chez le rat. *J. Physiol. (Paris)* 80:173–176, 1985.

Smith, R. E. Quantitative relations between liver mitochondria metabolism and total body weight in mammals. *Ann. N.Y. Acad. Sci.* 62:403–422, 1956.

Smith, R. J. Allometric scaling in comparative biology: problems of concept and method. *Am. J. Physiol.* 246:R152–R160, 1984.

Smith, R. E., and B. A. Horwitz. Brown fat and thermogenesis. *Physiol. Rev.* 49:330–425, 1969.

Smith, D., H. Green, J. Thomson, and M. Sharratt. Capillary and size interrelationships in developing rat diaphragm, EDL, and soleus muscle fiber types. *Am. J. Physiol.* 256:C50–C58, 1989.

Soppela, P., R. Sormunen, S. Saarela, P. Huttunen, and M. Nieminen. Localization, cellular morphology and respiratory capacity of "brown" adipose tissue in newborn reindeer. *Comp. Biochem. Physiol.* 101A:281–293, 1992.

Sørensen, S. C., and J. W. Severinghaus. Respiratory insensitivity to acute hypoxia persisting after correction of tetralogy of Fallot. *J. Appl. Physiol.* 25:221–223, 1968.

Sosenko, I. R. S., and L. Frank. Guinea pig lung development: antioxidant enzymes and premature survival in high O_2. *Am. J. Physiol.* 252:R693–R698, 1987.

———. Oxidants and antioxidants. In: *Basic Mechanisms of Pediatric Respiratory Disease*, edited by V. Chernick and R. B. Mellins. Philadelphia: Decker, ch. 23, pp. 315–327, 1991.

Sosenko, I. R. S., H. C. Nielsen, and L. Frank. Lack of sex differences in antioxidant enzyme development in the fetal rabbit lung. *Pediatr. Res.* 26:16–19, 1989.

Soust, M., A. M. Walker, and P. J. Berger. Blood flow to the respiratory muscles during hyperpnoea in the newborn lamb. *Respir. Physiol.* 76:93–105, 1989a.

———. Diaphragm O_2, diaphragm EMG, pressure-time product and calculated ventilation in newborn lambs during hypercapnic hyperpnoea. *Respir. Physiol.* 76:107–118, 1989b.

South, M., C. J. Morley, and G. Hughes. Expiratory muscle activity in preterm babies. *Arch. Dis. Child.* 62:825–829,1987.

Spells, K. E. Comparative studies in lung mechanics based on a survey of literature data. *Respir. Physiol.* 8:37–57, 1969/70.

Spiers, D. E. Nocturnal shifts in thermal and metabolic responses of the immature rat. *J. Appl. Physiol.* 64:2119–2124, 1988.

Stahl, W. R. Similarity and dimensional methods in biology. *Science* 137:205–212, 1962.

———. Scaling of respiratory variables in mammals. *J. Appl. Physiol.* 22:453–460, 1967.

Standaert, T. A., B. E. Wilham, D. E. Mayock, J. F. Watchko, R. L. Gibson, and D. E. Woodrum. Respiratory mechanics of the piglet during the first month of life. *Pediatr. Pulmonol.* 11:294–301, 1991.

Stanley, N. N., R. Alper, E. L. Cunningham, N. S. Cherniak, and N. A. Kefalides. Effects of a molecular change in collagen on lung structure and mechanical function. *J. Clin. Invest.* 55:1195–1201, 1975.

Stanton, A. N. Overheating and cot death. *Lancet* 2:1199–1201, 1984.

Stark, A. R., and I. D. Frantz, III. Prolonged expiratory duration with elevated lung volume in newborn infants. *Pediatr. Res.* 13:261–264, 1979.

Stenmark, K. R., J. Fasules, D. M. Hyde, N. F. Voelkel, J. Henson, A. Tucker, H. Wilson, and J. T. Reeves. Severe pulmonary hypertension and arterial adventitial changes in newborn calves at 4,300 m. *J. Appl. Physiol.* 62:821–830, 1987.

Stewart, J. H., R. J. Rose, and A. M. Barko. Respiratory studies in foals from birth to seven days old. *Equine Veter. J.* 16:323–328, 1984.

Stonehouse, B. *Sea Mammals of the World.* New York: Penguin, 1985.

Storrs, E. E. (photos: B. Lavies). The astonishing armadillo. *Nat. Geogr. Mag.* 161:820–830, 1982.

Stowe, C. M., and A. L. Good. Estimation of cardiac output by direct Fick technique in domestic animals, with observations on a case of traumatic pericarditis. *Am. J. Vet. Res.* 22:1093–1096, 1961.

Strang, L. B. *Neonatal Respiration. Physiological and Clinical Studies.* Oxford, U.K.: Blackwell Scientific Publications, p.69, 1977.

———. Fetal lung liquid: secretion and reabsorption. *Physiol. Rev.* 71:991–1016, 1991.

Suen, H. C., P. D. Losty, P. K. Donahoe, and J. J. Schnitzer. Accurate method to study static volume-pressure relationships in small fetal and neonatal animals. *J. Appl. Physiol.* 77:1036–1043, 1994.

Sugiura, T., M. Shunsuke, A. Morimoto, and N. Murakami. Regional differences in myosin heavy chain isoforms and enzyme activities of the rat diaphragm. *J. Appl. Physiol.* 73:506–509, 1992.

Sullivan, K. J., and J. P. Mortola. Effect of distortion on the mechanical properties of the newborn piglet lung. *J. Appl. Physiol.* 59:434–442, 1985.

———. Dynamic lung compliance in newborn and adult cats. *J. Appl. Physiol.* 60:743–750, 1986.

———. Age related changes in the rate of stress relaxation within the rat respiratory system. *Respir. Physiol.* 67:295–309, 1987.

Sundin, U., and B. Cannon. GDP-binding to the brown fat mitochondria of developing and cold-adapted rats. *Comp. Biochem. Physiol.* 65B:463–471, 1980.

Sutton, D., E. M. Taylor, and R. C. Lindeman. Prolonged apnea in infant monkeys resulting from stimulation of superior laryngeal nerve. *Pediatrics* 61:519–527, 1978.

Swift, P. G. F., and J. L. Emery. Clinical observations on response to nasal occlusion in infancy. *Arch. Dis. Child.* 48:947–951, 1973.

Tabachnik, E., N. L. Muller, A. C. Bryan, and H. Levison. Changes in ventilation and chest wall mechanics during sleep in normal adolescents. *J. Appl. Physiol.* 51:557–564, 1981.

Taeusch, H. W., Jr., I. Wyszogrodski, N. S. Wang, and M. E. Avery. Pulmonary pressure-volume relationships in premature fetal and newborn rabbits. *J. Appl. Physiol.* 37:809–813, 1974.

Taeusch, H. W., Jr., S. Carson, I. D. Frantz, III., and J. Milic-Emili. Respiratory regulation after elastic loading and CO_2 rebreathing in normal term infants. *J. Pediatr.* 88:102–111, 1976.

Takahashi, E., and H. Atsumi. Age differences in thoracic form as indicated by thoracic index. *Human Biol.* 27:65–74, 1955.

Talbot, F. B. Basal metabolism standards for children. *Am. J. Dis. Child.* 55:455–459,1938.

Talmadge, R. J., and R. R. Roy. Electrophoretic separation of rat skeletal muscle myosin heavy-chain isoforms. *J. Appl. Physiol.* 75:2337–2340, 1993.

Tamaki, N. Effect of growth on muscle capillarity and fiber type composition in rat diaphragm. *Eur. J. Appl. Physiol.* 54:24–29, 1985.

Tanswell, A. K., and B. A. Freeman. Pulmonary antioxidant enzyme maturation in the fetal and neonatal rat lung. I. Developmental profiles. *Pediatr. Res.* 18:584–587, 1984.

Tatsumi, K., C. K. Pickett, and J. V. Weil. Effects of haloperidol and domperidone on ventilatory roll off during sustained hypoxia in cats. *J. Appl. Physiol.* 72:1945–1952, 1992.

Taylor, A. The contribution of the intercostal muscles to the effort of respiration in man. *J. Physiol. (London)* 151:390–402, 1960a.

Taylor, P. M. Oxygen consumption in newborn rats. *J. Physiol. (London)* 154:153–169, 1960b.

Tchobroutsky, C., C. Merlet, and P. Rey. The diving reflex in rabbit, sheep and newborn lamb and its afferent pathways. *Respir. Physiol.* 8:108–117, 1969.

Teitel, D. F., H. S. Iwamoto, and A. M. Rudolph. Effects of birth-related events on central blood flow patterns. *Pediatr. Res.* 22:557–566, 1987.

———. Changes in the pulmonary circulation during birth-related events. *Pediatr. Res.* 27:372–378, 1990.

Teixeira, F. J., F. Aranda, and L. E. Becker. Postnatal maturation of the phrenic nerve in children. *Pediatr. Neurol.* 8:450–454, 1992.

Tenney, S. M., and D. F. Boggs. Comparative mammalian respiratory control. In: *Handbook of physiology: The Respiratory System, vol. II, Control of Breathing,* edited by N. S. Cherniack and J. G. Widdicombe. Bethesda, Md.: Am. Physiol. Soc., part 2, ch. 27, pp. 833–855, 1986.

Tenney, S. M., and L. C. Ou. Ventilatory response of decorticate and decerebrate cats to hypoxia and CO_2. *Respir. Physiol.* 29:81–92, 1977a.

———. Hypoxic ventilatory response of cats at high altitude: an interpretation of 'blunting.' *Respir. Physiol.* 30:185–199, 1977b.

Tenney, S. M., and J. E. Remmers. Comparative quantitative morphology of the mammalian lung: diffusing area. *Nature* 197:54–56, 1963.

Tepper, R. S., R. D. Pagtakhan, and L. M. Taussig. Noninvasive determination of total respiratory system compliance in infants by the weighted-spirometer method. *Am. Rev. Respir. Dis.* 130:461–466, 1984.

Thach, B. T., and H. W. Taeusch, Jr. Sighing in newborn human infants: role of inflation-augmenting reflex. *J. Appl. Physiol.* 41:502–507, 1976.

Thach, B. T., I. Wyszogrodski, and J. Milic-Emili. Effect of ether on control of rate and depth of breathing in newborn rabbits. *J. Appl. Physiol.* 40:281–286, 1976.

Thach, B. T., I. D. Frantz, III, S. M. Adler, and H. W. Taeusch, Jr. Maturation of reflexes influencing inspiratory duration in human infants. *J. Appl. Physiol.* 45:203–211, 1978.

Thach, B. T., I. F. Abroms, I. D. Frantz, III, A. Sotrel, E. N. Bruce, and M. D. Goldman. Intercostal muscle reflexes and sleep breathing patterns in the human infant. *J. Appl. Physiol.* 48:139–146, 1980.

Thach, B. T., A. P. Menon, and G. L. Schefft. Effects of negative upper airway pressure on pattern of breathing in sleeping infants. *J. Appl. Physiol.* 66: 1599–1605, 1989.

Thach, B. T., M. S. Jacobi, and W. M. Gershan. Control of breathing during asphyxia and autoresuscitation. In: *Developmental neurobiology of breathing,* Lung Biology in Health and Disease Series, vol. 53, edited by G. G. Haddad and J. P. Farber. New York: Marcel Dekker, ch. 23, pp. 681–699, 1991.

Thompson, G. E., and R. E. Moore. A study of newborn rats exposed to cold. *Can. J. Physiol. Pharmacol.* 46:865–871, 1968.

Thorburn, G. D. The placenta, PGE_2, and parturition. *Early Hum. Develop.* 29:63–73, 1992.

Thurlbeck, W. M. Structure of the lungs. In: *Respiratory Physiology II,* (Int. Rev. Physiol.), edited by J. G. Widdicombe. Baltimore: University Park Press, vol. 14, pp. 1–36, 1977.

Tod, M. L., and S. Cassin. Fetal and neonatal pulmonary circulation. In: *The Lung: Scientific Foundation,* 2nd ed., edited by R. G. Crystals, J. B. West, et al. Philadelphia: Lippincott-Raven, ch. 162, pp. 2129–2139, 1997.

Trang, T. T. H., N. Viires, and M. Aubier. *In vitro* functions of the rat diaphragm during postnatal development. *J. Dev. Physiol.* 17:1–6, 1992.

Trippenbach, T. Laryngeal, vagal and intercostal reflexes during the early postnatal period. *J. Develop. Physiol.* 3:133–159, 1981.

———. Chest wall reflexes in newborns. *Bull. Eur. Physiopathol. Respir.* 21: 115–122, 1985.

———. Ventilatory and metabolic effects of repeated hypoxia in conscious newborn rabbits. *Am. J. Physiol.* 266:R1584-R1590, 1994.

Trippenbach, T., and D. Flanders. Interaction between somatic and vagal afferent inputs in control of ventilation in 2-week-old rabbits. *Respir. Physiol.* 116:25–33, 1999.

Trippenbach, T., and G. Kelly. Phrenic activity and intercostal muscle EMG during inspiratory loading in newborn kittens. *J. Appl. Physiol.* 54:496–501, 1983.

Trippenbach, T., R. Zinman, and R. Mozes. Effects of airway occlusion at functional residual capacity in pentobarbital-anesthetized kittens. *J. Appl. Physiol.* 51:143–147, 1981.

Trippenbach, T., C. Gaultier, and L. Cooper. Effects of chest wall compression in kittens. *Can. J. Physiol. Pharmacol.* 60:1241–1246, 1982.

Trippenbach, T., G. Kelly, and R. Affleck. Effects of early hypoxia on breathing pattern in rabbit pups before and after vagotomy. *J. Appl. Physiol.* 58:1285–1290, 1985a.

Trippenbach, T., G. Kelly, and D. Marlot. Effects of tonic vagal input on breathing pattern in newborn rabbits. *J. Appl. Physiol.* 59:223–228, 1985b.

Tucker, P. S., and S. M. Horvath. Relationship between organ weight and blood flow in rats adapted to simulated high altitude. *Aerosp. Med.* 44:1036–1039, 1973.

Tucker, A., and D. G. Penney. Pulmonary vascular responsiveness in rats following neonatal exposure to high altitude or carbon monoxide. *Exp. Lung Res.* 19:699–713, 1993.

Tucker, A., N. Migally, M. L. Wright, and K. J. Greenlees. Pulmonary vascular changes in young and aging rats exposed to 5,486 m altitude. *Respiration* 46:246–257, 1984.

Turrens, J. F., B. A. Freeman, and J. D. Crapo. Hyperoxia increases H_2O_2 release by lung mitochondria and microsomes. *Arch. Biochem. Biophys.* 217:411–421, 1982.

Tyndale-Biscoe, H., and M. Renfree. *Reproductive Physiology of Marsupials.* Cambridge, U.K.: Cambridge University Press, 1987.

Van den Pol, A. N., and F. E. Dudek. Cellular communication in the circadian clock, the suprachiasmatic nucleus. *Neuroscience* 56:793–811, 1993.

Van Geijn, H. P., W. M. Kaylor, Jr., K. R. Nicola, and F. P. Zuspan. Induction of severe intrauterine growth retardation in the Sprague-Dawley rat. *Am. J. Obstet. Gynecol.* 137:43–47, 1980.

Van Weering, H. K., J. W. Wladimiroff, and P. J. Roodenburg. Effect of changes in maternal blood gases on fetal breathing movements. *Contr. Gynec. Obstet.* 6:88–91, 1979.

Vannucci, R. C., and T. E. Duffy. Carbohydrate metabolism in fetal and neonatal rat brain during anoxia and recovery. *Am. J. Physiol.* 230:1269–1275, 1976.

Várnai, I., M. Farkas, and S. Donhoffer. Thermoregulatory effects of hypercapnia in the newborn rat. Comparison with the effect of hypoxia. *Acta Physiol. Ac. Sci. Hung.* 38:225–235, 1970.

Vazquez, R., M. Daood, and J. F. Watchko. Regional distribution of myosin heavy chain isoforms in rib cage muscles as a function of postnatal development. *Pediatr. Pulmonol.* 16:289–296, 1993.

Velvis, H., P. Moore, and M. A. Heymann. Prostaglandin inhibition prevents the fall in pulmonary vascular resistance as a result of rhythmic distention of the lungs in fetal lambs. *Pediatr. Res.* 30:62–68, 1991.

Vilos, G. A., and G. C. Liggins. Intrathoracic pressures in fetal sheep. *J. Dev. Physiol.* 4:247–256, 1982.

Vince, M. A., and B. E. Tolhurst. The establishment of lung ventilation in the avian embryo: the rate at which lungs become aerated. *Comp. Biochem. Physiol.* 52A:331–337, 1975.

Visschedijk, A. H. J. The air space and embryonic respiration. 1. The pattern of gaseous exchange in the fertile egg during the closing stages of incubation. *Br. Poultry Sci.* 9:173–184, 1968a.

———. The air space and embryonic respiration. 2. The times of pipping and hatching as influenced by an artificially changed permeability of the shell over the air space. *Br. Poultry Sci.* 9:185–196, 1968b.

———. The air space and embryonic respiration. 3. The balance between oxygen and carbon dioxide in the air space of the incubating chicken egg and its role in stimulating pipping. *Br. Poultry Sci.* 9:197–210, 1968c.

Vizek, M., C. K. Pickett, and J. V. Weil. Biphasic ventilatory response of adult cats to sustained hypoxia has central origin. *J. Appl. Physiol.* 63:1658–1664, 1987.

Vyas, H., A. D. Milner, and I. E. Hopkin. Intrathoracic pressure and volume changes during the spontaneous onset of respiration in babies born by cesarean section and by vaginal delivery. *J. Pediatr.* 99:787–791, 1981.

Wagaman, M., J. G. Shutack, A. S. Moomjian, J. G. Schwartz, T. H. Shaffer, and W. W. Fox. Improved oxygenation and lung compliance with prone positioning of neonates. *J. Pediatr.* 94:787–791, 1979.

Waldron, M. A., and J. T. Fisher. Differential effects of CO_2 and hypoxia on bronchomotor tone in the newborn dog. *Respir. Physiol.* 72:271–282, 1988.

Waldschmidt, A., and E. F. Müller. A comparison of postnatal thermal physiology and energetics in an altricial (*Gerbillus perpallidus*) and a precocial (*Acomys cahirinus*) rodent species. *Comp. Biochem. Physiol.* 90A:169–181, 1988.

Walker, B. R., E. M. Adams, and N. F. Voelkel. Ventilatory responses of hamsters and rats to hypoxia and hypercapnia. *J. Appl. Physiol.* 59:1955–1960, 1985.

Wallace, M. J., S. B. Hooper, and R. Harding. Regulation of lung liquid secretion by arginine vasopressin in fetal sheep. *Am. J. Physiol.* 258:R104–R111, 1990.

Wallois, F., J.-M. Macron, N. Larnicol, and D. Rose. The sneeze: maturation of the reflex in kittens. *NeuroReport* 4:240–242, 1993.

Walters, D. V., and R. E. Olver. The role of catecholamines in lung liquid absorption at birth. *Pediatr. Res.* 12:239–242, 1978.

Walther, F. J., A. B. Wade, D. Warburton, and H. J. Forman. Ontogeny of antioxidant enzymes in the fetal lamb lung. *Exp. Lung Res.* 17:39–45, 1991.

Watanabe, T., P. Kumar, and M. A. Hanson. Effect of ambient temperature on respiratory chemoreflex in unanaesthetized kittens. *Respir. Physiol.* 106:239–246, 1996a.

———. Development of respiratory chemoreflexes to hypoxia and CO_2 in unanaesthetized kittens. *Respir. Physiol.* 106:247–254, 1996b.

———. Elevation of metabolic rate by pyrogen administration does not affect the gain of respiratory peripheral chemoreflexes in unanesthetized kittens. *Pediatr. Res.* 44:357–362, 1998.

Watchko, J. F., and G. C. Sieck. Respiratory muscle fatigue resistance relates to myosin phenotype and SDH activity during development. *J. Appl. Physiol.* 75:1341–1347, 1993.

Watchko, J. F., T. A. Standaert, and D. E. Woodrum. Distribution of diaphragmatic blood flow during inspiratory resistive loaded breathing in piglets (abstract). *Pediatr. Res.* 19:420A, 1985.

Watchko, J. F., D. E. Mayock, T. A. Standaert, and D. E. Woodrum. Postnatal changes in transdiaphragmatic pressure in piglets. *Pediatr. Res.* 20:658–661, 1986a.

Watchko, J. F., W. A. LaFramboise, T. A. Standaert, and D. E. Woodrum. Diaphragmatic function during hypoxemia: neonatal and developmental aspects. *J. Appl. Physiol.* 60:1599–1604, 1986b.

Watchko, J. F., T. A. Standaert, and D. E. Woodrum. Diaphragmatic function during hypercapnia: neonatal and developmental aspects. *J. Appl. Physiol.* 62:768–775, 1987.

Watchko, J. F., T. L. O'Day, B. S. Brozanski, and R. D. Guthrie. Expiratory abdominal muscle activity during ventilatory chemostimulation in piglets. *J. Appl. Physiol.* 68:1343–1349, 1990.

Watchko, J. F., B. S. Brozanski, T. L. O'Day, R. T. Guthrie, and G. C. Sieck. Contractile properties of the rat external abdominal oblique and diaphragm muscles during development. *J. Appl. Physiol.* 72:1432–1436, 1992a.

Watchko, J. F., M. J. Daood, R. L. Vazquez, B. S. Brozanski, W. A. LaFramboise, R. D. Guthrie, and G. C. Sieck. Postnatal expression of myosin isoforms in an expiratory muscle–external abdominal oblique. *J. Appl. Physiol.* 73:1860–1866, 1992b.

Waters, K. A., C. S. Beardsmore, J. Paquette, B. Meehan, A. Côté, and I. R. Moss. Respiratory responses to rapid-onset, repetitive vs. continuous hypoxia in piglets. *Respir. Physiol.* 105:135–142, 1996.

Webb, B., A. A. Hutchison, and P. W. Davenport. Vagally mediated volume-dependent modulation of inspiratory duration in the neonatal lamb. *J. Appl. Physiol.* 76:397–402, 1994.

Webster, S. H., and E. J. Liljegren. Organ : body weight ratios for certain organs of laboratory animals. II. Guinea pig. *Am. J. Anat.* 85:199–230, 1949.

Webster, S. H., E. J. Liljegren, and D. J. Zimmer. Organ : body weight ratios for liver, kidneys and spleen of laboratory animals. I. Albino rat. *Am. J. Anat.* 81:477–513, 1947.

Weil, J. V. Ventilatory control at high altitude. In: *Handbook of Physiology, Section 3: The Respiratory System, vol. II, Control of Breathing*, edited by N. S. Cherniack and J. G. Widdicombe. Bethesda, Md.: Am. Physiol. Soc., part 2, ch. 21, pp. 703–727, 1986.

Wennergren, G., and M. Wennergren. Respiratory effects elicited in newborn animals via the central chemoreceptors. *Acta Physiol. Scand.* 108:309–311, 1980.

————. Neonatal breathing control mediated via the central chemoreceptors. *Acta Physiol. Scand.* 119:139–146, 1983.

Wennergren, G., T. Hertzberg, J. Milerad, J. Bjure, and H. Lagercrantz. Hypoxia reinforces laryngeal reflex bradycardia in infants. *Acta Paediatr. Scand.* 78:11–17, 1989.

West, G. B., J. H. Brown, and B. J. Enquist. A general model for the origin of allometric scaling laws in biology. *Science* 276:122–126, 1997.

Whittaker, J. A. C., C. O. Trouth, Y. Pan, R. M. Millis, and D. G. Bernard. Age differences in responsiveness of brainstem chemosensitive neurons to extracellular pH changes. *Life Sci.* 46:1699–1705, 1990.

Widdicombe, J. G. Respiratory reflexes in man and other mammalian species. *Clin. Sci.* 21:163–170, 1961.

————. Reflexes from the upper respiratory tract. In: *Handbook of Physiology: The Respiratory System, vol. II, Control of Breathing,* edited by N. S. Cherniack and J. G. Widdicombe. Bethesda, Md.: Am. Physiol. Soc., part 1, ch. 11, pp. 363–394, 1986.

Williams, B. A., J. Smyth, A. W. Boon, M. A. Hanson, P. Kumar, and C. E. Blanco. Development of respiratory chemoreflexes in response to alternations of fractional inspired oxygen in the newborn infant. *J. Physiol. (London)* 442:81–90, 1991.

Williams, B. R., Jr., D. F. Boggs, and D. L. Kilgore, Jr. Scaling of hypercapnic ventilatory responsiveness in birds and mammals. *Respir. Physiol.* 99:313–319, 1995.

Wilson, T. A. Parenchymal mechanics at the alveolar level. *Federation Proc.* 38:7–10, 1979.

Wilson, S. L., B. T. Thach, R. T. Brouillette, and Y. K. Abu-Osba. Coordination of breathing and swallowing in human infants. *J. Appl. Physiol.* 50:851–858, 1981.

Winkler, G. C., and N. F. Cheville. Morphometry of postnatal development in the porcine lung. *Anat. Rec.* 211:427–433, 1985.

Wittenborg, M. H., M. T. Gyepes, and D. Crocker. Tracheal dynamics in infants with respiratory distress, stridor, and collapsing trachea. *Radiology* 88:653–662, 1967.

Wittmann, J., and J. Prechtl. Respiratory function of catecholamines during the late period of avian development. *Respir. Physiol.* 83:375–386, 1991.

Wolfson, M. R., J. S. Greenspan, K. S. Deoras, J. A. Allen, and T. H. Shaffer. Effect of position on the mechanical interaction between the rib cage and abdomen in preterm infants. *J. Appl. Physiol.* 72:1032–1038, 1992.

Wolsink, J. G., A. Berkenbosch, J. DeGoede, and C. N. Oliever. Ventilatory sensitivities of peripheral and central chemoreceptors of young piglets to inhalation of CO_2 in air. *Pediatr. Res.* 30:491–495, 1991.

Wong, K. A., A. Bano, A. Rigaux, B. Wang, B. Bharadwaj, S. Schürch, F. Green, J. E. Remmers, and S. U. Hasan. Pulmonary vagal innervation is required to establish adequate alveolar ventilation in the newborn lamb. *J. Appl. Physiol.* 85:849–859, 1998.

Wood, G. A., and R. Harding. The effects of pentobarbitone, diazepam and alcohol on oral breathing in neonatal and mature sheep. *Respir. Physiol.* 75:89–104, 1989.

Woodrum, D. E., T. A. Standaert, C. R. Parks, D. Belenky, J. Murphy, and W. A. Hodson. Ventilatory response in the fetal lamb following peripheral chemodenervation. *J. Appl. Physiol.* 42:630–635, 1977.

Woodrum, D. E., T. A. Standaert, D. E. Mayock, and R. D. Guthrie. Hypoxic ventilatory response in the newborn monkey. *Pediatr. Res.* 15:367–370, 1981.

Woźniak, W., R. O'Rahilly, and M. Bruska. Myelination of the human fetal phrenic nerve. *Acta Anat.* 112:281–296, 1982.

Wyszogrodski, I., B. T. Thach, and J. Milic-Emili. Maturation of respiratory control in unanesthetized newborn rabbits. *J. Appl. Physiol.* 44:304–310, 1978.

Yao, A. C., J. Lind, and V. Vuorenkoski. Expiratory grunting in the late clamped normal neonate. *Pediatrics* 48:865–870, 1971.

Yokoyama, E., and Z. Nambu. Age-dependence of V/P relationships and weight-related parameters in rat lungs. *Comp. Biochem. Physiol. A: Comp. Physiol.* 69:285–289, 1981.

Zamudio, S., S. K. Palmer, T. Droma, E. Stamm, C. Coffin, and L. G. Moore. Effect of altitude on uterine artery blood flow during normal pregnancy. *J. Appl. Physiol.* 79:7–14, 1995.

Zin, W. A., L. D. Pengelly, and J. Milic-Emili. Single-breath method for measurement of respiratory mechanics in anesthetized animals. *J. Appl. Physiol.* 52:1266–1271, 1982.

Index